DESIGN
AND MANUFACTURE
OF COMPOSITE
STRUCTURES

GEOFF ECKOLD

DESIGN AND MANUFACTURE OF COMPOSITE STRUCTURES

McGraw-Hill, Inc.

New York   San Francisco   Washington D.C.
Montreal   Toronto

Published by Woodhead Publishing Ltd,
Abington Hall, Abington,
Cambridge CB1 6AH, England

First published 1994, Woodhead Publishing Ltd

© Woodhead Publishing Ltd

Published in the United States of America and Canada by
McGraw-Hill, Inc.

Library of Congress Cataloging-in-Publication Data
A catalog record for this book is available from the Library of
Congress.

ISBN 0-07-018961-7

Printed in the United Kingdom

CONTENTS

PREFACE

Composite materials have experienced rapid development over the last 30 years and it is probably true to say that their use pervades virtually every industrial sector. From satellites to subsea, from automotive to artificial legs, there is likely to be some aspect of composite performance, be it light weight, corrosion resistance, unique mechanical characteristics, ease of fabrication or some other attribute which will prove to be of advantage. To capitalize on these characteristics, engineers and component designers are required to have a much broader appreciation of materials related issues than would be the case if options were restricted to simple monolithic systems. Heterogeneity, anisotropy and the concurrent manufacture of material *and* component can prove to be difficult concepts to grasp and must certainly be accounted for in the early stages of design. As such this book is primarily aimed at those in the component analysis and design arena and is presented with an engineering perspective. This is not to say that it is written solely for the benefit of engineers. Whilst it is fair to comment that the wider exploitation of composites has been inhibited by lack of awareness and experience of practising engineers, materials scientists also have an important part to play and this can be enhanced if they could have a clearer appreciation of engineering concerns.

The design art for all material systems is multifaceted and this is especially true for composites where an appreciation of property characteristics, mechanics and manufacture is key to successful execution of a product development cycle. In this book, each part of the design process is considered with chapters describing materials and properties, mechanics and material behaviour, the analysis of basic structural forms such as beams and shells, aspects of component design and manufacture. Although these various topics are presented in sequence, the need for strong interaction between each element of the material/design/manufacture equation is emphasised. Of course no single volume can attempt to cover in detail every aspect of composite performance or do justice to the wealth of experience established in the materials community at large and the following chapters do not attempt to do so. The prime intention is to provide those whose interest in composites is to convert materials promise into component reality with information in sufficient depth such that they can gain an appreciation of the key issues involved.

The book has four main themes. The first three chapters have a materials focus providing an overview of the fibres and matrices in current use and the mechanics of how they can be combined to form a structural material of the required stiffness and strength. The discussion is centred around continuously reinforced systems and this is continued throughout the text, the emphasis being on the application of composites for the more structurally demanding duties. Materials considerations are continued in Chapter 7 where the use of ceramics and metals as matrices is reviewed. Whilst these are still in the development phase they offer future potential in extending the options available to the designer. Component design and performance provide the basis of Chapters 4 and 5. In the first the analysis of beams, plates and shells is presented. Although at first sight the mathematical nature of the presentation appears daunting, time invested in coming to terms with the derivations outlined will pay dividends. The components concerned form the basic elements of virtually every engineering structure and an understanding of their behaviour and how it differs from those manufactured from conventional materials is fundamental. Even if it is intended to use a numerical method such as finite elements the ability to carry out scoping calculations must feature in the design exercise. Other aspects of design are described in Chapter 5. Once a method is established for the evaluation of basic laminate configurations consideration must be given to issues such as jointing, free edge effects and environmental conditions. The intention is not to be exhaustive or provide a design manual approach but to give an introduction to these important topics. Manufacture methods are reviewed in Chapter 6. In addition to descriptions of the processing techniques themselves the discussion is extended to include machining and non-destructive testing. The final chapter is concerned with examples of component application where the materials are currently being used or under investigation for the future. Sections are concluded with an example component, each of which varies with respect to materials of construction, motivating factors which determined the choice of composites in the first instance, and degree of maturity in the product development cycle.

A book of this type is possible only because of the support and assistance given by my many co-workers in the materials field and their contribution is gratefully acknowledged. Particular thanks are due to my friends and colleagues, past and present, at the Harwell Laboratory and at Plastics Design and Engineering where I learned what engineering with composites was all about. Finally, I would like to thank Marian Gibson who persevered with the text against great adversity and Catherine and children who can now have the dining room back.

Geoff Eckold

1

INTRODUCTION

Materials and their application are the fundamental constituents of any engineering design. Regardless of whether or not the calculations for stress and strain are accurate, the fabrication is performed according to proper quality-assured procedures, and installation and commissioning are completed as specified, the incorrect selection of materials will ultimately lead to the component or structure not achieving its potential in terms of performance or lifetime. As a consequence, the concept of tailoring materials at a fundamental level to meet specific design requirements has intrigued engineers for many years. Whilst the principle is by no means new, it is only comparatively recently that the use of 'materials design' has been fully incorporated in the overall product development process. Examples range from the macroscopic, such as steel reinforcing bars in concrete for tensile strength, through the microscopic and molecular, for example whisker reinforcement of ceramics for enhanced toughness and the copolymerization of polypropylene and polyethylene for good mechanical behaviour below ambient temperatures, to the atomic, such as ion implantation of surfaces to increase wear properties.

This book is concerned with one family of designer materials – fibre reinforced composites. In many respects they represent excellent examples of the principles of material design where the performance of the whole is greater than the sum of its parts. Fibres, lightweight, immensely strong and stiff (in some cases not too distant from the theoretical maximum), but easily damaged and in a form of limited engineering application, and the matrix, comparatively weak and often brittle and not usually attractive for structural load-bearing applications. Together, however, they offer a vast array of materials and great scope for optimization in the true sense of the word. This is not only limited to mechanical characteristics but also extends to thermal properties, acoustic and electromagnetic response, creep and fatigue, ballistic performance and chemical resistance. Indeed, for the great majority of parameters in which an engineer may be interested, fibre reinforced composites provide alternatives often only limited by the imaginations of the designers themselves. Of course, it would be misleading to suggest that these systems supply answers to all

1

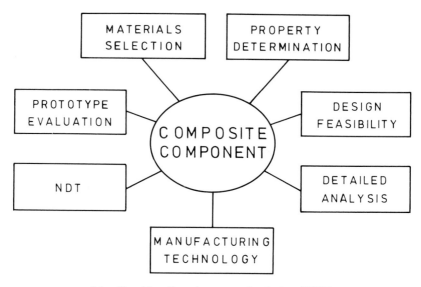

1.1   Considerations in composite design (NDT,
non-destructive testing).

problems: where there are advantages, there are also disadvantages. It is for this reason that this new class of materials needs a complement – a new breed of engineers, not only familiar with the principles of mechanics and design, but also conversant with fabrication science, chemistry, materials physics, and new test and inspection procedures (Fig. 1.1). Only with an appreciation of all these facets of the composites' equation will the required levels of structural efficiency and reliability be achieved. It is the purpose of the following chapters to introduce these issues in a design-orientated context and, while claiming not to be exhaustive, to cover the main points of the engineering process from conceiving the materials to completing the component.

Historical perspective

Amongst the earliest applications of the use of particulate or fibrous material to reinforce another material are natural fibres such as grass or animal hair which were used to improve the strength and to alleviate shrinkage of pottery prior to firing.[1] In a similar way potters were able to modify the porosity of their finished artefacts to produce receptacles that would provide cooling by evaporation. Modified polymeric materials also featured in ancient times. Evidence dating to around 3000 BC suggests reinforced bitumen/pitch building products, and bitumen embedded with papyrus reeds for boat building were in widespread application. An example of what must be one of the first exercises in surface modification is in the use of stone or fired clay cones hammered into

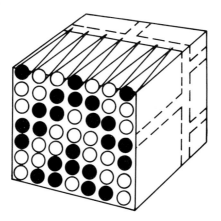

1.2   Reinforcement of soft mud brick walls.

building walls (Fig. 1.2).[2] This would have had the effect of improving erosion and weathering resistance.

Applications can also be found in early armaments. In this period powerful composite bows were fabricated from wood and horn lamallae or even more exotic combinations such as animal tendons, wood and silk bonded together using adhesive.[2] The makers of armour made use of the perforation resistance of laminated structures fabricated from alternate layers of iron and steel. In one example[3] of the many variants of compound blade, three types of steel were used. Edge steel was folded up to 20 times, core steel 8 times and the three layers of skin steel folded by a similar amount. All were hammered out and welded together into a blade containing many thousands of layers. Special heat treatment utilizing quenching and slow cooling of different parts of the structure converted the edge into a very hard material indeed. Similarly, oriental gunsmiths utilized different kinds of iron and steel fashioned into strips, wound into a helix and then welded to form the barrel of a gun.

In terms of the modern exploitation of composite materials, this was essentially pioneered following the work of Griffiths' in the 1920s;[4] who reported strengths of freshly drawn glass fibres of up to 900 000 psi (6000 MPa). Since this time research and development have been intensive. The initial work centred on the defence uses of glass reinforced plastics (GRPs) such as aircraft radomes, boat hulls and seaplane floats.[5] Up to that time the cost of the basic resin and glass materials precluded general application and it was not until the late 1940s with the advent of cold setting resin systems and cheaper forms of reinforcement that the industry expanded beyond specialist uses. Translucent sheeting, boat hulls, chemical-resistant process equipment and car bodies all became widespread.

Perhaps the single most important event in the advanced composites calendar was the development of carbon fibre.[6] Although not new, Edison

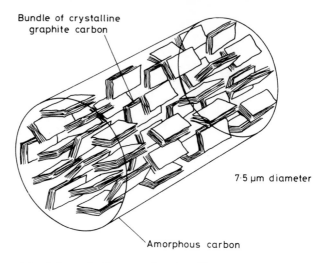

Bundle of crystalline
graphite carbon

7·5 µm diameter

Amorphous carbon

1.3   Schematic diagram of carbon fibre structure.

employed fine carbon filaments in his early electric lamps, the apparent
simplicity of the material,[7] (Fig. 1.3) and its exciting range of mechanical
properties captured the imagination. The initial developments at the Royal
Aircraft Establishment in the early 1960s used highly orientated polyac-
rylonitile (PAN) as a precursor. The key steps in the process (Fig. 1.4) were as
follows:[7]

- Tows of continuous PAN were wound on to frames and heated in an air
  oven at 220–230 °C. During this time an oxidation reaction occurred
  which changed the fibre from its original white to black.
- The fibres were cut from the frames and placed in a graphite crucible and
  heated from 400 to 1600 °C in a nitrogen atmosphere.
- Further heating to around 2300 °C for about an hour depending on the
  grade of material required.

It was soon discovered that the properties of the fibre obtained could be
controlled by the thermal cycle adopted during processing (Fig. 1.5).[8] Two
types of fibre were defined:

- Type 1: high stiffness (modulus $\sim$ 350 GPa, strength $\sim$ 2.0 GPa).
- Type 2: high strength (modulus $\sim$ 240 GPa, strength $\sim$ 2.4 GPa).

From this starting point there is now a whole array of carbon fibre materials
from which a selection can be made with moduli and strengths approaching
800 GPa and 6.5 GPa respectively. Variations are now manufactured from
different precursors using a variety of processing conditions, and with different
surface and size treatments all of which are honed to give a specific range of
properties.

Stage (1)

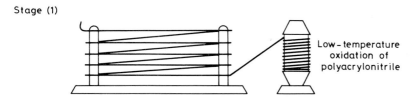

Low-temperature
oxidation of
polyacrylonitrile

Stage (2) Carbonisation of nitrogen containing ring polymer from stage (1)

Inert
gas atmosphere

Stage (3) Heat treatment of orientated carbon fibre from stage (2)

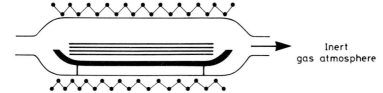

1.4   Preparation of carbon fibre from polyacrylonitrile
precursor.

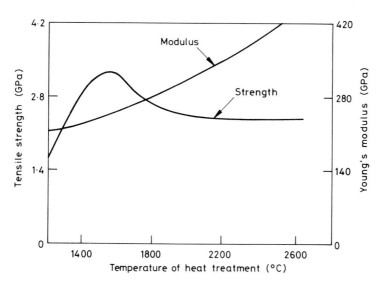

1.5   Properties of carbon fibre against temperature.

1.6   Structure of aramid polymer.

The development of aramid fibres, 'Kevlar', is also a case of 'designer' materials science. In this case an aromatic polymer (Fig. 1.6), similar in generic form to nylon, is spun and drawn into fibre form. The combination of the regular pattern of benzene rings and hydrogen bonding leads to a highly crystalline polymer with a significant level of molecular orientation. The net result is fibres having both high strength and high stiffness.

Although it is probably fair to say that the pace of developments has been dictated by progress in fibre properties, they are only part of the system; the matrix is of key importance. Great strides have been made in synthetic polymer chemistry and the choice available extends much further than the phenol–formaldelyde and melamine–formaldelyde systems available many years ago. Polyesters and epoxies form the basic resin range, but these are augmented by a whole range of specialist materials including polyimides, polybismaleimides (BMI) and multifunctional epoxies amongst the thermosetting polymers, and polyphenylene sulphide (PPS), polyether sulfone (PES) and polyether ether ketone (PEEK) amongst the thermoplastic systems. Many of these systems have been formulated for specialist applications, notably high temperature capability. The choice is not limited to polymers and both ceramics and metal matrices can feature in the materials engineer's armoury.

As to the future, it is clear that the development of property enhancement will continue, be it stiffness, strength, toughness or temperature capability. The next milestone is likely to be the advent of the so-called smart materials. By incorporating sensors and actuators (see, for example, Fig. 1.7) into the material during fabrication, a structure can be produced which not only reacts in a prescribed way to external loading but also senses its environment and modifies its behaviour to suit. On-line health monitoring, self-damping, damage assessment and intelligent processing are all within reach, so in addition to becoming a 'chemist', the design engineer will need to become familiar with concepts of control and signal processing as well!

Composites: advantages and disadvantages

The potential advantages for considering composites for any one application are manifold, but it is likely for the majority of cases that interest will lie in one

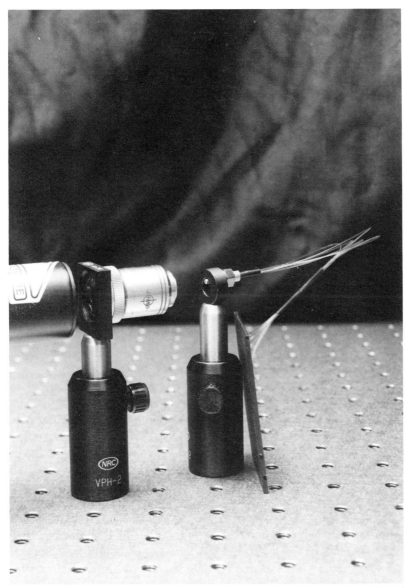

1.7   Fibre optic sensors incorporated in a carbon fibre
laminate.

of two areas: light weight manifested in the form of high specific properties, or corrosion resistance. Figure 1.8 shows the specific properties (stiffness or strength divided by specific gravity) of composites compared with those of other engineering materials. The benefits are self-evident with even glass reinforced materials having properties equivalent to, or greater than, steel. Carbon reinforced materials would actually be outside the scale of the figure.

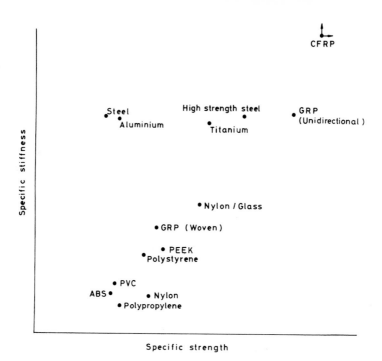

1.8   Specific properties of composites (CFRP, carbon
fibre reinforced plastic).

One note of caution, however, is that the properties shown are for those in the direction of reinforcement. Stiffnesses and strengths in other directions can be markedly lower. Therein lies one of the complications with composites – anisotropy. This arises when the properties of a material vary with direction and although it is possible to put fibres precisely where they are needed, i.e. in the direction of applied loading, this may be difficult in some cases. Indeed, there may be insufficient knowledge regarding the loads in the first instance. Design with composites therefore requires much more property data and is inherently more complex than design with more traditional materials.

Corrosion resistance is another outstanding attribute, particularly those with a polymeric matrix. Table 1.1 indicates the range of media where the materials have been used successfully.[9] To be successful, special attention must be given to the details of the design to preclude ingress of fluids into the composite laminate where it may have a deleterious effect on the interface between matrix and reinforcement and on the properties of the reinforcement itself.

Flexibility of manufacture is also a unique characteristic. Large, complex structures can be fabricated in one piece minimizing tooling costs and obviating the need for joints and fastenings. The fact that during manufacture

Table 1.1. Chemical resistance of GRP composites

Acids	Sulphuric
	Hydrochloric
	Nitric
	Phosphoric
Alkalis	Sodium hydroxide
	Potassium hydroxide
Aqueous salts	Ammonium, calcium, sodium, potassium salts
Oxidizing agents	Chlorine
	Sodium hypochloride
Organics	Alcohols, some acids, glycols, Aviation and diesel fuels

Tabulated information is indicative only. Reference must be made to technical data sheets pertaining to individual resin systems.

the material itself is being made at the same time as the component gives rise to tremendous scope for innovative design without the constraints of conventional metal-forming or machining processes.

Other significant advantages may include:

- **Thermal properties**. It is possible to design components with very low, even zero, thermal expansivity, a key factor in controlling dimensional tolerances. Also thermal conductivities are generally low which may be of significance in certain applications.
- **Fatigue**. Certain composites, particularly those with carbon reinforcement have excellent fatigue characteristics – a clear benefit in many areas.
- **Wear resistance**. Again, carbon fibre materials behave well due to the formation of a graphitic layer as wear begins.
- **Electrical properties**. GRP is an excellent electric insulator. Electromagnetic characteristics can be especially valuable for a certain range of structures.
- **Finish**. Because of the flexibility available during manufacture finishes such as pigmentation and textures can be easily incorporated.

In terms of disadvantages, there are two that require particular attention: lack of ductility and inspectability. By and large whenever a ductile material is filled with a reinforcement, be it in particulate or fibre form, a loss in toughness will result. This leads to a stress/strain curve which is essentially linear to failure. Composites, therefore, are less forgiving materials than, say, metals, where stress concentrations and peak loadings can often be accommodated by redistribution as a result of local yielding. Great care must be taken with composites to ensure that all loadings are accounted for in the initial design. Coping with this can be made more difficult by the effects of anisotropy.

The difficulty with inspectability is not so much associated with the technique, although the heterogeneity of many composites can cause difficulties, but the scale of the required task. Metals can be procured in part finished form, e.g. plate, rod or tube, and supplied with a certificate of conformity which provides evidence of quality. Subsequent component inspection, over and above that required to check manufacturing tolerance, etc., can be limited to welds and joints. Composite components on the other hand require inspection over the whole surface which can lead to a different scale of problem altogether. Whilst this issue is by no means insurmountable, quality assurance and inspection procedures must feature in the initial feasibility assessment.

It should be emphasized that composites are not a panacea to all problems. An advantage for one application may be a disadvantage for another and often a compromise may be the most appropriate course of action. The point which can be made, however, is that the added flexibility offered by composites, together with the inherent advantages of light weight, corrosion resistance, etc., they possess over other materials, ensures that they must feature in the first rank of materials' options available to the engineer.

Design methods

In all design assessments there are a number of stages that must be undertaken in order to derive a satisfactory structure both in terms of performance and cost. Inevitably there will be a number of options for the engineer to consider which will result in the design being an iterative procedure. Where a metal is the chosen material of construction it is likely that, owing to the extensive knowledge of material properties, analysis methods and fabrication techniques that are available, the number of stages in the design will be reduced to a minimum. In composites, however, this will not be the case and, with the level of technology as it now stands, due attention must be given to each stage of the design process:

- **Functional specification**: the first step in the design of any component is to define clearly the requirements of the application. For composites it is rarely satisfactory to declare that the component must simply perform as a direct replacement for an existing structure, say manufactured from a metal, as this may result in a design that is not the best achievable. Particular attention should not only be given to the magnitude of the primary mechanical and thermal loads, but also to the direction of the load path. Loads are usually combined and secondary loading such as through thickness components of stress, impact or thermal transients should also be considered.
- **Materials of construction**: for metals there is a wealth of information concerning behaviour under a variety of conditions, much of which is

available in formal documentation such as standards and codes of practice. This greatly eases the material selection procedure. With composites there is a general lack of property data and hence it is often necessary to test materials in conjunction with the design. Another important consideration with composites is the ability to tailor properties or materials, for example by changing fibre orientation or type of reinforcement, thereby allowing the material itself to be designed to suit operating conditions.

- **Design and analysis**: the two most pronounced features of the behaviour of composites are anisotropy and heterogeneity. These affect elastic, strength and thermal properties and must be considered in design.
- **Fabrication limitations**: it is well known that the method of fabrication and the processing conditions employed can have a marked effect on composite properties. Clearly the availability of a suitable production route for the artefact of concern must be an important feature of the design.
- **Reliability requirements**: the essence of any design exercise is the assessment of reliability and the potential risk of component failure. This is most important as the engineer must take these into account when assessing the factors of safety which should be applied to the design. To a certain extent the selection of factors of safety is a matter of judgement based on experience, and this can present difficulties with new material systems such as composites. The relative brittleness of these materials can also have ramifications on reliability-related calculations.
- **Cost considerations**: at the end of the design process the resulting component must be competitive both in terms of performance and cost. Structures fabricated from composites can be expensive compared with those of traditional materials, particularly on the basis of initial procurement cost, and should be viewed in terms of total life costs and improved performance.

There are a number of approaches that may be used for design; the choice of which depends on the experience, data and resources available at the time at which the assessment is made.

Empirical design

Empirical design is a trial and error technique that employs an iterative approach and entails a fabrication and testing programme. A minimum of technical analysis is employed. Although this approach can be satisfactory when supplied property data for an analysis are not available, it is rarely possible to simulate the full range of operating conditions in the laboratory. A consequence of assumptions on scaling, superposition or extrapolation of results can be the imposition of high safety factors on the design.

Deterministic design

In this approach the stress distribution within the component is calculated using analytical or numerical techniques. A failure criterion, for example Von Mises, maximum stress or Tresca,[10] is normally used in conjunction with experimentally determined strength data. From this a margin of safety can be determined and a judgement made on the acceptability of a candidate material. This technique is often used for metals where the relatively low scatter in property values means that the material can be employed close to its ultimate strength. Indeed, factors of safety as low as 1.5 are not uncommon. For brittle materials, however, the wide distribution in measured strength values can lead to high factors of safety if a low risk of failure is to be assured. Another factor which cannot be taken into account with deterministic design is the dependence of strength on geometry and scale, a feature of brittle materials.

Probabilistic design

For brittle materials where strength measurements exhibit wide variability and a marked dependence on specimen size a statistical approach is often employed to assist design. The most commonly used method is Weibull analysis.[11] From this it can be shown that the probability of failure at a particular stress is given by:

$$P_f = 1 - \exp\left[-\int_v \frac{(\sigma - \sigma_u)^m}{\sigma_0} dv \right] \qquad [1.1]$$

where P_f is the probability of failure, σ_u is the threshold stress (lowest value of stress for which the failure probability is finite), σ_0 is the normalizing parameter (dimensions of stress), m is the Weibull modulus (this is associated with the variability of failure stresses) and v represents volume. Equation 1.1 can be rewritten in a more useful form and in terms of physically significant parameters as follows:[12]

$$P_f = 1 - \exp\left[-\int_v \left(\frac{1}{m!}\right)^m (\sigma/\bar{\sigma}_f)^m dv \right] \qquad [1.2]$$

where $\bar{\sigma}_f$ is the mean value of failure stress and $(1/m)!$ is the gamma function of $(1/m) + 1$. Figure 1.9 shows equation 1.2 as a function of different values of Weibull modulus. It provides a useful guide to the variability of strength values, the lower the value of m, the greater the variation.

When applying this method to structures, further information is required concerning the stress distribution. The level of stress in each element within the component needs to be compared with material strength data and the

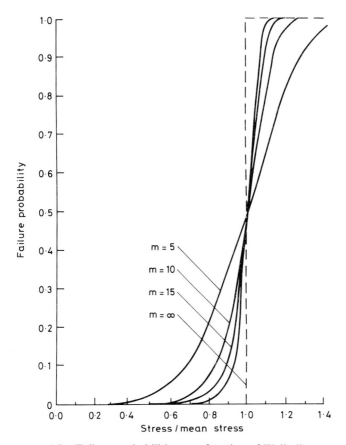

1.9   Failure probabilities as a function of Weibull
modulus.

probability of failure determined. These values are then integrated over the structure (stress volume integral) to derive a probability of failure for the component as a whole.

The main advantage of a probabilistic method is that it allows areas of high localized stress to be treated more pragmatically. For example, a material with a low Weibull modulus could still be considered acceptable for a component with a stress concentration, because the volume of the element of concern is such that it would not provide a major contribution to the total probability of failure.

As well as advantages there are also a number of problems with this design technique which should be considered:

• The determination of stress distributions within components must be carried out with a high degree of accuracy. With metals the inherent ductility of the materials, which allows some local redistribution of stress, means that the need for accuracy is not as severe.

- A large number of specimens (> 20) must be tested to determine the Weibull modulus. Large uncertainties in the value of *m* employed in the calculation can negate the value of the analysis.
- The Weibull relationship assumes a uniform random flaw distribution. Should the population of flaws take a different form the use of a single modulus value may not be appropriate.

Fracture mechanics

Fracture mechanics is a widely employed technique where critical defects within the material are considered in the assessment of structural integrity. For any particular section of a component, defects of various sizes will be present and from a knowledge of applied stress distribution the stress intensity factor or strain release rate at each flaw can be determined. By examining the propagation of defects during service an assessment can be made of the end of life flaw sizes and these can then be compared with critical values. Fracture mechanics methods have been used successfully in the design of a range of components, for example rotors and pressure vessels,[13] where they have been used in conjunction with non-destructive testing (NDT) methods for crack detection.

The approach most often adopted to structures is known as linear elastic fracture mechanics (LEFM) and is based on the Griffith theory of fracture.[4] The basis of the derived principle is that a crack cannot grow unless the stored elastic energy released during crack growth is equal to, or greater than, the thermodynamic surface energy of the material. For an elastically isotropic solid it can be shown that:

$$\sigma_f = Y . \sqrt{\frac{E\gamma_f}{c}} \qquad [1.3]$$

where σ_f is the fracture stress, Y is a geometrical constant depending on specimen shape and dimensions, γ_f is the energy absorbed during the creation of a unit area of fracture surface, $2c$ is the crack length and E is Young's modulus. Table 1.2 shows typical values of γ_f and, as can be seen, the ranking of materials corresponds to that which would be expected intuitively on the basis of toughness or brittleness.[2]

Equation 1.3 can be rearranged to

$$K_c^2 = Y'\sigma_f^2 c \qquad [1.4]$$

where K_c is known as the stress intensity factor.

Table 1.2. Typical fracture surface energies

Material	Fracture surface energy γ_f (J/m^2)
Dural	1.4×10^5
Copper	5×10^4
Cast iron	4×10^3
Polystyrene	1×10^3
Perspex	5×10^2
Epoxy resin	3.3×10^2
Polyester resin	2.2×10^2
Graphite	50–100
Glass	4

The position with regard to the application of fracture mechanics to composites, which are often anisotropic, is somewhat more complex and this needs to be reflected in the above equations. A particularly interesting application of fracture mechanics in terms of design is in the development of strength/probability/time (SPT) diagrams.[14,15] This technique, developed with brittle ceramic materials in mind, attempts to relate time of failure, failure load and the probability of failure. For example, consider a material which when loaded by a given stress breaks in a short time. A similar section of the same material when stressed to, say, 0.5 of the original load would still fracture, but only when subjected to load for a considerably longer period of time. This time dependence of strength is clearly important in design and is due to sub-critical crack growth occurring under stress. Derivation of a SPT diagram requires a combination of data relating to fracture mechanics properties and the statistical variation of strength. An example of the end result of the analysis is shown in Fig. 1.10, in this case for alumina specimens in bending. The survival probability of a component is plotted as a function of applied stress and time. To demonstrate its application consider a requirement of 95% survival probability and a component lifetime of 10^4 s. It can be seen that the permissible working stress for this system is approximately 205 MPa as compared to a mean short term strength of about 380 MPa.

Nomenclature

Inevitably as an emerging technology develops it will be accompanied by its own jargon. In many cases the terminology will not be new, but only employed in a context that is unfamiliar. Composite materials are not exceptions to this rule and today's designers and engineers have had to become accustomed to a number of new concepts.

In almost all cases where design is carried out using traditional materials, conditions of *isotropy* are assumed. An *isotropic* body is one with uniform properties throughout that are not functions of position or orientation. In

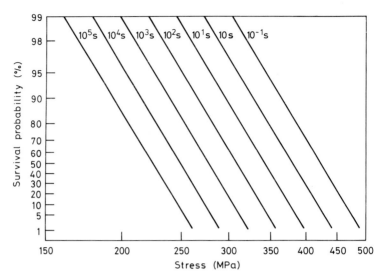

1.10   SPT diagram for 95% alumina in bending.

such a material elastic behaviour can be characterized by two independent elastic constants and the following relationships apply:

$$G = \frac{E}{2(1 + v)} \qquad [1.5]$$

$$- 1 \leq v \leq 0.5 \qquad [1.6]$$

where G is the shear modulus, E the tensile modulus and v Poisson's ratio.

An *anisotropic* body on the other hand has material properties that are different in all directions with no planes of symmetry. There are 21 independent elastic constants. In composites there are two particularly important conditions of symmetry: *orthotropy* and *transverse isotropy*. In the former case there are three orthogonal planes of symmetry and in the latter there is a plane at each point in the material in which properties are equal in all directions. The stress/strain relations have nine and five independent constants, respectively. For conditions of plane stress this reduces to just four constants. In terms of a comparison with isotropic materials, the main difference with the plane stress case is the independence of the shear modulus; there is no equivalent expression to equation 1.5. Also values for Poisson's ratio can be difficult to visualize, ranging from nearly 0 to, in some cases, well over unity.

The study of composites can be divided into two levels. The first, *micromechanics*, is based on the consideration of the interaction of the constituent materials at a microscopic level. The determination of constitutive properties from those of fibre and matrix is one such example. At the second

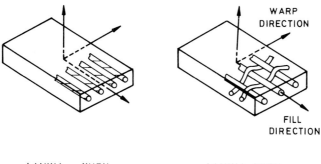

WARP
DIRECTION

FILL
DIRECTION

LAMINA WITH
UNIDIRECTIONAL FIBRES

LAMINA WITH
WOVEN FIBRES

1.11 Typical types of lamina.

level, *macromechanics*, the materials are assumed to be homogeneous with 'effective' composite properties being used that are not specifically related to those of the individual constituents.

The relative quantity of the various constituents in the composite is known as the *volume fraction* and is normally expressed as the ratio of volume of reinforcement and the total volume of composite. Expressing contents in terms of volume is often convenient as this can then be directly related to concepts of load share within the material. However, during manufacture it is more difficult to measure the volume of a material than its weight, so the use of *weight fraction* is also common. Except in the rare occasions where all constituents have the same density the two measures will have different values, but can be interrelated through the following expressions:

$$w_f = \frac{v_f \rho_f}{v_f \rho_f + (1 - v_f)\rho_m} \tag{1.7}$$

$$v_f = \frac{w_f}{w_f + (1 - w_f)(\rho_f/\rho_m)} \tag{1.8}$$

where v_f and w_f are volume and weight fractions of fibre and ρ_f and ρ_m are densities of fibre and matrix. In the event of the composite containing more than just the fibre and matrix, equations 1.7 and 1.8 can be easily extended to accommodate the additional phases.

For most engineering applications of composites the materials are used in the form of *lamina* which are combined in various ways to form a *laminate*. A lamina (Fig. 1.11) is essentially the repeating unit of the construction and can take a variety of forms. In high performance uses it usually comprises a directional array of fibres within a matrix. The laminate (Fig. 1.12), is then a sequence of such layers bonded with reinforcement in appropriate orientations as dictated by the design. Clearly an infinite array of laminates is

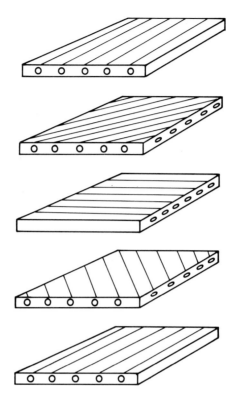

1.12   Laminate construction.

possible, but there are a number of technically important configurations. A *symmetrical laminate* is one where there is a plane of symmetry of lamina in terms of thickness, orientation and properties with respect to the centre-line. Another form of construction is the *balanced laminate* where the number of layers in orientations $\theta_1, \theta_2, \theta_3, \ldots$, are matched by an equal number in the $-\theta_1, -\theta_2, -\theta_3, \ldots$, directions. A balanced laminate can be, but is not necessarily, symmetrical. One form of balanced construction which regularly occurs in practice is the *angle-ply*. Here there are equal numbers of $+\theta$ and $-\theta$ layers. The *cross-ply* is also commonly found and here the layers are orientated at 90° with respect to one another. It is possible to have a laminate where the lamina are so arranged that, to an engineering approximation, the properties do not vary with direction. Such laminates are known as *quasi-isotropic* and are often used where the applied loadings are such that no one direction predominates. The significance of lamina type is not just limited to simple considerations of stiffness and strength but, as will be seen in subsequent chapters, symmetry (or lack of it) and lamina stacking sequence can have a profound effect on the performance of materials.

References

1 A Kelly and S T Mileiko (Eds), *Fabrication of Composites*, Elsevier, Amsterdam, 1983.

2 M O W Richardson (Ed), *Polymer Engineering Composites*, Applied Science, London, 1977.

3 G H Dietz, (Ed), *Composite Engineering Laminates*, MIT Press, Cambridge, Massachusetts, 1969.

4 Griffith A A, 'The phenomena of rupture and flow in solids' *Trans R Soc (London) ser. A*, **221**, 163, 1921.

5 B Parkyn (Ed), *Glass Reinforced Plastics*, Butterworth, London, 1970.

6 Johnson W, Phillips L N, and Watt W, British Patent 1110791, 1968.

7 L N Phillips (Ed), *Design with Advanced Composite Materials*, Springer-Verlag, London, 1989.

8 Watt W, 'Production and properties of high modulus carbon fibres', *Proc. R. Soc. (London) ser A*, **319**, 5, 1970.

9 'Maximum working temperatures for crystic polyester resins', Scott Bader Technical Leaflet No. 145, 1986.

10 Timoshenko S, *Strength of Materials*, Van Nostrand, New York, 1956.

11 Weibull W, 'Statistical distribution function of wide applicability', *J Appl Mech*, **18**, 293, 1951.

12 Stanley P, Fessler H and Sivill A D, 'An engineer's approach to the prediction of failure probability of components', *Proc Br Ceram Soc*, **22**, 453, 1973.

13 Irvine W H and Quirk A, 'Stress concentration approach to fracture mechanics', Conference on Practical Application of Fracture Mechanics to Pressure Vessel Technology, I Mech E, London, pp 76–84, 1971.

14 Davidge R W, *Mechanical Behaviour of Ceramics*, Cambridge University Press, Cambridge, 1979.

15 Davidge R W, McLaren J R and Tappin G, 'Strength probability time (SPT) relationships in ceramics', *J Mat Sci*, **8**, 1699, 1973.

2

MATERIALS AND PROPERTIES

One of the first steps in any design exercise is the consideration of candidate materials. For fibre composites this requires an assessment of each constituent phase and usually focuses on the selection of matrix and reinforcement. In addition to characteristic mechanical properties, equal attention must be given to compatibility and processability. Clearly these are interrelated as the processes for forming the material and fabricating the structure occur concurrently. Other materials may also feature at this stage depending on the application. These may include fillers, stabilizers, pigments, fire retardants, etc. In almost all cases of structural application the fibre acts as the primary load bearing constituent and the matrix serves as a medium of load transfer on to those fibres. These are the basic mechanics of a composite material. The interface between fibre and matrix is therefore vitally important as this transfer occurs through shear at this connection. A further function of the matrix is to protect the interface and the fibres from the action of any environmental effects. For many components the role of the matrix may not be entirely non-structural, however, as it cannot always be assumed that all loads will act in the direction of the fibres. In the non-fibre directions properties become strongly influenced by matrix characteristics.

During fabrication the materials can be handled in one of two ways: on-line impregnation or use of preimpregnated tows. In the former case the matrix is used in a liquid form, or converted into such by the use of a solvent or by melting, and the fibres are wetted out during component manufacture. Preimpregnated tows (prepregs) on the other hand are fibres that have been combined with matrix in a preliminary processing operation which can then be fabricated into a final component form. This has certain advantages as it eliminates much of the chemistry from the component fabrication, but can limit flexibility and is usually more expensive.

Whatever choice of matrix and reinforcement is made, the key point is that they are selected as a system. It is only by considering constituents in this way that the full design potential can be realised.

Matrices

In general terms a matrix can take the form of almost any material. However, those that have attracted most interest are those based on polymeric systems. New materials based on metals and ceramics are becoming available, but are still in their formative stages of development (see Chapter 7). Polymers used as matrices can be one of two types. The first, and most common, are of thermosetting character where solidification from the liquid phase takes place by the action of an irreversible chemical cross-linking reaction. This usually occurs as a result of the addition of other chemicals to initiate and accelerate the reaction and may involve the application of heat and pressure. The second type of polymers are thermoplastic in nature and forming can be carried out as a result of the physical processes of melting and freezing. Generally speaking, these reactions are reversible.

The type of resin matrix will govern the details of the manufacture technique employed. For example, some thermosets are sufficiently fluid to allow processing without further modification, whilst others require the application of heat or the use of diluents to lower viscosity levels. Prepreg materials, where the resin which is already incorporated and has been allowed to react to an initial stage, can be handled as a solid feedstock which is then consolidated through the action of pressure and temperature. The fabrication of thermoplastics is carried out primarily through the action of heating and cooling. Whatever the application, the selection of matrix cannot be divorced from either design or processing.

Unsaturated polyester resins

Polyesters are, certainly in tonnage terms, probably the most commonly used of polymeric resin materials. One of their major advantages is the ability for cure at room temperature. This allows large and complex structures to be fabricated where an oven cure would not be practical. Essentially they consist of a relatively low molecular weight unsaturated polyester dissolved in styrene (Table 2.1). Curing occurs by the polymerization of the styrene which forms cross-links across unsaturated sites in the polyester. A good degree of chemical resistance gives them wide application. A point to note is that the curing reaction is strongly exothermic (Fig. 2.1), and this can affect processing rates as excessive heat can be generated which can damage the final laminate.[1] Shrinkage on cure (approx 7–8%) can also be a problem.

Because of the popularity of these systems, a family of resins has been developed to offer specific properties. Amongst the most important of these are those tailored for chemical resistance. For example, alkali resistance can be enhanced through the use of the so-called bisphenol modified resins where the number of sites for alkali hydrolysis is reduced to a minimum. A further system

Table 2.1. Structure of common thermosetting resin systems

Polymer	Structure
Polyester	$\left[\begin{array}{c} R - O - \underset{\underset{O}{\\|}}{C} - CH = CH - \underset{\underset{O}{\\|}}{C} - O \end{array}\right]$
Vinyl Ester	$\left[\begin{array}{c} \overset{\overset{O}{\mid}}{-CH_2\,CH\,CH_2O} - C - \underset{\underset{O}{\\|}}{\overset{\overset{R'}{\mid}}{C}} = CH_2 \end{array}\right]$
Epoxy	$R \left[\begin{array}{c} CH_2 - CH \overset{O}{-} CH_2 \end{array}\right]$
Phenolic	$\left[\begin{array}{c} \overset{O}{} \quad CH_2 \quad \overset{OH}{} \end{array}\right]$

related to the polyesters, in that diluents such as styrene are used, is the vinyl ester family. Here the unsaturation occurs at the ends of the polymer chain, giving good chemical resistance and comparatively large strains to failure.

In all these cases additives may be used to impart certain characteristics. The most notable are those for fire and flame retardance, and ultraviolet absorbers to improve weathering resistance.

Epoxy resins

These resins are those most often used for advanced structural applications. They are generally two-part systems consisting of an epoxy resin and a hardener which is either an amine or anhydride. A wide variety of formulations is available to give a broad spectrum of properties after cure and to meet a diverse range of processing conditions. The higher performance epoxies require the application of heat during a controlled curing cycle to achieve the best properties. Prepregs can be made with epoxies where the fibres are impregnated with resin which is then partially cured. The objective of prepreg manufacture is to derive a material that can then be used in a fabrication environment, has an acceptable shelf life and good tack. This last property is important as it permits good adhesion between adjacent layers during

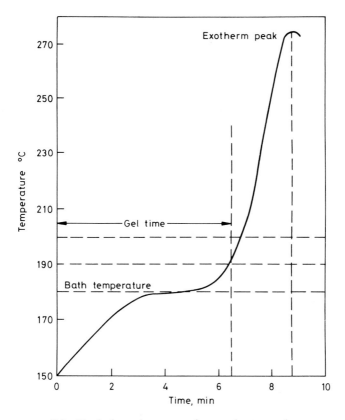

2.1 Typical exotherm cure for a polyester resin.

assembly operations. Compared with polyesters, epoxies tend to demonstrate better mechanical properties, better performance at elevated temperature (depending on cure cycle) and a lower degree of shrinkage (typically 2–3%); Table 2.2.[1]

Phenolic resins

Phenolics are of particular interest in structural applications owing to their inherent fire resistance properties. This is accomplished without the use of fillers which, although effective in inhibiting flame spread, tend to increase smoke generation. The relatively recent development of ambient temperature cure systems has provided a stimulus to their use in a range of applications. They possess two significant disadvantages, however: low toughness and a curing reaction that involves the generation of water. This latter effect can cause problems since if it remains trapped within the composite, steam can be generated during a fire which can then damage the structure of the material.

Table 2.2. Typical properties of cast resin systems

Property	Polyester	Epoxy
Specific gravity	1.1–1.5	1.2–1.3
Rockwell hardness	M70–M115	M100–M110
Impact strength (Izod) (J/m)	16–32	8–80
Thermal conductivity (W/m/°C)	–	0.17–0.21
Thermal expansivity (°C^{-1} × 10^5)	–	5–8
Specific heat (kJ/kg/°C)	–	1.25–1.80
Volume resistivity (Ω cm)	10^{15}	10^{17}
Dielectric constant (at 60 Hz)	3.0–4.4	2.5–4.5
Tensile strength (MPa)	40–90	55–130
Flexural strength (MPa)	60–160	125
Tensile modulus (GPa)	20–44	28–42

High temperature thermosetting resins

Because of the increasing level of interest for the use of composites at ever higher operating temperatures, there has been a continuing programme to develop organic matrices with good performance in this respect. There are now a number of options with reported survivable temperature capability in the region of 200 °C.[2] Examples include:

- Multifunctional epoxies.
- Polyimides.
- Bismaleimides.
- Polystyryl pyridenes.
- PMR (*in situ* polymerization with monomeric reactants).

All of these systems have glass transition temperatures in the range 180–400 °C.

Most epoxy resin systems in common usage tend to have upper working temperatures around the 100–120 °C range. Those rated for higher temperature applications are usually based on an epoxy novolac or a tetrafunctional resin with additions to control viscosity and toughness. Polyimides are a family of polymers with arguably the best upper working temperature attainable in a readily available polymer system ($\sim$ 300 °C). They are, however, expensive, somewhat temperamental to fabricate owing to the chemistry involved and, when cured, prone to microcracking. Bismaleimides are a class of thermosetting resins of somewhat complex chemistry related to polyimides.[3] Generally, these systems have a better upper working temperature than epoxies, though the cured resins are more brittle. Fabrication difficulties can also arise with the material in prepreg form as there is little tack. The polystyryl pyridine polymer was originally developed as an ablative system and hence has excellent high temperature properties with a glass transition temperature in the region of 280 °C.[4] Again, fabrication is difficult

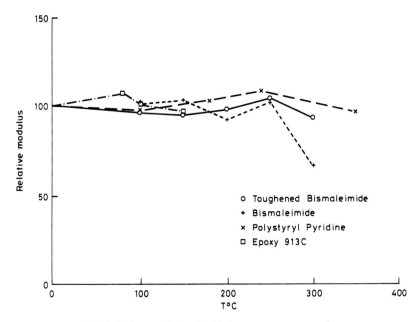

2.2 Relative modulus for high temperature matrices.

as, in this case, solvents must be used in a high temperature curing cycle. The last system cited, PMR, is particularly interesting as it involves a different approach in its formulation. As the term PMR suggests, a composite material is formed by impregnating fibrous reinforcement with monomers dissolved in a low boiling point alcohol. The monomers are unreactive at low temperatures, but react *in situ* at elevated temperatures to form a stable polyimide. The performance of some of the higher temperature matrices is summarized in Fig. 2.2 and 2.3.[2]

The processing of high temperature resin systems is not straightforward as they inevitably involve curing at high temperatures for extended periods of time. The processing cycle can be further complicated by the need to remove the solvent used to ease processing or the requirement to heat the material to reduce viscosity. Also, as a rule of thumb, increased temperature capability results in a loss of toughness which can affect performance or even give rise to cracking because of the residual stresses generated at high temperature curing.

Thermoplastic matrices

The thermoplastic resin systems are fundamentally different from the thermosets in that they do not undergo irreversible cross-linking reactions, but instead melt and flow on application of heat and pressure. Table 2.3 shows details of thermoplastics in current use (or undergoing evaluation).[5] They

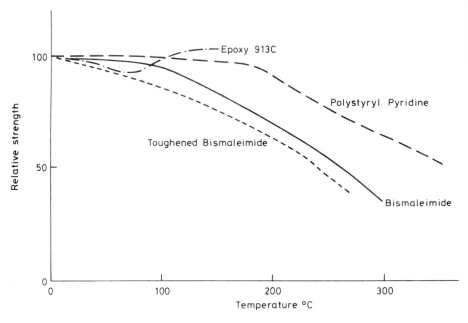

2.3   Relative strength for high temperature matrices.

range from common materials, for example nylon and polypropylene, to those that have been specially formulated with particular regard to elevated temperature performance such as those based on phenylene (PEEK, PES, PEK – poly(ether ketone)) and aromatic groups (PPS). Properties of the more commonly considered thermoplastics are given in Table 2.4.[5,6]

The main advantages of thermoplastics over thermosets are indefinite shelf life, good toughness and the fact that processing is concerned with physical transformations only. There is no chemistry involved and therefore extended cure cycles are not necessary. As a consequence there is potential for rapid, low cost fabrication with simplified quality control procedures. Post-forming operations which provide scope for added flexibility in component design are also possible. However, there are attendant difficulties as the temperatures required for processing can be high (Table 2.5), resulting in expensive tooling requirements where particularly complex shapes are required. Higher temperature performance is also seen as a potential advantage for thermoplastics. Figures 2.4 and 2.5 show the effect of temperature on mechanical property retention for a range of systems.[5]

Thermoplastic composites are almost always processed in the form of prepreg materials. Impregnation of the fibres to form the prepreg can be difficult owing to high viscosities of the melt thermoplastic or the requirement to use high boiling point, polar solvents. As an alternative post-impregnated forms are available where the basic feedstock does not contain fully wetted

Table 2.3. Thermoplastics matrix materials

Polymer	Glass Transition Temp. °C	Structure
Poly (Phenylene Sulphide)	85	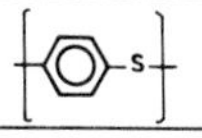
Poly (Ether Ether Ketone)	143	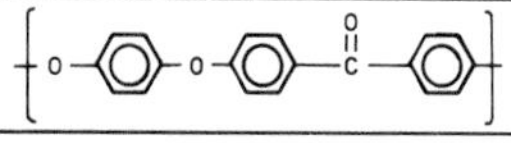
Poly (Ether Ketone)	165	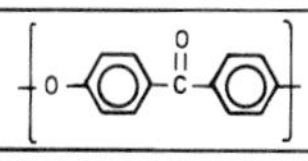
Poly (Sulphone)	190	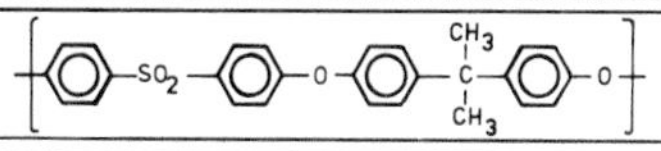
Poly (Ether Imide)	216	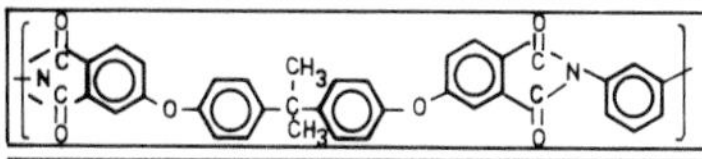
Poly (Phenyl Sulphone)	220	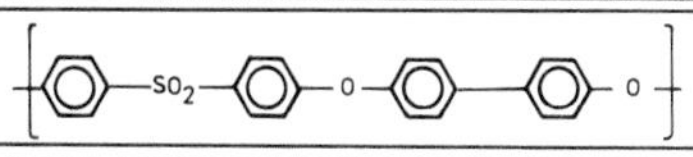
Poly (Ether Sulphone)	230	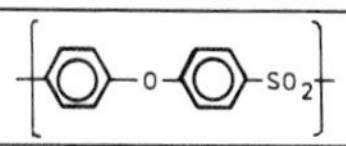
Poly (Amide Imide)	249–288	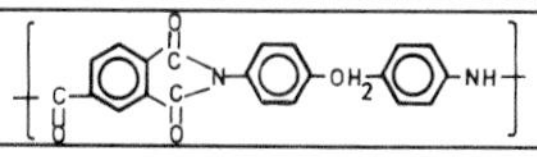
Poly (Imide)	256	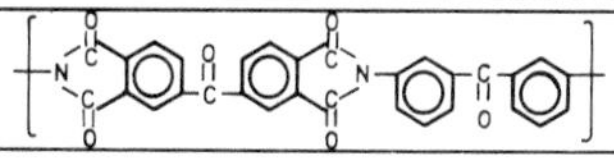

fibres, but a system which is only physically mixed (Fig. 2.6). Full impregnation is achieved on subsequent processing.[5]

Thermoplastics continue to be the subject of much development activity, particularly in the areas of fabrication science and property characterization, since there is some considerable progress required before the potential indicated by basic polymer properties is achieved in a composite. Should these developments be successful, a significant increase in the use of thermoplastics for a variety of applications can be expected.

Fibres

Fibres are the dominant constituent of most composite systems, and one of the main objectives of any design should be to place the fibres in positions and orientations so that they are able to contribute efficiently to load-carrying capability. The most widely available fibre form for advanced structural applications is continuous tows. These produce highly anisotropic materials of very high stiffness and strength in the direction of the reinforcement. Away

Table 2.4. Properties of thermoplastic matrices

	Specific gravity	Modulus (GPa)	Tensile strength (MPa)	Elongation (%)	Fracture toughness	Izod
Polypropylene (PP)	0.90	1.1–1.6	29–37	200–700	–	–
Nylon	1.15	1.2–2.9	61–82	60–300	–	–
Polyphenylene sulphide (PPYS)	1.29	2.1	72	60	3.5	–
PPS	1.36	3.4	79	2–20	–	21
PEEK	1.27–1.32	3.7	92	50	4.8	83
PEK	1.3	4.0	105	5	–	–
Polysulphone (PS)	1.27	3.6	70	50–100	2.5	69
Polyether imide (PEI)	1.29	3	103	30	–	40
Polyether sulphone (PES)	1.37	2.6	84	40–80	3	84
Polyamide imide (PAI)	1.3–1.4	3.8–4.8	152–193	10–15	3.2–3.9	80
Polyimide (PI)	1.3–1.4	3.5	101–166	9–14	–	21

Table 2.5. Thermal properties of thermoplastic matrices

	Glass transition temperature $T_g(^\circ C)$	Melting temperature $T_m(^\circ C)$	Processing temperature $T_p(^\circ C)$
PP	−27	200–280	200–280
Nylon	80–100	180–270	220–310
PPYS	220	*	330
PPS	85	285	320–340
PEEK	143	334	380
PEK	162–167	360–365	400
PS	190	*	300
PEI	216	*	360
PES	230	*	350
PAI	249–288	*	400
PI	250–261	270–340	34–350

* Amorphous (therefore there is no apparent T_m)

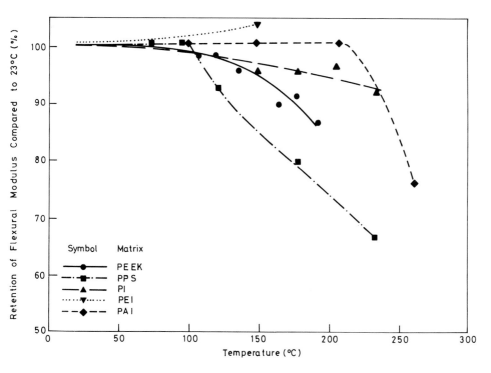

2.4   Flexural modulus of thermoplastic composites at
elevated temperatures.

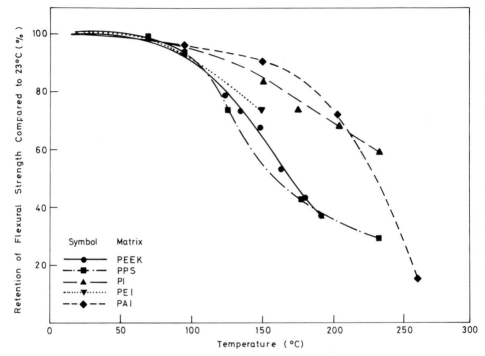

2.5   Flexural strengths of thermoplastic composites at
elevated temperatures.

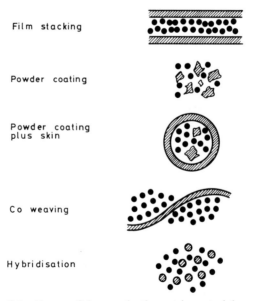

Film stacking

Powder coating

Powder coating
plus skin

Co weaving

Hybridisation

2.6   Forms of thermoplastic matrix material.

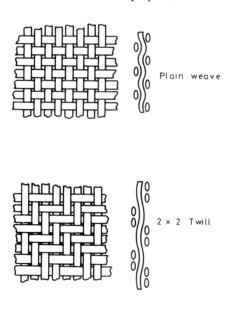

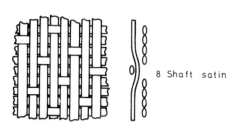

2.7   Forms of woven cloth.

from this direction, properties tend to fall away rapidly until at the orientation perpendicular to the fibres they become similar to those of the matrix. It is common for individual plies of unidirectional material to be combined to form a more complex construction. Such a laminate may contain many individual layers, each at different orientations with respect to one another, the sequence of plies being determined by design considerations.

Fibre systems are also available in a number of other forms. Most common are woven cloths. These may have equal numbers of fibres in warp and weft directions and therefore equal properties in those directions, or have a degree of bias giving rise to a level of anisotropy. The type of weave can also be important as this determines aspects of processability, such as cloth drape over a curved surface (Fig. 2.7).[4] Both cloths and tows can be preimpregnated, processed and then cured as a prepreg.

Table 2.6. Typical glass fibre properties

Glass type	SG	Thermal expansivity ($°C^{-1}$)	Tensile modulus (GPa)	Strength (GPa)	
				Undamaged	Strand from roving
A-glass	2.46	7.8×10^{-6}	72	3.5	–
E-glass	2.54	4.9×10^{-6}	72	3.6	1.7–2.7
AR-glass	2.7	7.5×10^{-6}	70–75	3.6	1.5–1.9
S/R-glass	2.5	–	85	4.5	2.0–3.0

Discontinuous fibres can also be usefully employed to provide reinforcement to a matrix. Typical of these is 'chopped strand mat' which consists of chopped fibres about 30–50 mm long distributed in a random manner in a plane and held together with a resin binder. Such an arrangement provides relatively low enhancement of stiffness and strength, but the resulting laminate is effectively isotropic in the plane of the reinforcement.

When considering fibre properties it is important to recognize that these may not be fully achieved within a composite. Fibre damage, misalignment and variations in volume fractions can all be deleterious to property values. Strength levels in particular can be dramatically reduced. In design, composite properties as opposed to fibre properties should be used wherever possible.

Glass fibres

Glass fibres account for around 90% of the reinforcement used in structural reinforced plastic applications. There are a number of different compositions all based on silicate glasses. The mechanical properties are not strongly dependent on composition, but chemical behaviour which is reflected in terms of durability and strength retention in corrosive environments is influenced by the details of the chemistry. Most continuous fibre for reinforcement purposes is manufactured from E-glass which is of low alkali content. Table 2.6 gives typical property values for different glass types.[7] In addition to the types shown, a variant of E-glass – E-CR glass – is becoming increasingly important. This has been specially developed to meet the requirements of acid corrosion.[8] S-glass is noted for its high strength and is used in preference to E-glass where its added cost is deemed acceptable. Typical properties for glass fibre composites are given in Tables 2.7 and 2.8.[5,9]

An essential feature of glass fibre production is the application of size which coats the surface of the fibre. The primary purpose of the size is to protect the fibre from abrasive damage and to enhance the strength of the fibre matrix interfacial bond. The nature of the size can be tailored to ensure compatibility with the proposed matrix.

Table 2.7. Mechanical properties of glass fibre/thermoplastic matrix composites

	Fibre volume (%)	Tensile modulus (GPa)	Tensile strength (MPa)	Compressive modulus (GPa)	Compressive strength (MPa)	Flexural modulus (GPa)	Flexural strength (MPa)	Short beam shear strength (MPa)	Interlaminar fracture toughness (kJ/m^2)
Unidirectional E-glass									
PEEK	62	55	1126	–	1044	53	1213	95	3.3
PPS	57	52	835	–	800	42	1145	43	1.3
Woven E-glass									
PPS	60	24	248	–	297	21	366	–	–
PEI	53	21	338	24	572	21	414	70	–

Table 2.8. Typical mechanical properties of composites

	Fibre volume (%)	Tensile strength 0° (MPa)	Tensile strength 90° (MPa)	Compressive strength 0° (MPa)	Compressive strength 90° (Mpa)	Shear strength (MPa)
UD E-glass/epoxy	53	1190	73	1001	159	67
UD carbon/epoxy	57	2040	90	1000	148	49
UD aramid/epoxy	60	1379	30	276	138	60
0/90 woven E-glass/epoxy	33	360	360	240	205	98
±45 woven E-glass/epoxy	33	185	185	122	122	137
0/90 woven carbon/epoxy	50	625	625	500	500	130
±45 woven carbon/epoxy	50	240	240	200	200	–
0/90 woven aramid/epoxy	50	517	517	172	172	110
CSM E-glass/polyester	19	108	108	148	148	85

	Modulus 0° (GPa)	Modulus 90° (GPa)	Shear modulus (GPa)	ILSS (MPa)	Poisson's ratio	Density, g/cm^3
UD E-glass/epoxy	39	15	4	90	–	1.92
UD carbon/epoxy	134	11	5	94	0.26	1.57
UD aramid/epoxy	76	5	2	83	0.34	1.38
0/90 woven E-glass/epoxy	17	17	5	60	0.24	1.92
±45 woven E-glass/epoxy	10	10	8	48	0.7	1.92
0/90 woven carbon/epoxy	70	70	5	57	–	1.53
±45 woven carbon/epoxy	18	18	27	57	–	1.53
0/90 woven aramid/epoxy	31	31	2	70	–	1.33
CSM E-glass/polyester	8	8	2.75	–	0.32	1.45

Note: UD unidirectional, CSM chopped strand mat, ILSS interlaminar shear strength.

Macroscopically, glass fibre behaviour at ambient temperature can be described with a linear stress/strain curve to failure. Under continuous loading a comparatively small (< 5% of elastic strain) viscoelastic deformation can also be measured. At higher temperatures creep effects can become more prominent. As would be expected from a brittle material the strength values are sensitive to flaws and stress concentrations and are subject to a significant size effect. This is the reason why fibres taken from rovings which have been mechanically handled have strengths much less than 'undamaged' filaments. Strength effects can be treated using statistical concepts. Considering equation 1.1 given in the previous chapter, the form of the expression can be simplified to:

$$P_f = 1 - \exp\left(-\lambda\sigma^m\right) \tag{2.1}$$

where λ is a scale factor dependent on volume which in the case of fibres is governed by length and m is again the Weibull modulus.

As can be seen the probability of failure, P_f, is strongly dependent on fibre length, i.e. the longer the fibre, the greater the chance of a critical flaw being present and hence the greater the likelihood of failure. At the mean failure stress, $\bar{\sigma}_f, P_f = 0.5$. Therefore:

$$\bar{\sigma}_f = \left[\frac{\log_e 2}{\lambda}\right]^{1/m} \tag{2.2}$$

A tow of nominally identical fibres will most likely fail below the mean failure stress of individual fibres because the load capacity of the bundle gradually deteriorates as a result of failure of the weaker fibres as the load increases. Catastrophic failure will occur when the load is reached where the breakage of just one more fibre reduces the effective cross-sectional area below that which can just support the current level of load. This is then the tow strength, σ_t, and can be shown to be:

$$\sigma_t = (m\lambda)^{1/m}\exp\left(-1/m\right) \tag{2.3}$$

The ratio of tow strength to single filament mean strength is therefore:

$$\frac{\sigma_t}{\bar{\sigma}_f} = \frac{m\lambda^{-1/m}\exp(-1/m)}{(\log_e 2/\lambda)^{1/m}} \tag{2.4}$$

Figure 2.8 shows a plot of equation 2.4.[10] As can be seen, the strength of a fibre bundle approaches the mean strength of individual fibres with increasing Weibull modulus.

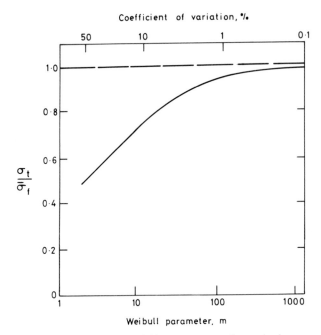

2.8   Dependence of fibre tow strength on single
filament strength values.

Carbon fibres

Carbon fibres are typified by a combination of low density, high strength and high stiffness. The characteristics of an individual fibre are determined by the degree of orientation of graphite planes achieved during manufacture. Currently there is a great range of materials available each with a different combination of properties (Fig. 2.9).[11] Current developments are focusing on the requirement for increasing strength coupled with a capability for high strain to failure. As with glass, carbon fibres are susceptible to damage and require a size for protection and surface treatment to promote matrix adhesion. Properties for a range of composites, both carbon and glass reinforced, are given in Table 2.8.[5]

One aspect of the behaviour of carbon fibres which is rarely reported is the nonlinear form of the stress/strain curve. All fibre types show a marked rise in stiffness with increasing strain. Figure 2.10 shows typical data. In some cases the difference between initial and final modulus may be as much as 25%. The shape of the stress/strain curve can be described by:

$$E_c = E_0(1 + f\varepsilon) \qquad\qquad [2.5]$$

where E_0 is the modulus at zero strain E_c that at strain ε, and f is an empirical

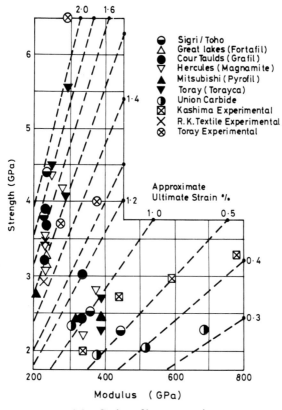

2.9   Carbon fibre properties.

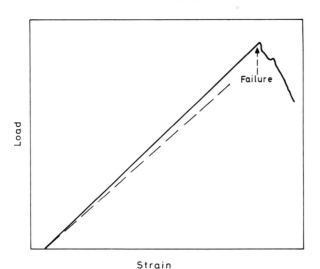

2.10   Typical nonlinear stress/strain curve for
carbon fibre.

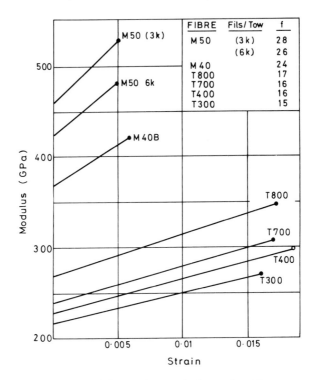

2.11   Modulus against strain for carbon fibres
(Toryaca fibres).

factor dependent on fibre type. Figure 2.11 shows the variation in modulus
with strain for a variety of carbon types. In general, the value of the parameter f
increases with fibre modulus.[11] Perhaps the more important issue arising from
this effect is not so much the use of equation 2.5 in a design – often design
strains are low in any case – but in the interpretation of values quoted in the
literature. Values based on, say, a secant modulus may be higher than those
that should be used.

As already indicated current developments in carbon fibre technology are
providing ever-higher values of stiffness and strength and, by and large, this
can be replicated within a unidirectional lamina. A key point to note, however,
is that as properties in the fibre direction improve, those transverse to the fibre
tend to deteriorate. As most applications involve transverse loading to some
extent, it may be difficult to utilize fibre properties fully as failure may occur in
other directions due to secondary loading or even residual stresses generated
in the fabrication process.

Table 2.9. Typical properties for aramid, boron and UHMPE fibres

Fibre	SG	Tensile strength (GPa)	Young's modulus (GPa)	Elongation at break (%)
Aramid (low modulus)	1.44	2.65	59	4
Aramid (high modulus)	1.45	2.65	127	2.4
Boron	2.63	2.76–3.45	380–400	0
UHMPE	0.96	1.5	70	4–18*

* Dependent on strain rate.

Aramid fibres

Aramid fibres are available in two forms: low and high modulus. The difference between the two results from basic structural and orientation variations at the polymer chain level. Low modulus can be converted to high modulus material through an annealing process. The main advantage of aramids is their very low density (lower than glass or carbon), giving high values of specific strength and stiffness combined with excellent toughness. As a result of this latter characteristic the material is frequently used in applications where impact resistance and ballistic energy absorption are important. However, there are some difficulties with the material. Because of the details of the polymer morphology the fibre has low longitudinal shear modulus, poor transverse properties and low axial compression strength. These effects are reflected in resulting composite properties where, in bending and compression the response is nonlinear with relatively low ultimate strengths. As a result they are frequently used in a hybrid construction with other fibres such as carbon with the aramid providing toughness characteristics to the laminate. Some example property data for aramid composites and aramid fibres are given in Tables 2.8 and 2.9.

Other fibre systems

The vast majority of applications using fibre reinforced polymers employ either glass, carbon or aramid fibres. However, two other fibres are of interest:

- **Boron.** Boron fibres were amongst the first fibres specially developed for advanced composites.[1] They have a density similar to glass, but a tensile modulus up to six times greater. The production route is rather complex, involving chemical vapour deposition (CVD) onto a tungsten wire substrate and the high cost of this process means that the fibres are relatively expensive. Because of their large size and stiffness boron filaments cannot be woven into cloths or handled like other fibres, so they are usually processed in parallel arrays of single thickness preimpregnated sheets or tapes.

- **Polyethylene fibres**. Fibres produced from ultra high molecular weight polyethylene (UHMWPE) are now available and have similar tensile behaviour to that of aramids.[12] A very low density ($\sim$ 30% lower than aramids) means that specific properties are considerably higher. Polyethylene has good chemical, abrasion resistance and low moisture absorption, but like aramids its composites display poor compression and shear behaviour. A further disadvantage is their low melting point ($\sim$ 150 °C) which limits applications to 130 °C or less.

Table 2.9 shows typical property values for aramid, boron and UHMWPE fibres.[1,9,12]

Property measurement

Owing to the complexity of the structure of composites and the unusual effects that can arise as a consequence, it is essential that properly formulated test methods are employed to measure property data. There is enormous potential for generating poor quality information which is misrepresentative of actual material behaviour. Many of the documented procedures available for composites are derived from those developed from monolithic polymers. Examples include:

- ASTM D3410 – compression.
- ASTM D3039; BS 2782 Pt 10 Method 1003 – tension.
- BS 2782 Pt 10 Method 1005 – flexure.
- BS 2782 Pt 3 Method 341; ASTM D2344 – interlaminar shear strength.
- ASTM D3518 – in plane shear.
- ASTM D3479 – tensile fatigue.

In tensile tests it is common to use tabs bonded to the ends of specimens of either aluminium or GRP ($\pm$ 45°), the latter being of particular usefulness for testing at elevated temperature to minimize thermal expansion mismatch. Wasting of specimens may be found to be necessary where strength values are to be obtained. Bending tests are also commonly used normally in the form of three-point bending on parallel-sided specimens. Figure 2.12 shows details of a typical tensile test specimen.

There is a variety of methods for measurement of in-plane shear properties. One test consists of tensile loading of a $\pm$ 45° laminate; the shear modulus being determined by a subsequent analysis procedure. Perhaps the most difficult property values to determine are those in compression. The low modulus of many composite systems means there can be an overriding tendency for specimens to buckle under load. This has dictated the development of a range of devices to support material samples whilst under load (Fig. 2.13).

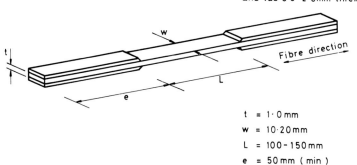

End tab 0·5 - 2·0mm thick

Fibre direction

t = 1·0 mm
w = 10·20 mm
L = 100 - 150 mm
e = 50 mm (min)

2.12   Test specimen for unidirectional tensile strength
and modulus.

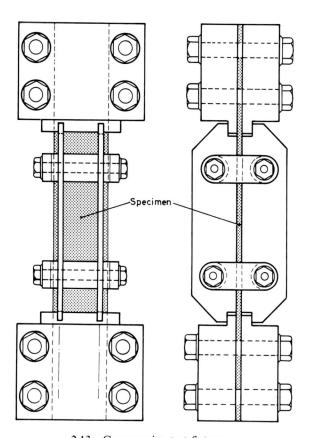

Specimen

2.13   Compressive test fixture.

Accurate determination of fibre volume is vitally important in the characterization of composites and there are two methods available:

- ASTM D3171 – acid digestion.
- ISO 1172 – combustion.

It should be noted that the latter technique cannot be used for carbon- or aramid-based fibre materials owing to the possible destruction of the fibre phase during test.

The development of test methods is continuing and there are a number of initiatives underway to establish a consistent set of procedures which can be applied to a range of materials.[13] Typical of these are the CRAG procedures (Composites Research Advisory Group – MOD, UK) and these are awaiting the results of a validation programme before formal adoption by the BSI.[14] A list of CRAG standard property test methods is given in Table 2.10.

As examples of the results which can be obtained from a typical test programme, Table 2.11 shows elastic and strength data for carbon and glass reinforced materials.[15] Samples were prepared from E-glass rovings and a high strength carbon fibre. The resin used throughout was a standard di-functional liquid epoxide filament winding system. Fibre volume fractions were measured using gravimetric and acid digestion methods. Typically, fibre volume fractions were in the region of 60%. Tensile coupons were cut from plates using a water-cooled diamond wheel and care was exercised over minimizing edge damage. Woven GRP end tabs were bonded to the coupons using a room temperature curing epoxide paste adhesive. Specimen dimensions conformed closely with BS 2782: Part 3: Method 320E: 1976 and recommended CRAG procedures were followed.

All tests were carried out under displacement control on a universal testing machine with a calibrated 25 mm $\pm$ 2.5 mm extensometer attached to each specimen. In addition a 0/90 strain gauge rosette was used to collect axial and transverse strains and therefore Poisson's ratio as a function of applied load.

Figures 2.14–2.19 show derived stress/strain curves. The unidirectional T300 samples exhibited a gradual increase in modulus with strain, and comparing this with equation 2.5 yields a value of $f = 15$. This is of similar magnitude to data obtained from single fibre properties (Fig. 2.11). For $[0,90]_s$ cross-ply laminates, transverse ply cracking was observed at approximately 0.5%. Failure occurred with little evidence of longitudinal splitting or delamination. The $[0,90]_s$ cross-ply GRP failure process was easier to observe than that for the carbon and began with transverse ply cracking at approximately 0.5% strain and this continued until a uniform crack density was established. A noticeable decrease in stiffness was observed after transverse cracking (first ply failure), unlike the carbon fibre reinforced plastic (CFRP) sample. A series of longitudinal splits was also observed to grow from the region of the end tabs until they covered the gauge length. These splits

Table 2.10. CRAG test methods

Shear test methods	
Method 100	Interlaminar shear strength
Method 101	In-plane shear strength and modulus
Method 102	Lap shear strength
Flexural test methods	
Method 200	Flexural strength and modulus
Tensile test methods	
Method 300	Longitudinal tensile strength and modulus of unidirectional composites
Method 301	Transverse tensile strength and modulus of unidirectional composites
Method 302	Tensile strength and modulus of multidirectional composites
Method 303	Notched tensile strength of multidirectional composites
Compression test methods	
Method 400	Longitudinal compression strength and modulus of unidirectional composites
Method 401	Longitudinal compression strength and modulus of multidirectional composites
Method 402	Notched compression strength of multidirectional fibre composites
Method 403	Residual compression strength after impact of multidirectional composites
Methods of test for fatigue properties	
Method 500	Test specimens for the measurement of fatigue properties
Methods of test for toughness	
Method 600	Interlaminar fracture toughness
Methods of test for bearing properties	
Method 700	Bearing properties of multidirectional composites
Physical test methods	
Method 800	Density
Method 801	Coefficient of linear thermal expansion
Method 802	Outgassing properties
Environmental effects	
Method 900	Background information on environmental effects
Method 901	Diffusivity properties
Method 902	Conditioning under hot/wet environments
Miscellaneous tests	
Method 1000	Fibre volume fraction
Method 1001	Void volume fraction by ultrasonic scanning

appeared to be less densely spaced than the transverse cracks (approximately 2–5 mm for the axial splits compared to 1–2 mm for the transverse cracks). Failure appeared to be precipitated from edge delamination growth, usually in the region of the end tabs, which grew rapidly. The ensuing fracture resulted in the characteristic 'brush-like' appearance, typical for this type of material.

The $[\pm 75]_s$ angle-ply CFRP failed at a very low strain compared with the

Table 2.11. Compilation of mechanical property data

Lay-up	Fibre	Modulus (MPa)		Strength (MPa)*		Ult. strain (%)		Volume fraction (%)		Thermal expansion ($\times 10^{-6}$ m/m/°C)	
		Mean	Standard deviation	Mean	Standard deviation	Mean	Standard deviation	Archem	Chem. Dig.	45–110 °C	45–145 °C
0°	T300	131 120	3711	1632	92	1.19	0.07	60.91	58.78	0.00	0.00
0°	E-glass	46 584	629	663	129	1.43	0.31	56.65	48.37	5–7	5–7
90°	T300	7935	394	47	5	0.59	0.06	60.91	58.78	38–39	48–54
90°	E-glass	10 360	499	47	5	0.44	0.06	56.65	48.37	31–32	42–46
[0, 90]$_s$	T300	68 642	2595	924	63	1.28	0.08	60.00	57.05	–	–
[0, 90]$_s$	E-glass	30 028	1370	598	47	2.46	0.32	63.16	53.40	–	–
[±15]$_s$	T300	78 098	8017	–	–	–	–	59.20	49.01	–	–
[±15]$_s$	E-glass	39 605	2999	–	–	–	–	61.60	53.33	–	–
[±75]$_s$	T300	8956	456	36	6	0.41	0.06	59.20	49.01	–	–
[±75]$_s$	E-glass	13 040	546	47	4	0.40	0.04	61.60	53.33	–	–

* Strength values of unidirectional GRP low; specimens underwent longitudinal splitting before fibre failure.

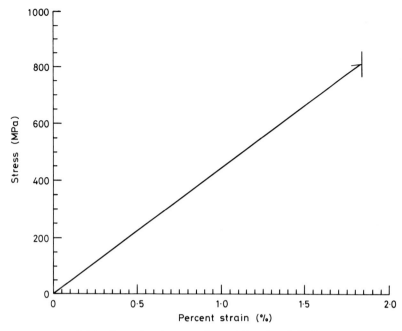

2.14   Stress/strain curve for unidirectional GRP.

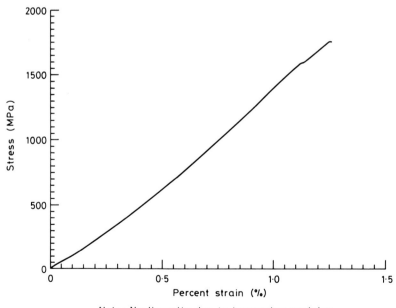

Note: Nonlinearity due to increasing modulus

2.15   Stress/strain curve for unidirectional CFRP.

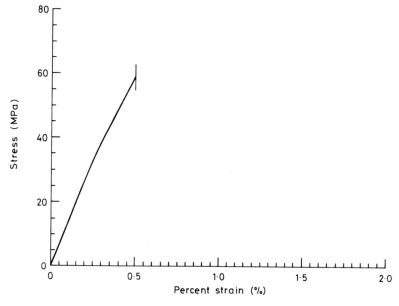

2.16   Stress/strain curve for transverse GRP.

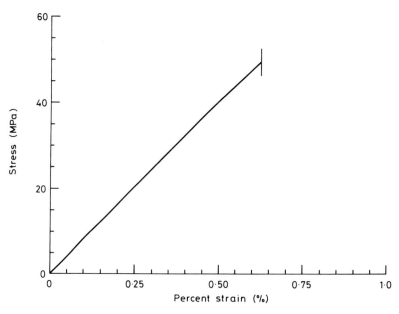

2.17   Stress/strain curve for transverse CFRP.

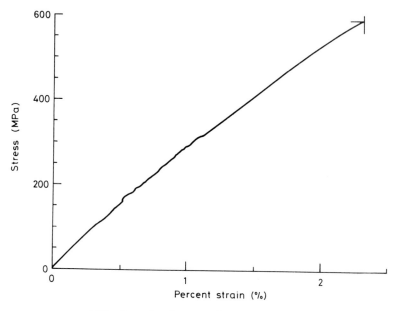

2.18 Stress/strain curve for cross-ply GRP.

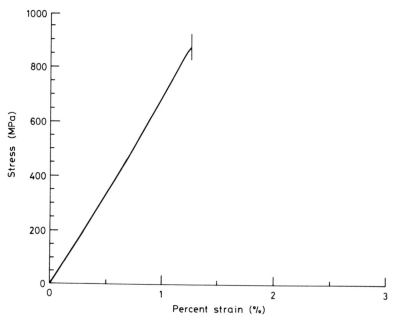

2.19 Stress/strain curve for cross-ply CFRP.

unidirectional laminate. Typically the failure strain was less than 0.5% with a certain degree of non-linearity. Stress/strain graphs indicated a slowly decreasing modulus. Failure occurred with little evidence of longitudinal splitting or delamination. The [± 75]$_s$ angle-ply GRP failed at a very similar strain, although at slightly increased stress. Failure strain was again less than 0.5% with a slight degree of nonlinearity.

References

 1 Lubin G, *Handbook of Composites*, Van Nostrand, New York, 1982.
 2 Hancox N L, 'Elevated temperature polymer composites', *Materials and design*, **12**, 317, 1991.
 3 McLaren J R and Hancox N L, 'The production and properties of bismaleimide composites', Institute of Physics Conference, Ser. No. 89: Session 1, Warwick, 1987.
 4 Hancox N L, 'Carbon fibre reinforced bismaleimide and polystyryl' *Plastics Rubber Proc. Appl.*, **10**, 131, 1988.
 5 Partridge I K, *Advanced Composites*, Elsevier, London, 1989.
 6 Billmeyer F W, *Textbook of Polymer Science*, Wiley, New York, 1971.
 7 Kelly A, *Concise Encyclopedia of Composite Materials*, Springer-Verlag, Oxford, 1989.
 8 Hogg P J and Arslanian Y, *Future Trends in Non Metallic Process Equipment in Chemical and Process Industries*, I. Mech. E., Manchester, 1989.
 9 Phillips L N, *Design with Advanced Composite Materials*, Springer-Verlag, London, 1989.
10 Harris B, *Engineering Composite Materials*, Institute of Metals, London, 1986.
11 Hughes J D H, 'Strength and modulus of current carbon fibres', *Carbon*, **25**, 551, 1986.
12 Ward I M, *Fibre Reinforced Composites*, I. Mech E., Liverpool, 1986.
13 Sims G D, 'Development of standards for advanced polymer matrix composites', *Composites*, **22**, 267, 1991.
14 Curtis P T, 'CRAG test methods for measurement of engineering properties of fibre reinforced plastics', RAe Tech. report TR 88012, 3rd edn, 1988.
15 Lee R J, AEA Technology, private communication, 1992.

3

MECHANICS OF MATERIAL BEHAVIOUR

The properties of composites are strongly dependent on many parameters such as fibre volume fraction, fibre length, packing arrangement and fibre orientation. For high performance applications the reinforcement consists of continuous aligned or woven fibres in discrete layers, with typical fibre volume fractions being in the range 50–70%. To design effectively with these systems it is essential that the material can be characterized in terms of its thermal and mechanical properties. In structural composites the reinforcement of high stiffness and strength provides the load-bearing constituent, whilst the matrix is generally of low modulus and modest strength. Its contribution, however, is significant in that it acts as the medium of load transfer into the fibres as well as offering a degree of protection. In deriving expressions to predict mechanical behaviour, the contribution of each constituent needs to be taken into account.

Stiffness properties

The analysis of the mechanics of composite material response can take place on a number of levels. On a micromechanical basis assumptions can be made regarding the nature of the interaction between constituents and expressions derived to relate the behaviour of fibre and matrix directly to that of the composite. With a larger scale an assessment based on macroscopic homogeneity can be made to relate properties of a structural form to be compiled from those of individual lamina layers. Each level of analysis has its own strengths and weaknesses, the relative magnitude of which depends largely on the extent and quality of property data available for the design exercise of concern.

Micromechanics

For a material such as a unidirectional laminate the conditions of transverse anisotropy apply. The stress/strain relationship can be written as:

$$\varepsilon_i = S_{ij}\sigma_j \qquad [3.1]$$

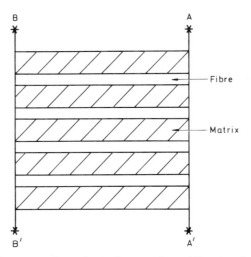

3.1   Representative volume element of a unidirectional
composite.

where ε_i and σ_j are components of strain and stress and the stiffness matrix S_{ij} is given by:

$$S_{ij} = \begin{bmatrix} 1/E_1 & -v_{12}/E_1 & 0 \\ -v_{12}/E_1 & 1/E_2 & 0 \\ 0 & 0 & 1/G_{12} \end{bmatrix}$$

To describe material behaviour it is therefore necessary to derive expressions for the four independent constants in the stress/strain equation: E_1, the modulus in the fibre direction; E_2, the modulus in the transverse direction; v_{12}, the value of Poisson's ratio; and G_{12} the in-plane shear modulus.

The simplest approach is that known as the 'rule of mixtures'. Figure 3.1 shows a representative volume element of a unidirectional composite. Applying a uniform displacement on boundary AA′ and fixing the boundary at BB′ yields an expression for the modulus in the fibre direction:

$$E_1 = E_f v_f + E_m(1 - v_f) \tag{3.2}$$

where E_f and E_m are the moduli of fibre and matrix respectively, and v_f is the fibre volume fraction. The physical assumption that forms the basis of the derivation of this expression is that of compatibility of strain between fibre and matrix. This results in the load share between constituent phases being proportional to the modulus and volume fraction of each component.

Using a similar approach, except assuming continuity of stress between fibre and matrix (Fig. 3.1), expressions can also be derived for the transverse

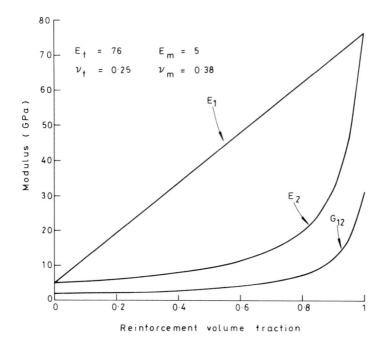

3.2   Variation of modulus values with volume
fraction.

and shear modulus values:

$$E_2 = \frac{E_f E_m}{E_m v_f + E_f (1 - v_f)} \qquad [3.3]$$

$$G_{12} = \frac{G_f G_m}{G_m v_f + G_f (1 - v_f)} \qquad [3.4]$$

where G_f and G_m are the shear moduli of fibre and matrix.

In terms of Fig. 3.1, the boundary conditions appropriate to the two cases
are:

- Transverse modulus:
    boundaries AB′ and AB′: uniform stress;
    boundaries AA′ and BB′: stress free.
- Shear modulus:
    all boundaries: uniform shear stress.

In Fig. 3.2 the variation of moduli as a function of fibre volume fraction is
shown. As can be seen, unlike the longitudinal modulus where values increase
in proportion to volume fraction, there is only modest improvement of
transverse and shear moduli with increasing fibre fraction. The data used for

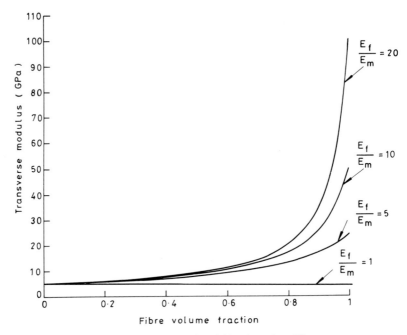

3.3   Effect of the ratio of fibre/matrix stiffness on
composite transverse modulus.

these calculations are typical values for GRP. The domination of the response
of the matrix for these modes of deformation is clear. Figure 3.3 demonstrates
the same issue in a different way. Here the transverse modulus is plotted with
respect to volume fraction and the ratio of fibre and matrix stiffness. Over the
range of typical volume fractions (50–70%) the calculated property values for
the composite are comparatively insensitive to increasing fibre stiffness.

The major Poisson's ratio, v_{12}, can also be determined using a similar
approach:

$$v_{12} = v_f v_f + v_m(1 - v_f) \qquad [3.5]$$

where v_f and v_m are those values of fibre and matrix respectively.

To obtain the minor Poisson's ratio, v_{21}, it can be demonstrated that, as the
stiffness matrix is symmetric, the following elastic identity may be used:

$$v_{21} = \frac{E_2 v_{12}}{E_1} \qquad [3.6]$$

Clearly, these expressions can only be regarded as being approximate since the
physical model that they describe is a highly idealized representation. For
example, in the derivation of the equation for transverse modulus, the
condition that the boundaries AA′ and BB′ are stress-free is not accurate as the

modulus and Poisson's ratio of the matrix and fibre are markedly different. Accommodating this into the derivation yields:[1]

$$\frac{1}{E_2} = \frac{v_f}{E_f} + \frac{(1 - v_f)}{E_m} - v_f(1 - v_f) \cdot \frac{v_f^2 E_m/E_f + v_m^2 E_f/E_m - 2v_f v_m}{E_f v_f + E_m(1 - v_f)} \quad [3.7]$$

There has been much work, both of continuum mechanics and numerical natures, to develop and refine models which represent composite behaviour more accurately. Early work in this area focused on the derivation of upper and lower bounds for materials of arbitrary internal geometry.[2-5] The resulting expressions for axial, shear and bulk moduli are as follows:

- Longitudinal modulus:
  (a) upper bound

$$E_1 = v_f E_f + (1 - v_f)E_m + \frac{4v_f(1 - v_f)(v_f - v_m)^2}{v_f/k_m + (1 - v_f)/k_f + 1/G_f} \quad [3.8]$$

  (b) lower bound

$$E_1 = v_f E_f + (1 - v_f)E_m + \frac{4v_f(1 - v_f)(v_f - v_m)^2}{v_f/k_m + (1 - v_f)/k_f + 1/G_m}$$

- Poisson's ratio:
  (a) upper bound

$$v_{12} = v_f v_f + (1 - v_f)v_m + \frac{v_f(1 - v_f)(v_f - v_m)(1/k_m - 1/k_f)}{v_f/k_m + (1 - v_f)/k_f + 1/G_f} \quad [3.9]$$

  (b) lower bound

$$v_{12} = v_f v_f + (1 - v_f)v_m + \frac{v_f(1 - v_f)(v_f - v_m)(1/k_m - 1/k_f)}{v_f/k_m + (1 - v_f)/k_f + 1/G_m}$$

- Shear modulus:
  (a) upper bound

$$G_{12} = G_f + \frac{1 - v_f}{1/(G_m - G_f) + v_f/2G_f}$$

$$G_{21} = G_f + \frac{1 - v_f}{1/(G_m - G_f) + v_f(k_f + 2G_f)/2G_f(k_f + G_f)}$$

  (b) lower bound

$$G_{12} = G_m + \frac{v_f}{1/(G_f - G_m) + v_f/2G_m} \quad [3.10]$$

$$G_{21} = G_m + \frac{v_f}{1/(G_f - F_m) + (1 - v_f)(k_m + 2G_m)/2G_m(k_m + G_m)}$$

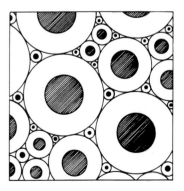

3.4   Composite cylinder assemblage model.

- Bulk modulus:
  (a) upper bound

$$k = k_f + \frac{1 - v_f}{1/(k_m - k_f) + v_f/(k_f + G_f)} \qquad [3.11]$$

  (b) lower bound

$$k = k_m + \frac{v_f}{1/(k_f - k_m) + (1 - v_f)/(k_m + G_m)}$$

To improve these bounds more information on an internal geometry must be specified. One such model is that based on an assemblage of composite cylinders.[6] This consists of an infinite number of concentric cylinders of different diameter packed to fill the total volume (Fig. 3.4). The inner cylinders have elastic properties of the fibre, whilst the outer cylinders represent the matrix. The cylinder diameters are arranged so that the fibre volume fraction of each is equal to the overall average for the composite. The analysis results in closed-form solutions for longitudinal, shear and bulk moduli and Poisson's ratio:

$$E_1 = v_f E_f + (1 - v_f)E_m + \frac{4v_f(1 - v_f)(v_f - v_m)}{v_f/k_m + (1 - v_f)/k_f + 1/G_m}$$

$$v_{12} = v_f v_f + (1 - v_f)v_m + \frac{v_f(1 - v_f)(v_f - v_m)(1/k_m - 1/k_f)}{v_f/k_m + (1 - v_f)/k_f + 1/G_m}$$

$$[3.12]$$

$$G_{12} = G_m \left[\frac{G_m(1 - v_f) + G_f(1 + v_f)}{G_m(1 + v_f) + G_f(1 - v_f)} \right]$$

$$k = \frac{k_m(k_f + G_m)(1 - v_f) + v_f k_f(k_m + G_m)}{(1 - v_f)(k_f + G_m) + v_f(k_m + G_m)}$$

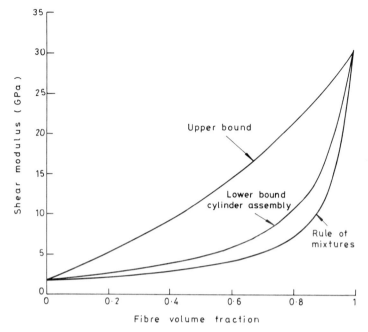

3.5   Calculation of shear modulus by different
micromechanical models.

Figures 3.5 and 3.6 show the results of calculation of elastic properties on each
of the different models.

Another approach that has been considered is the case where the fibres are
modelled as parallel, cylindrical inclusions with varying degrees of contact, as
expressed by a contiguity factor C.[7] When all fibres are isolated within the
matrix, C is given the value 0; when all fibres are in contact, $C = 1$. In practice,
the true value will lie somewhere in between these two extremes. The resulting
relationships are:

$$E_2 = 2[1 - v_f + (v_f - v_m)v_m]$$

$$\times \left[(1 - C)\frac{K_f(2K_m + G_m) - G_m(K_f - K_m)(1 - v_f)}{(2K_m + G_m) + 2(K_f - K_m)(1 - v_f)} \right.$$

$$\left. + C\frac{K_f(2K_m + G_f) + G_f(K_m - K_f)(1 - v_f)}{(2K_m + G_f) - 2(K_m - K_f)(1 - v_f)} \right]$$

$$v_{12} = (1 - C)\frac{K_f v_f(2K_m + G_m)v_f + K_m v_m(2K_f + G_m)(1 - v_f)}{K_f(2K_m + G_m) - G_m(K_f - K_m)(1 - v_f)}$$

$$+ C\frac{K_m v_m(2K_f + G_f)(1 - v_f) + K_f v_f(2K_m + G_f)v_f}{K_f(2K_m + G_f) + G_f(K_m - K_f)(1 - v_f)} \qquad [3.13]$$

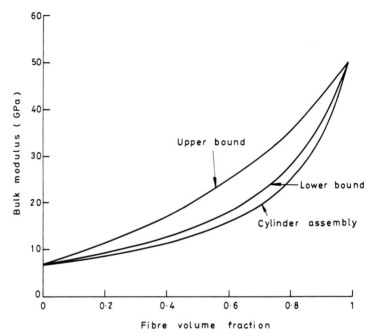

3.6   Calculation of bulk modulus by different
micromechanical models.

$$G_{12} = (1 - C) G_m \frac{2G_f - (G_f - G_m)(1 - v_f)}{2G_m + (G_f - G_m)(1 - v_f)}$$

$$+ CG_f \frac{(G_f + G_m) - (G_f - G_m)(1 - v_f)}{(G_f + G_m) + (G_f - G_m)(1 - v_f)}$$

where K_m and K_f are $E_m/2(1 - v_m)$ and $E_f/2(1 - v_f)$ respectively and the similar expressions for G_m and G_f are $E_m/2(1 + v_m)$ and $E_f/2(1 + v_f)$.

To apply these analyses, the contiguity factor, C, is determined from one data set and it is then applied to other properties for a given specimen. In this respect, the approach must be considered as semi-empirical.

Whilst these approaches have a certain attraction in terms of analytical rigour, their general applicability has not been established. A more pragmatic approach is one that employs the simple rule of mixtures expressions and modifies them by the use of empirical constants. For those properties that are dominated by the presence of the fibre, e.g. longitudinal modulus and the major Poisson's ratio, equations 3.2 and 3.5 are generally adequate, provided there are no problems of fibre misalignment. For the resin-dominated

Table 3.1. Halpin–Tsai approximation, $P = \dfrac{1}{v_f + \eta(1 - v_f)} \cdot [v_f P_f + \eta(1 - v_f)P_m]$

P	P_f	P_m	η
E_1	E_1	E_f	E_m	1
v_{12}	v_{12}	v_f	v_m	1
E_2	$1/E_2$	$1/E_f$	$1/E_m$	η_y
k	$1/k$	$1/k_f$	$1/k_m$	η_k
G_{12}	$1/G_{12}$	$1/G_f$	$1/G_m$	η_s

properties, Halpin and Tsai adopted the following expression:[8]

$$\frac{P_c}{P_m} = \frac{(1 + \xi \eta v_f)}{(1 - \eta v_f)} \qquad [3.14]$$

where $\eta = (P_f/P_m - 1)/(P_r/P_m + \xi)$, P is the composite modulus of concern (E_2 or G_{12}), P_f is the corresponding fibre modulus (E_f or G_f), and P_m is the corresponding matrix modulus (E_m or G_m). The constant ξ is a reinforcement geometry factor which depends on prevailing boundary conditions and is determined by comparing the rule of mixtures with those solutions employing a more formal elasticity theory. Developing the method further, thereby removing the need to obtain solutions to complex elasticity problems, results in the series of expressions shown in Table 3.1.[9] The parameters, η, given in the table are a measure of the relative magnitude of average stress in the fibre and matrix. If the values of these coefficients are set to 1, the equations in Table 3.1 equate to those for the rule of mixtures. Ideally, values for η_y, η_s and η_k should be determined from correlation with experiment, but approximate values can be determined from:

$$\eta_y = \tfrac{1}{2}$$

$$\eta_s = \tfrac{1}{2} \qquad [3.15]$$

$$\eta_k = \frac{1}{2(1 - v_m)}$$

As can be seen, the coefficients are independent of volume fraction and therefore, once determined for a given fibre/matrix combination, should be generally applicable. Figures 3.7–3.9 show a comparison between experimental modulus values and the rule of mixtures and Halpin–Tsai approximation.[7,10,11] As expected for longitudinal properties, correlation is good but for matrix–dominated properties the rule of mixtures equations are comparatively inadequate.

The micromechanical expressions developed above are concerned with composites with continuous reinforcement. For uniaxially aligned fibres of finite length the rule of mixtures can be modified by the inclusion of a length

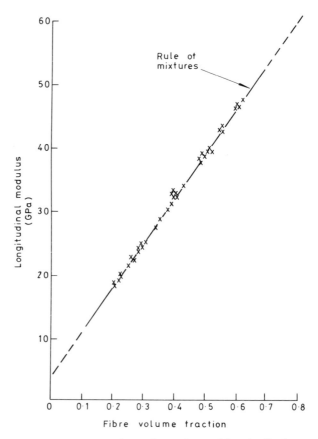

3.7   Comparison of experimental longitudinal
modulus with rule of mixtures.

correction factor, η_L,[12] such that:

$$E = \eta_L E_f v_f + E_m v_m \qquad [3.16]$$

It can be shown that:[13]

$$\eta_L = 1 - \frac{\tanh(\beta L/2)}{(\beta L/2)}$$

where

$$\beta = \frac{G_m 2\pi}{E_f A_f \ln(R/R_f)}$$

where G_m is the shear modulus of the matrix, L is the fibre length, A_f is the cross-sectional area of the fibre, R_f is the radius of the fibre, and R is the mean separation of the fibres.

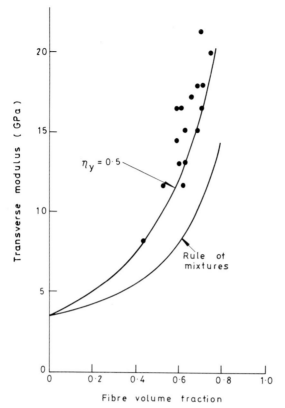

3.8 Comparison of experimental transverse modulus with rule of mixtures and Halpin–Tsai approximations.

The effect of a fibre length distribution can be described as:[12]

$$E_1 = \sum_{j=i}^{M} E_f v_{fj} \left[1 - \frac{\tanh (\beta L_j/2)}{(\beta L_j/2)} \right] + E_m (1 - v_f) \qquad [3.17]$$

where M is the number of intervals in the distribution, and v_{fj} is the volume fraction of fibres with length L_j.

In many respects, hygrothermal properties may be treated in the same manner as elastic constants. For thermal expansivities:[14]

$$\alpha_1 = \frac{E_f \alpha_f v_f + E_m \alpha_m (1 - v_f)}{E_f v_f + E_m (1 - v_f)}$$

$$[3.18]$$

$$\alpha_2 = \alpha_f v_f + \alpha_m (1 - v_f) + v_f \alpha_f v_f + v_m \alpha_m (1 - v_f) - [v_f v_f + v_m (1 - v_f)]\alpha_1$$

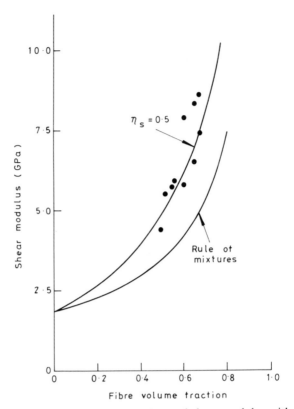

3.9   Comparison of experimental shear modulus with
rule of mixtures and Halpin–Tsai approximations.

where α_1 and α_2 are the longitudinal and transverse thermal expansivities of
the composite and α_f and α_m are those for fibre and matrix respectively. Figure
3.10 shows the variation of thermal expansivity of a typical GRP laminate
with fibre volume fraction.

Similarly for thermal conductivities:[15]

$$\lambda_1 = \lambda_f v_f + \lambda_m (1 - v_f)$$

[3.19]

$$\frac{1}{\lambda_2} = \frac{v_f}{\lambda_f} + \frac{1 - v_f}{\lambda_m}$$

where λ_1, λ_2 and λ_f, λ_m are conductivity values of composite and constituent
materials, and for swelling coefficients:[9]

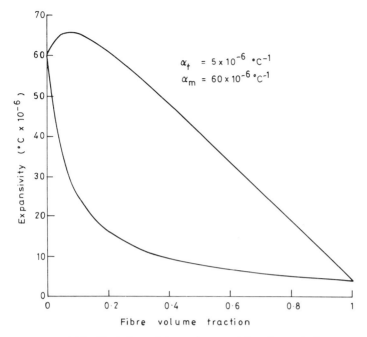

$$\alpha_f = 5 \times 10^{-6} \, {}^\circ C^{-1}$$
$$\alpha_m = 60 \times 10^{-6} \, {}^\circ C^{-1}$$

3.10   Variation of thermal expansivity of a typical
GRP laminate with fibre volume fraction.

$$\beta_1 = \frac{E_f c_{fm} \beta_f v_f + E_m \beta_m (1 - v_f)}{[E_f v_f + E_m (1 - v_f)][s_m (1 - v_f) + s_f c_{fm} v_f]} \cdot s$$

[3.20]

$$\beta_2 = \frac{c_{fm} \beta_f v_f (1 + v_f) + \beta_m (1 - v_f)(1 + v_m)}{s_m (1 - v_f) + s_f c_{fm} v_f} \cdot s - (v_f v_f + v_m (1 - v_f) \beta_1$$

where β_1, β_2 and β_f, β_m are swelling coefficients for composite, and fibre and matrix, s is the composite specific gravity and c_{fm} is the ratio of moisture concentrations in fibre and matrix.

Lamina properties

The micromechanical equations that have been presented relate to properties in the primary material directions, i.e. parallel and perpendicular to the fibre direction (Fig. 3.11). For practical purposes and in design it is necessary to compute values in other orientations. Equation 3.1 shows the constitutive relationship for a unidirectional laminate in plane stress and indicates how the terms in the compliance matrix can be related to engineering constants. This

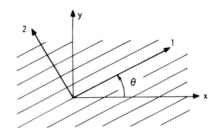

3.11   Material and laminate axis systems.

equation can be inverted to obtain the stress/strain relationships:

$$\begin{bmatrix} \sigma_1 \\ \sigma_2 \\ \tau_{12} \end{bmatrix} = \begin{bmatrix} Q_{11} & Q_{12} & 0 \\ Q_{12} & Q_{12} & 0 \\ 0 & 0 & Q_{66} \end{bmatrix} \begin{bmatrix} \varepsilon_1 \\ \varepsilon_2 \\ \gamma_{12} \end{bmatrix}$$

[3.21]

where the terms of Q_{ij} are:

$$Q_{11} = \frac{E_1}{1 - v_{12}v_{21}}$$

$$Q_{12} = \frac{v_{12}E_2}{1 - v_{12}v_{21}}$$

$$Q_{22} = \frac{E_2}{1 - v_{12}v_{21}}$$

$$Q_{66} = G_{12}$$

To determine lamina properties with respect to other coordinate axes, the stiffness and compliance matrices must be transformed. For rotation by angle θ to axes x, y, it can be shown that:

$$\begin{bmatrix} \sigma_x \\ \sigma_y \\ \tau_{xy} \end{bmatrix} = [\bar{Q}] \begin{bmatrix} \varepsilon_x \\ \varepsilon_y \\ \gamma_x \end{bmatrix}$$

[3.22]

where $[\bar{Q}]$ is transformed stiffness matrix given by:

$$[\bar{Q}] = [T]^{-1}[Q][T]^{-T}$$

and for compliance:

$$\begin{bmatrix} \varepsilon_x \\ \varepsilon_y \\ \gamma_{xy} \end{bmatrix} = [\bar{S}] \begin{bmatrix} \sigma_x \\ \sigma_y \\ \tau_x \end{bmatrix}$$

[3.23]

where S is the transformed compliance matrix given by:

$$[\bar{S}] = [T]^T [S] [T]$$

where $[T]$ is the transformation matrix for a second-order tensor. In expanded form it may be written as:

$$[T] = \begin{bmatrix} m^2 & n^2 & 2mn \\ n^2 & m^2 & -2mn \\ -mn & mn & m^2 - n^2 \end{bmatrix}$$

where $m = \cos \theta$ and $n = \sin \theta$.

Combining equations 3.21 and 3.22 allows the engineering constants to be expressed as functions of the off-axis angle:

$$\frac{1}{E_x} = \frac{1}{E_1} \cos^4 \theta + \left(\frac{1}{G_{12}} - \frac{2v_{12}}{E_1} \right) \sin^2 \theta \cos^2 \theta + \frac{1}{E_2} \sin^4 \theta$$

$$v_{xy} = E_x \left[\frac{v_{12}}{E_1} (\sin^4 \theta + \cos^4 \theta) - \left(\frac{1}{E_1} + \frac{1}{E_2} - \frac{1}{G_{12}} \right) \sin^2 \theta \cos^2 \theta \right] \quad [3.24]$$

$$\frac{1}{E_y} = \frac{1}{E_1} \sin^4 \theta + \left(\frac{1}{G_{12}} - \frac{2v_{12}}{E_1} \right) \sin^2 \cos^2 \theta + \frac{1}{E_2} \cos^4 \theta$$

$$\frac{1}{G_{xy}} = 2 \left(\frac{2}{E_1} + \frac{2}{E_2} + \frac{4v_{12}}{E_1} - \frac{1}{G_{12}} \right) \sin^2 \theta \cos^2 \theta + \frac{1}{G_{12}} (\sin^4 \theta + \cos^4 \theta)$$

In Fig. 3.12 the variation of elastic properties with reinforcement angle is shown. The values shown are typical for a GRP material. The key point to note is the rapid reduction in properties away from the longitudinal direction. On transformation of the stiffness matrix in equation 3.21, $[Q]$, to that for an off axis lamina, $[\bar{Q}]$, the matrix becomes fully populated. As a result, the application of, for example, a normal load to an off-axis lamina will generate a shear strain. Figure 3.13 shows the effect and indicates the magnitude of the coupling for the material properties used to calculate the stiffness data in Fig. 3.12. The coupling coefficient is 0 at 0° and 90° and reaches a maximum at a fibre angle of approximately 20°.

When considering the design of lamina for stiffness applications it is sometimes useful to consider combinations of stiffness terms which are in themselves invariant with orientation. Tables 3.2 and 3.3 show these combinations of stiffness and their use in the calculation of off-axis properties. By expressing stiffness in this way an understanding of the variation of properties with orientation is rapidly gained. For example, it can be seen that the stiffness in the x direction, $\bar{Q}_{11}$, consists of a component that is constant,

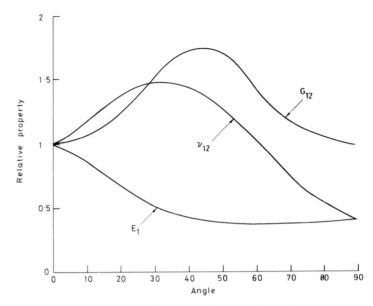

3.12   Variation of elastic properties with
reinforcement angle.

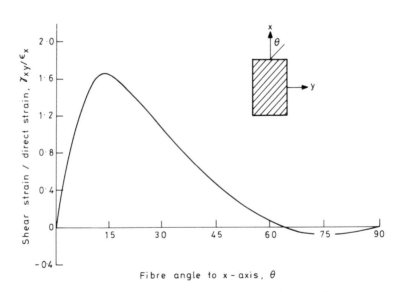

3.13   Effect of coupling between normal forces and
shear strain.

Table 3.2. Calculation of invariant properties

Invariant term	Q_{11}	Q_{22}	Q_{12}	Q_{33}
U_1	3/8	3/8	1/4	1/2
U_2	1/2	−1/2	0	0
U_3	1/8	1/8	−1/4	−1/2
U_4	1/8	1/8	3/4	−1/2
U_5	1/8	1/8	−1/4	1/2

Combinations $(U_4 + 2U_5)$, $(U_1 - 2U_5)$ and $\{(U_1 - U_4)/2\}$ are also invariant.

Table 3.3. Calculation of off-axis properties using invariants

Stiffness term	Constant	cos 2θ	cos 4θ	sin 2θ	sin 4θ
$\bar{Q}_{11}$	U_1	U_2	U_3	0	0
$\bar{Q}_{22}$	U_1	$-U_2$	U_3	0	0
$\bar{Q}_{12}$	U_4	0	$-U_3$	0	0
$\bar{Q}_{66}$	U_5	0	$-U_3$	0	0
$\bar{Q}_{16}$	0	0	0	$1/2U_2$	U_3
$\bar{Q}_{26}$	0	0	0	$1/2U_2$	$-U_3$

U_1, a component that varies slowly with orientation, $U_2 \cos 2\theta$, and a component that varies rapidly with orientation, $U_3 \cos 4\theta$.

A further advantage of the technique is that it enables calculation of quasi-isotropic properties. These are the characteristics of a laminate where layers are disposed in such a way that the in-plane properties are effectively isotropic. By definition the invariants are themselves the elastic constants of a composite in quasi-isotropic form. Therefore:

$$[Q]_{iso} = \begin{bmatrix} U_1 & U_4 & 0 \\ U_4 & U_1 & 0 \\ U & U & U_5 \end{bmatrix} \qquad [3.25]$$

which leads to:

$$E_{iso} = (U_1^2 - U_4^2)/U_1$$

$$\nu_{iso} = U_4/U_1 \qquad [3.26]$$

$$G_{iso} = U_5$$

where E_{iso}, ν_{iso} and G_{iso} are the elastic properties of the quasi-isotropic material.

Laminate properties

In practice, it is common for plies of different angles to be used in the form of a multi-directional laminate, the number and disposition of plies being dictated

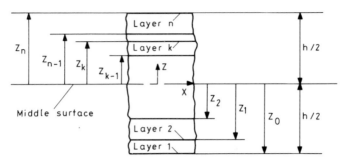

3.14 Laminate geometry.

by the applied loading conditions. Figure 3.14 shows the geometry of a layered laminate where the x–y plane corresponds to the geometric mid-plane. For a plate comprising such a system of layers the strain components can be written as follows:

$$\begin{bmatrix} \varepsilon_x \\ \varepsilon_y \\ \gamma_{xy} \end{bmatrix} = \begin{bmatrix} \varepsilon_x^0 \\ \varepsilon_y^0 \\ \gamma_{xy}^0 \end{bmatrix} + z \begin{bmatrix} \kappa_x \\ \kappa_y \\ \kappa_{xy} \end{bmatrix} \qquad [3.27]$$

where ε_x^0, ε_y^0, γ_{xy}^0 are mid-plane strains and κ_x, κ_y, κ_{xy} are the mid-plane curvatures.

The force and moment resultants (Fig. 3.15) acting on the laminate are:

$$\begin{bmatrix} N_x \\ N_y \\ N_{xy} \end{bmatrix} = \int_{-h/2}^{h/2} \begin{bmatrix} \sigma_x \\ \sigma_y \\ \tau_{xy} \end{bmatrix} dz \qquad [3.28]$$

$$\begin{bmatrix} M_x \\ M_y \\ M_{xy} \end{bmatrix} = \int_{-h/2}^{h/2} \begin{bmatrix} \sigma_x \\ \sigma_y \\ \tau_{xy} \end{bmatrix} z \, dz$$

The stress/strain relationship for the laminate may now be written:

$$\begin{bmatrix} [N] \\ [M] \end{bmatrix} = \begin{bmatrix} [A] & [B] \\ [B] & [D] \end{bmatrix} \begin{bmatrix} [\varepsilon] \\ [\kappa] \end{bmatrix} \qquad [3.29]$$

where the laminate stiffness matrices can be calculated as follows:

$$[A] = \Sigma [\bar{Q}]_k (z_k - z_{k-1})$$

$$[B] = \tfrac{1}{2} \Sigma [\bar{Q}]_k (z_k^2 - z_{k-1}^2)$$

$$[D] = \tfrac{1}{3} \Sigma [\bar{Q}]_k (z_k^3 - z_{k-1}^3)$$

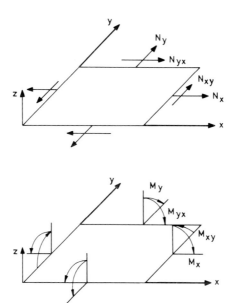

3.15 Force and moment resultants.

where the summation is carried out through the laminate thickness, $k = 1$ to n.

In equation 3.29, $[A]$ are extensional stiffnesses, $[B]$ are coupling stiffnesses and $[D]$ are bending stiffnesses, the value of z is defined by reference to Fig. 3.14. The coupling stiffness terms relate normal forces and extensions to moments and curvatures. Generally, if the laminate is symmetrical about its mid-plane, the terms of $[B]$ will reduce to zero.

As with the stiffness properties for single plies, $[\bar{Q}]$, these matrices simplify considerably if the laminate possesses symmetry. The most simple form of laminate is a symmetrical cross-ply with equal numbers of $0°$ and $90°$ layers of constant thickness. For such a material the elastic constants are defined by:

$$A_{11} = A_{22} = \tfrac{1}{2}(Q_{11} + Q_{22}) \cdot t$$
$$A_{12} = Q_{12} \cdot t$$
$$A_{16} = Q_{16} \cdot t \qquad\qquad [3.30]$$
$$A_{16} = A_{26} = 0$$

In this case all coefficients in the coupling matrix $[B]$ are zero due to symmetry. Similar simplifications are available for other laminate geometries.[9,16] Angle ply laminates where individual layers have alternating $+\theta$ and $-\theta$ orientations are amongst the most common types of layup used in design.

For this case examination of equation 3.29 shows that:

$$A_{11} = t \bar{Q}_{11}$$
$$A_{12} = t \bar{Q}_{12}$$
$$A_{22} = t \bar{Q}_{22} \qquad [3.31]$$
$$A_{66} = t \bar{Q}_{33}$$

The values of the other constants depend on the details of the laminate construction. For a symmetric laminate where each layer is of equal thickness:

$$A_{16} = \frac{t}{n} \bar{Q}_{16}$$

$$A_{26} = \frac{t}{n} \bar{Q}_{26} \qquad [3.32]$$

and

$$B_{ij} = 0$$

where n is the number of layers, and in this case n will be an odd number.

Due to the non-zero values of the A_{16} and A_{26} terms there will be coupling between normal loads and shear strains, but symmetry means that the other coupling terms, B_{ij}, are zero. For a similar laminate, except that n is an even number, symmetry about the mid-plane will no longer be present, e.g. a four-layer laminate will consist of a $+ \theta, - \theta, + \theta, - \theta$ configuration. Here the terms A_{16} and A_{26} are zero, but the constants B_{16} and B_{26} can become significant. They are given by:

$$B_{16} = \frac{t^2}{2n} \bar{Q}_{16}$$

$$[3.33]$$

$$B_{26} = \frac{t^2}{2n} \bar{Q}_{26}$$

The presence of these terms means that the laminate will undergo twisting deformation on the application of a normal load. Figure 3.16 shows the effect. Figure 3.17 shows the magnitude of the term B_{16} for an unsymmetric angle-ply GRP laminate as a function of fibre orientation and the number of plies. In the case shown, the maximum degree of coupling occurs at a fibre orientation of approximately 20°. A feature of the calculation for B_{16} is the sensitivity of its magnitude with the number of layers. Increasing the ply count to achieve a given thickness is one means of minimizing unwanted deformation.

A method of eliminating coupling effects altogether is to modify the disposition of layers from the true angle-ply disposition $(+ \theta, - \theta)_n$ to one

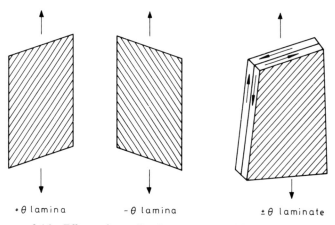

+θ lamina   −θ lamina   ±θ laminate

3.16   Effects of coupling between normal forces and
twisting deformation.

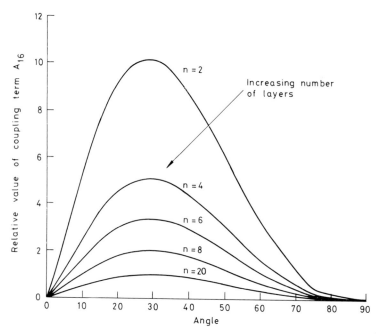

3.17   Coupling for an unsymmetrical angle-ply GRP
laminate.

which is both balanced, i.e. equal number of $+\theta$ and $-\theta$ plies and
symmetrical. For a four-layer material this would be of the form $+\theta$,
$-\theta$, $-\theta$, $+\theta$. In such a laminate, shear strain will be eliminated and this will
be associated with a shear stress in individual layers. The uncomplicated

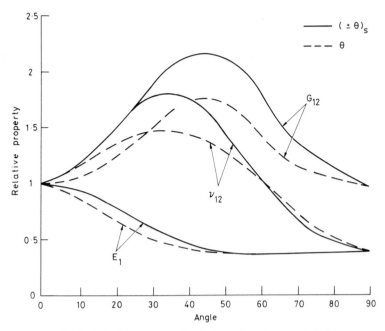

3.18 Elastic properties for balanced and symmetrical
angle-ply laminates.

response of this type of configuration to applied loads makes it an effective choice for design. In Fig. 3.18 typical properties for balanced and symmetrical GRP angle-ply laminates are shown and compared with the single ply material.

In carrying out design assessments for laminates it is often more convenient to invert equation 3.29 to derive the strain/stress relationship:

$$\begin{bmatrix} [\varepsilon^0] \\ [\kappa] \end{bmatrix} = \begin{bmatrix} [A^*] & [B^*] \\ [B^*]^T & [D^*] \end{bmatrix} \begin{bmatrix} [N] \\ [M] \end{bmatrix} \qquad [3.34]$$

Determination of the terms of the compliance matrix can be achieved through relatively straightforward manipulation of equation 3.29. The first step is to solve for $[\varepsilon^0]$, viz.:

$$[\varepsilon^0] = [A]^{-1}[N] - [A]^{-1}[B][\kappa]$$

Substitution into the expression for $[M]$ yields:

$$[M] = [B][A]^{-1}[N] + (-[B][A]^{-1}[B] + [D])[\kappa]$$

Repeating the process firstly by solving for $[\kappa]$ and then substituting into the expression for $[\varepsilon^0]$ gives rise to the following:

$$[A^*] = [A]^{-1} + [A]^{-1}[B][C]^{-1}[B][A]^{-1}$$
$$[B^*] = [A]^{-1}[B][C]^{-1} \qquad\qquad\qquad [3.35]$$
$$[D^*] = [C]^{-1}$$

where

$$[C] = [D] - [B][A]^{-1}[B]$$

Although as presented the evaluation of the stiffness and/or compliance matrices appears complex, the matrix calculations are easily performed using standard computer software packages which are widely available.

The stress analysis of a composite laminate can proceed in a straightforward manner given the stiffness properties and details of the applied loads. The steps in the assessment can be listed as follows:

- Calculation of mid-plane strains and curvatures arising due to the application of design loads and moments (equation 3.34).
- Calculation of strains for each layer within the laminate from mid-plane values (equation 3.27).
- Calculation of stresses for each layer within the laminate from strains (equation 3.22).
- Conversion of stresses (or strains) from the laminate axis to the primary material directions for each layer using the transformation matrix $[T]$ (defined in equation 3.23).

Once the stress system in each laminate layer is available it is then possible to compare results with an acceptance criterion which would be established for each component or application. In the simplest form it may be just a series, or some combination, of strength values for the material. More sophisticated approaches may involve design margins, degradation factors to account for the effect of the environment such as temperature and chemical media or parameters to cater for issues such as cyclic loading, creep and transients.

Strength properties

As with stiffness properties it is possible to provide estimates of lamina strength values from basic considerations of fibre and materials. Whilst strength predictions are more tentative than, say, those for elastic modulus, the approximate results which are available can often be used as an initial basis for design.

Micromechanics

From a simple rule of mixtures approach it can be shown that the composite

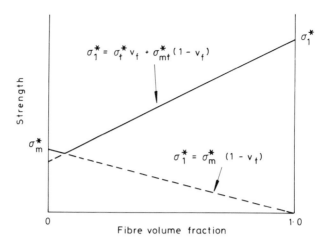

3.19   Effect of fibre volume fraction on longitudinal
tensile strength.

tensile strength, σ_1^*, in the fibre direction is given by:

$$\sigma_1^* = \sigma_f^* v_f + \sigma_{mf}^* (1 - v_f) \qquad [3.36]$$

where σ_f^* is the fibre tensile strength and σ_{mf}^* is the matrix stress at a matrix strain equal to the strain in the fibres. In the majority of cases, $\sigma_f^* >> \sigma_{mf}^*$ and therefore equation 3.36 can be effectively reduced to the first term only.

For the fibres to have a reinforcing influence, i.e. $\sigma_1^* > \sigma_m^*$, the fibre volume fraction must be:

$$v_f \geq \frac{\sigma_m^* - \sigma_{mf}^*}{\sigma_f^* - \sigma_{mf}^*} \qquad [3.37]$$

For very low volume fractions there may not be sufficient fibres to control matrix elongation and in this situation the composite strength would be given by:

$$\sigma_1^* = \sigma_m^* (1 - v_f) \qquad [3.38]$$

Equations 3.36 and 3.38 are plotted on Fig. 3.19. When the volume fraction is less than v_{min}, as indicated in Fig. 3.19, the strength is controlled by the matrix and, indeed, is less than the matrix strength. Only when the proportion of fibres exceeds v_{min} do the fibres contribute to material strength. It should be noted that as σ_f^* is generally much larger than σ_m^*, v_{min} is very small. The expressions given above are based on a deterministic approach to strength and assume all fibres are of equal strength. Clearly this is not the case, particularly for brittle materials such as glass, and a more rigorous approach would be to consider the statistical variations inherent within the material. This aspect of behaviour is considered briefly in Chapter 2.

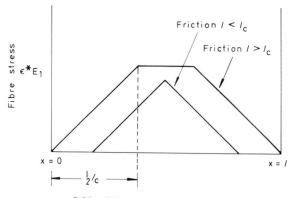

3.20   Fibre/matrix load transfer.

For aligned, discontinuous fibres considerations of fibre length must be taken into account. Load is transferred from the matrix to the reinforcement through a shear mechanism where stress builds up from the fibre ends and is at a maximum at the centre. There is, as a result, a critical fibre length, l_c, required for the stress to develop to such an extent to reach the fibre failure stress:

$$l_c = r\frac{\sigma_f^*}{\tau} \qquad [3.39]$$

where r is the fibre radius and τ is the interfacial shear stress.

At values of $l < l_c$ load transfer into the fibre is not maximized (Fig. 3.20). For the case when the fibre length is greater than the critical value, consideration of the geometry of the stress distribution in Fig. 3.20 allows the average stress $\bar{\sigma}_f$ at fibre failure to be calculated:

$$\bar{\sigma}_f = (1 - l_c/2l)\sigma_f^* \qquad [3.40]$$

Combining equations 3.40, 3.39 and 3.36 yields:

$$\sigma_1^* = \left(1 - \frac{\sigma_f^* r}{2\tau l}\right)\sigma_f^* v_f + \sigma_{mf}^*(1 - v_f) \qquad [3.41]$$

For the converse case where the fibre length is less than the critical value, failure will be determined by the matrix strength. At the matrix failure stress the average fibre stress is given by:

$$\bar{\sigma}_f = \frac{l\tau}{2r} \qquad [3.42]$$

and the strength of the composite is given by:

$$\sigma_1^* = \left(\frac{l\tau}{2r}\right)v_f + \sigma_m^*(1 - v_f) \qquad [3.43]$$

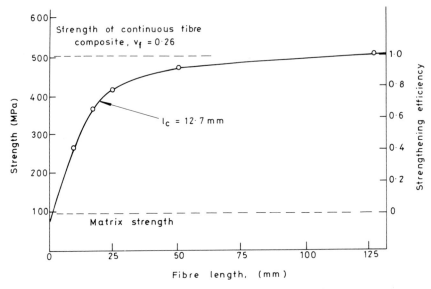

3.21   Effect of fibre length on strength of
unidirectional GRP.

Figure 3.21 shows the effect of fibre length on unidirectional GRP.[17] The value
for l_c ($\sim$ 12.7 mm) was calculated by curve fitting to the experimental points.
The full strength potential of the composite is achieved for fibre lengths of
about 10 l_c but the majority of the equivalent unidirectional strength is
developed when $l \approx 4l_c$. It is worth noting that the fibre lengths in structural
reinforcing mats are of the order of 50 mm[18] and this is as a consequence of
considerations of critical lengths. In the majority of cases fibre length will be
variable within a discontinuously reinforced composite. Damage and breakage during processing can be particularly important in reducing fibre length
and therefore reinforcing efficiency. Where all fibres are such that $l < l_c$ the
composite strength is given by:

$$\sigma_1^* = \sum \frac{l_i \tau}{2r} v_f + \sigma_m^* (1 - v_f) \qquad [3.44]$$

For the case where the distribution of lengths is either side of l_c the value of
σ_1^* can be shown to be:

$$\sigma_1^* = \sum_{l_i < l_c} \frac{l_i \tau}{2r} v_i + \sum_{l_j > l_c} \left(1 - \frac{l_c}{2l_j}\right) \sigma_f^* v_j + \sigma_{mf}^* (1 - v_f) \qquad [3.45]$$

where

$$\sum_i v_i + \sum_j v_j = v_f$$

Table 3.4. Formulae for in-plane strengths

Longitudinal Tension	$\sigma_1^* = V_f\,\sigma_f^*$	
Longitudinal Compression • Fibre compression • Shear • Microbucking	$\sigma_{1c}^* = V_f\,\sigma_{fc}^*$ $\sigma_{1c}^* = 10\,\tau_{12}^* + 2\cdot5\,\sigma_m^*$ $\sigma_{1c}^* = \dfrac{G_m}{1 - V_f\,(1 - \frac{G_m}{G_{12}})}$	
Transverse Tension	$\sigma_2^* = [\,1 - (V_f^{1/2} - V_f)(1 - \frac{E_m}{E_2})\,]\,\sigma_m^*$	
Transverse Compression	$\sigma_{2c}^* = [\,1 - (V_f^{1/2} - V_f)(1 - \frac{E_m}{E_2})\,]\,\sigma_{mc}^*$	
Shear	$\tau_{12}^* = [\,1 - (V_f^{1/2} - V_f)(1 - \frac{G_m}{G_{12}})\,]\,\tau_m^*$	

As can be seen from equations 3.44 and 3.45 the composite strength remains a strong function of fibre volume fraction and this will be a linear relationship if the distribution in l is constant from one material to another.

Strength values in other directions and for other modes are also important in design. In these instances strengths are usually dominated by matrix behaviour, stress concentrations due to the presence of fibres and voids, and the interfacial properties of the fibre/matrix bond.[19] Micromechanical expressions which attempt to address these factors with the aim of providing a means of calculating strengths are at best somewhat tentative and can only be regarded as being approximate. Tables 3.4 and 3.5 show a series of formulae that can be used to estimate strength values of various form.[20,21] As a guide, matrices that have brittle characteristics, and where the effect of stress concentrations can be severe, result in strengths less than the matrix, whilst

Table 3.5. Formulae for out-of-plane strengths

Interlaminar Shear	$\tau_{13}^* = [\, 1 - (\, V_f^{1/2} - V_f \,)(1 - {}^{G_m}/G_{12})\,]\, \tau_m^*$ $\tau_{23} = \dfrac{1 - V_f^{1/2}\,(1 - {}^{G_m}/G_{23})}{1 - V_f\,(1 - {}^{G_m}/G_{23})}\; \tau_m^*$
Short Bean Shear	$(\tau_{13}^*)_{beam} = 1 \cdot 5\;\tau_{13}^*$ $(\tau_{23}^*)_{beam} = 1 \cdot 5\;\tau_{23}^*$
Flexural	$\mathfrak{I}_{1f}^* = \dfrac{3\,V_f\;\sigma_1^*}{1 + (\sigma_1^*/\sigma_{1c}^*)}$ $\mathfrak{I}_{2f}^* = \dfrac{3\,[1 - (V^{1/2} - V)(1 - {}^{E_m}/E_2)]\;\sigma_m^*}{1 - (\sigma_m^*\;\sigma_{mc}^*)}$

those that are more ductile can have higher strengths. In this latter case, redistribution of stresses due to relaxation is of benefit. Failure in longitudinal compression occurs by a range of mechanisms, the most significant of which are fibre buckling and delamination initiated by a shear failure. Both of these can result in compression strengths significantly less than their tensile counterparts. For transverse compression, failure is usually by a shear mode on the 45° plane and as with monolithic materials compression strengths can be high and certainly greater than those in tension.

Failure theories

As it is not possible to obtain strengths in all possible lamina orientations or for all combinations of lamina, a means must be established by which these characteristics can be determined from basic layer data. Theories of failure are hypotheses concerning the limit of load-carrying ability under different load combinations. Using expressions derived from these theories it is possible to construct failure envelopes or, if in three dimensions, failure surfaces, which represent the limit of usefulness of the material as a load-bearing component, i.e. if a given loading condition is within the envelope the material will not fail. The suitability of any proposed criterion is determined by a number of factors, the most important of which being concerned with the nature of the failure mode, e.g. do criteria represent the limit of proportionality (yield), or do they represent the limit of load-carrying ability (ultimate strength)? Additional issues feature, such as fracture, fatigue, creep and corrosion. As a result it is important that proposed failure criteria are accompanied by a definition of material behaviour.

The most important criteria for isotropic materials are:[22]

- Tresca theory. Failure of a material subject to a stress system occurs when the maximum shearing stress at any point reaches a particular value for that material:

$$|\sigma_1 - \sigma_2| = |\sigma^*|$$
$$|\sigma_2 - \sigma_3| = |\sigma^*|$$
$$|\sigma_3 - \sigma_1| = |\sigma^*| \qquad [3.46]$$

where σ_1, σ_2 and σ_3 are the principal stresses and σ^* is the yield or ultimate stress of the material.

- Von Mise Theory. Failure occurs when the internal strain energy of distortion in the stressed material reaches a particular value for that material:

$$\sigma_1^2 + \sigma_2^2 + \sigma_3^2 + \sigma_1\sigma_2 + \sigma_2\sigma_3 + \sigma_1\sigma_3 = 2(\sigma^*)^2 \qquad [3.47]$$

- Maximum stress theory. Failure occurs when the maximum normal stress reaches a value at which failure occurs in a simple tensile test, i.e.:

$$\begin{vmatrix} \sigma_1 \\ \sigma_2 \\ \sigma_3 \end{vmatrix} = \begin{vmatrix} \sigma^* \\ \sigma^* \\ \sigma^* \end{vmatrix} \qquad [3.48]$$

where σ^* is the ultimate strength.

With composite materials, the effects of anisotropy must be catered for in the analysis and this has led to a number of developments based on the expressions for isotropic systems. The distortional energy criterion in general form can be written as:

$$(\sigma_2 - \sigma_3)^2 + (\sigma_3 - \sigma_1)^2 + (\sigma_1 - \sigma_2)^2 + 6(\tau_{23} + \tau_{13} + \tau_{12}) = 2(\sigma^*)^2 \qquad [3.49]$$

To allow for orthotropy, equation 3.49 can be generalized by the inclusion of six constants, each of which is a function of either the tensile or yield strength in the principal material directions i.e. the 1, 2 and 3 directions where the axis denoted 1 corresponds to the fibre direction:[23]

$$F(\sigma_2 - \sigma_3)^2 + G(\sigma_3 - \sigma_1)^2 + H(\sigma_1 - \sigma_2)^2 + (2L\tau_{23}^2 + 2M\tau_{13}^2 + 2N\tau_{12}^2) = 1 \qquad [3.50]$$

where by substitution of uniaxial loading conditions:

$$2F = \frac{1}{(\sigma_2^*)^2} + \frac{1}{(\sigma_3^*)^2} - \frac{1}{(\sigma_1^*)^2}$$

$$2G = \frac{1}{(\sigma_3^*)^2} + \frac{1}{(\sigma_1^*)^2} - \frac{1}{(\sigma_2^*)^2}$$

$$2H = \frac{1}{(\sigma_1^*)^2} + \frac{1}{(\sigma_2^*)^2} - \frac{1}{(\sigma_3^*)^2}$$

[3.51]

$$2L = \frac{1}{(\tau_{23}^*)^2}$$

$$2M = \frac{1}{(\tau_{13}^*)^2}$$

$$2N = \frac{1}{(\tau_{12}^*)^2}$$

where σ_1^*, σ_2^* and σ_3^* are the uniaxial strengths in the 1, 2 and 3 directions respectively and τ_{23}^*, τ_{13}^* and τ_{12}^* are the shear strengths in the 2–3, 1–3 and 1–2 planes respectively. This approach makes no attempt to allow for differing tensile and compressive strengths and it is assumed that hydrostatic pressure has no effect on yield. This theory was originally developed for materials with preferential orientation (about 10% difference between maximum and minimum strengths), e.g. cold-rolled metals.

Reducing equation 3.48 to cater for the conditions of plane stress yields:

$$\frac{\sigma_1^2}{(\sigma_1^*)^2} - \frac{\sigma_1\sigma_2}{(\sigma_1^*)^2} + \frac{\sigma_2^2}{(\sigma_2^*)^2} + \frac{\tau_{12}^2}{(\tau_{12}^*)^2} = 1$$

[3.52]

This is widely known as the Tsai–Hill criterion.[24,25] For a single unidirectional lamina it can be shown that equation 3.52 has three turning points for the conditions of a uniaxial load. The first two, as would be expected, correspond to the transverse and longitudinal directions, while the third point is defined by the equation:

$$\theta = \tan^{-1}\left[\frac{[(\sigma_1^*)^2 - 3(\tau_{12}^*)^2](\sigma_2^*)^2}{(\sigma_2^*)^2[(\sigma_1^*)^2 - (\tau_{12}^*)^2] - 2(\sigma_1^*)^2(\tau_{12}^*)^2}\right]$$

[3.53]

The significance of this third point is dependent on the relative values of σ_2^* and τ_{12}^*. Low shear strengths lamina of intermediate orientation may have a slightly lower strength than the transverse value. Differing tensile and compressive strengths can be allowed for in equation 3.52 by inserting the appropriate values, depending on the sign of the stresses, into the equation. There have been a great many attempts to derive a criterion based on the

Table 3.6. Distortional energy failure criteria

Tsai–Hill

$$\left(\frac{\sigma_1}{\sigma_1^*}\right)^2 - \frac{\sigma_1\sigma_2}{(\sigma_1^*)^2} + \left(\frac{\sigma_2}{\sigma_2^*}\right)^2 + \left(\frac{\tau_{12}}{\tau_{12}^*}\right)^2 = 1$$

Norris

$$\left(\frac{\sigma_1}{\sigma_1^*}\right)^2 - \left(\frac{\sigma_1\sigma_2}{\sigma_1^*\sigma_2^*}\right) + \left(\frac{\sigma_2}{\sigma_2^*}\right)^2 + \left(\frac{\tau_{12}}{\tau_{12}^*}\right)^2 = 1$$

Puppo–Evensen

$$\left(\frac{\sigma_1^*}{X^*}\right)^2 - \gamma\frac{\sigma_1^*\sigma_2^*}{(Y^*)^2} + \gamma\left(\frac{\sigma_2^*}{Y^*}\right)^2 + \left(\frac{\tau_{12}^*}{S^*}\right)^2 = 1$$

$$\gamma\left(\frac{\sigma_1^*}{X^*}\right)^2 - \gamma\frac{\sigma_1^*\sigma_2^*}{(X^*)^2} + \left(\frac{\sigma_2^*}{Y^*}\right)^2 + \left(\frac{\tau_{12}^*}{S^*}\right)^2 = 1$$

where

$$\gamma = \frac{3(T^*)^2}{X^*Y^*}$$

Stresses and strengths shown refer to the principal strength axes defined as when the parameter γ is a minimum with respect to rotation.

Puck–Schneider

$$\left(\frac{\sigma_1}{\sigma_1^*}\right)^2 = 1$$

$$\left(\frac{\sigma_1}{\sigma_{mf}^*}\right)^2 + \left(\frac{\sigma_2}{\sigma_2^*}\right)^2 + \left(\frac{\tau_{12}}{\tau_{1/2}^*}\right)^2 = 1$$

where $\sigma_{mf}^* = E_1\varepsilon_m^*$

original distortional energy concept and a list of the more prominent variants is given in Table 3.6.[26–29]

As an alternative to the distortional energy approach a number of phenomenological criteria have been proposed that link experimental constants but do not attempt to describe any physical occurrence. The first such analysis expressed in tensor notation can be written as:[30]

$$(F_i\sigma_i)^\alpha + (F_{ij}\sigma_i\sigma_j)^\beta + (F_{ijk}\sigma_i\sigma_j\sigma_k)^\gamma + \ldots = 1 \qquad [3.54]$$

where α, β, γ are material parameters and F_i, F_{ij} and F_{ijk} are strength tensors of the appropriate rank.

Subsequently, Tsai and Wu postulated that a failure surface can be described in the form:[31,32]

$$F_i\sigma_i + F_{ij}\sigma_i\sigma_j = 1 \qquad [3.55]$$

which in expanded form for plane stress is:

$$F_1\sigma_1 + F_2\sigma_2 + F_{66}\tau_{12} + F_{11}\sigma_1^2 + F_{22}\sigma_2^2 + F_{66}\tau_{12}^2 + 2F_{12}\sigma_1\sigma_2 = 1 \quad [3.56]$$

The constants F_1, F_2, F_{11}, F_{22} and F_{66} can be defined in terms of uniaxial tensile, compressive and shear strengths as follows:

$$F_1 = \frac{1}{\sigma_1^*} + \frac{1}{\sigma_{1c}^*}$$

$$F_2 = \frac{1}{\sigma_2^*} + \frac{1}{\sigma_{2c}^*}$$

$$F_{11} = -\frac{1}{\sigma_1^*\sigma_{1c}^*} \qquad\qquad [3.57]$$

$$F_{22} = -\frac{1}{\sigma_2^*\sigma_{2c}^*}$$

$$F_{66} = \frac{1}{(\tau_{12}^*)^2}$$

where the presence of the subscript c denotes a compressive strength.

The determination of constant F_{12} is not straightforward as it requires results from a biaxial loading experiment, for example, for the condition $\sigma_1 = \sigma_2 = \sigma$:

$$F_{12} = \frac{1}{2\sigma^2}\left[1 - \sigma\left(\frac{1}{\sigma_1^*} + \frac{1}{\sigma_{1c}^*} + \frac{1}{\sigma_2^*} + \frac{1}{\sigma_{2c}^*}\right) + \sigma^2\left(\frac{1}{\sigma_1^*\sigma_{1c}^*} + \frac{1}{\sigma_2^*\sigma_{2c}^*}\right)\right] \quad [3.58]$$

Alternatively by using the fact that the condition $\sigma_1 = \sigma_2 = \sigma_{12} = \sigma$ is equivalent to a uniaxial tension test on a sample in which the fibres are orientated 45° yields:

$$F_{12} = \frac{2}{(\sigma_{45}^*)^2}\left[1 - \frac{(\sigma_{45}^*)^2}{2}\left(\frac{1}{\sigma_1^*} + \frac{1}{\sigma_{1c}^*} + \frac{1}{\sigma_2^*} + \frac{1}{\sigma_{2c}^*}\right)\right.$$

$$\left. + \frac{(\sigma_{45}^*)^2}{2}\left(\frac{1}{\sigma_1^*\sigma_{1c}^*} + \frac{1}{\sigma_2^*\sigma_{2c}^*} + \frac{1}{(\tau_{12}^*)^2}\right)\right] \qquad [3.59]$$

where σ_{45}^* is the tensile strength of the 45° material.

The criteria which are simplest in concept are the maximum stress and strain theories and these are direct extensions of those for isotropic materials. They assume that failure will occur when the stress or strain in the primary material direction exceeds the ultimate tensile, compressive or shear values in that direction.

Maximum stress:

$$\frac{\sigma_1}{\sigma_1^*} = 1 \quad or \quad \frac{\sigma_1}{\sigma_{1c}^*} = 1$$

$$\frac{\sigma_2}{\sigma_2^*} = 1 \quad or \quad \frac{\sigma_2}{\sigma_{2c}^*} = 1 \qquad [3.60]$$

$$\frac{\tau_{12}}{\tau_{12}^*} = 1$$

Maximum strain:

$$\frac{\varepsilon_1}{\varepsilon_1^*} = 1$$

$$\frac{\varepsilon_2}{\varepsilon_2^*} = 1 \qquad [3.61]$$

$$\frac{\gamma_{12}}{\gamma_{12}^*} = 1$$

To apply the failure criteria to off-axis lamina it is first necessary to transform the stress system generated by the applied loading to the primary material directions. Once in this form stresses can be used directly with the strength values appropriate to the material of concern. For example, the maximum stress criteria for an off-axis lamina subject to a uniaxial stress, σ, can be written as:

$$\sigma \cdot \frac{\cos^2 \theta}{\sigma_1^*} = 1$$

$$\sigma \cdot \frac{\sin^2 \theta}{\sigma_2^*} = 1 \qquad [3.62]$$

$$\sigma \cdot \frac{\sin \theta \cos \theta}{\tau_{12}^*} = 1$$

Figure 3.22 shows the correlation between the maximum stress criteria for off-axis CFRP material.[33] Also shown is the Tsai–Hill criterion, and for both approaches agreement with the data is good. The tensor criteria postulated by Tsai and Wu also provides an effective means of predicting strength values (Fig. 3.23).[34] This is found to be generally the case for uniaxial tensile loading and for unidirectional composites including those of glass, carbon and boron reinforcement.[35,36] The use of failure theories with combined stresses is shown

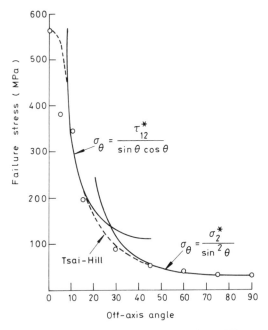

3.22   Failure stresses for off-axis CFRP.

in Fig. 3.24, in this case for GRP tubular specimens tested with a combination of axial tension and shear.[37] For this configuration axial stress is equivalent to loading in the transverse material direction and the Tsai–Hill criterion reduces to:

$$\left(\frac{\sigma_2}{\sigma_2^*}\right)^2 + \left(\frac{\tau_{12}}{\tau_{12}^*}\right)^2 = 1 \qquad\qquad [3.63]$$

A good level of agreement is achieved between experimental data and equation 3.63. However, in this situation the utility of the maximum stress approach is poor. This is not too surprising as the expressions comprising the maximum stress criterion do not allow for any interaction between normal and shear effects. The same is true for isotropic materials where in the shear/tension quadrant of the failure envelope, the Tresca criterion is usually preferred.

The application of failure criteria to laminates follows the same procedure as with single layers, with calculations first being carried out to evaluate stresses in the primary material directions; this needs to be evaluated on a layer-by-layer basis. The choice of this coordinate system as a reference is somewhat arbitrary as the principal stresses do not necessarily lie on these axes.[38] A major advantage of using such a coordinate system for failure calculations, however, is that only one set of allowable stresses or strains

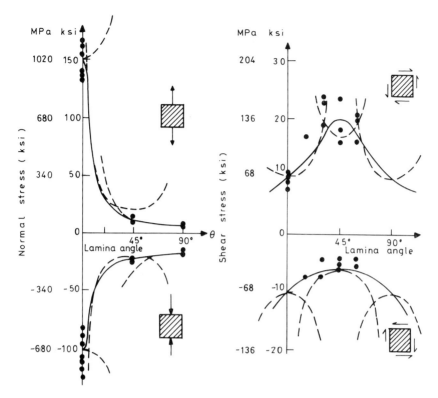

3.23 Application of Tsai–Wu criteria for off-axis
CFRP unidirectional lamina.

require determination and can be used for any combination of layers of fibre angles. Any other system, e.g. one based on principal planes, would require a different set of strength values for every lamina sequence as the orientation of the coordinate system would vary for each. The primary material directions also tend to correspond to those of lowest strength, particularly shear strength.[39,40]

Examination of the response of an angle-ply laminate in terms of primary material directions to simple unidirectional loads is shown in Fig. 3.25. These strains are those generated in one layer of the laminate. For angles between 0° and ± 17° the maximum strain is parallel to fibres, between ± 18° and ± 68° the maximum strain is shear and between ± 69° and ± 90° it is perpendicular to the fibres. The effect of combinations of load applied simultaneously to a laminate may be treated in a similar fashion. Figure 3.26 shows the strain distribution within a laminate of fibre orientation ± 55° to the axis for different combinations of direct load applied in the x and y directions. The strains plotted are those generated in the primary material directions and are normalized with respect to the maximum applied strain. This example is

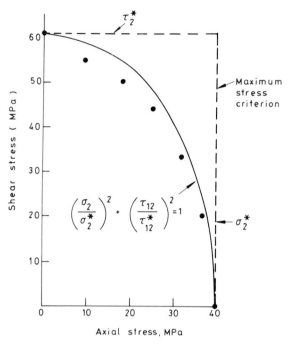

3.24   Failure curve of unidirectional GRP in
        combined tension and shear.

analogous to a cylinder in which the fibres are wound with an angle of $\pm\,55°$ to its longitudinal axis. For the loading condition $\sigma_x > 0$ and $\sigma_y = 0$ ($\tan^{-1}$ $\sigma_y/\sigma_x = 0$) the cylinder would be under uniaxial longitudinal tension; for $\sigma_x = 0$ and $\sigma_y > 0$ ($\tan^{-1}\,\sigma_y/\sigma_x = 90$) the cylinder would be under uniaxial circumferential tension, i.e. internal pressure without end closure; whilst $\sigma_x > 0$ and $\sigma_y = 0$ ($\tan^{-1}\,\sigma_y/\sigma_x = 180$) corresponds to the case where the cylinder is under longitudinal compression.

By selecting a maximum allowable strain and then comparing it with the strains generated within a laminate, it is possible to construct a biaxial failure envelope for the material. Figure 3.27 shows such an envelope for a laminate of winding angle $\pm\,55°$. In the case of a cylinder, loads plotted on the abscissa would give rise to circumferential stresses whilst those plotted on the ordinate would cause longitudinal stresses. For the purposes of presentation the loads shown are plotted relative to the uniaxial longitudinal strength. Also shown on Fig. 3.27 is the nature of the strain, i.e. in-plane shear (γ_{12}), transverse (ε_2) or longitudinal (ε_1), which determines the maximum allowable load values. Structures that are subjected to load systems located within the failure envelope will be below the allowable strain limit, but it should be noted that the loads must be applied simultaneously. For example, consider point 'A'.

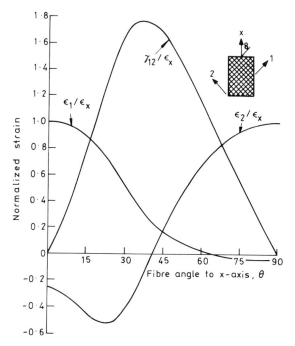

3.25　Normalized direct and shear strain for the
application of a unidirectional load.

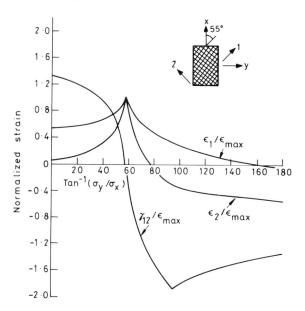

3.26　Normalized direct and shear strain for different
combinations of load in a laminate of fibre
angle $\pm 55°$.

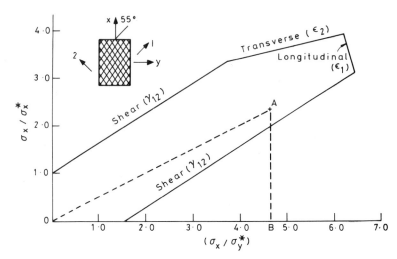

3.27   Biaxial failure envelope.

Following the load path denoted by 'OA' would be satisfactory, but following the route 'OBA' would cause the laminate to exceed the design limits.

A very simple approach to the prediction of strength under combined loads is provided by netting analysis.[41] In this method it is assumed that all of the load is carried by the reinforcement. For a simple angle-ply laminate allowable combinations of normal stresses σ_x and σ_y are given by:

$$\sigma_x = v_f \sigma_f^* \cos^2 \theta$$
$$\sigma_y = v_f \sigma_f^* \sin^2 \theta$$

[3.64]

These expressions can, however, only be used for the load case where the stresses combine to act in the fibre direction and for an angle ply laminate this ratio of σ_x and σ_y is:

$$\frac{\sigma_x}{\sigma_y} = \tan^2 \theta$$

[3.65]

Although very simplistic the calculation method can be used to obtain a crude form of biaxial envelope in conjunction with measured uniaxial data.[38] Figure 3.28 shows an example. Uniaxial strengths in tension and compression are measured values and the envelope is constructed by joining these points and those calculated by equation 3.64. Netting analysis can also be used for more complex layups.[42,43] For example, a cylinder with a laminate that is constructed from a combination of hoop and $\pm \theta$ laminae can be treated as shown in Fig. 3.29. The basis of the analysis is that of static equilibrium with no consideration of strain compatibility. Summing forces in the axial direction

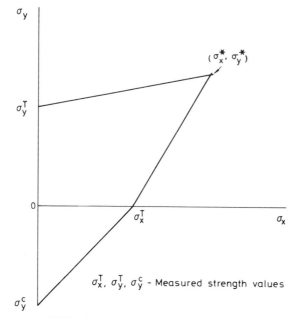

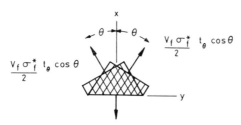

3.28 Approximate biaxial failure envelope.

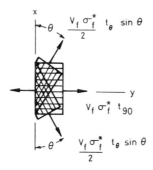

3.29 Netting analysis applied to hoop/helical winding.

gives:

$$v_f \sigma_f^* t_\theta \cos^2 \theta = \frac{pr}{2} \qquad [3.66]$$

where t_θ is the thickness of the helical layers, p is the applied pressure and r is the cylinder radius.

The thickness, t_θ, is therefore:

$$t_\theta = \frac{pr}{2v_f \sigma_f^* \cos^2 \theta} \qquad [3.67]$$

Using equation 3.66 and summing forces in the circumferential direction gives:

$$t_{90} = \frac{pr}{2v_f \sigma_f^*}(2 - \tan^2 \theta) \qquad [3.68]$$

Owing to the assumption that it is only the reinforcement that carries load, the use of this type of expression is limited either to single load cases (equation 3.59) or where there are two independent angles in the laminate.

The application of failure criteria generally to multiply laminates follows the steps already outlined. Given the applied loads, the stresses in each layer are examined in the context of the selected failure criterion. In some circumstances the point at which the first ply exceeds the allowable loading will be deemed to be the failure point. However, this may or may not be the ultimate load of the material. For example, in a 0/90 cross-ply laminate subject to a uniaxial tensile load the stress/strain curve will be linear to failure of the transverse plies, but the arrangement will still be able to accept increasing load due to the capabilities of the 0° layers. The stress initially carried by the 90° ply will be transferred on to the other layers. Such events can be handled in an analysis by setting the elastic properties of the failed layer to zero, recalculating the stiffness matrix for the modified laminate and repeating the analysis as before. This procedure can be completed until the final layer has failed. Comparison of the failure analysis of multi-layer laminates with experimental results generally leads to a disappointing degree of correlation. This is especially true with conditions of combined loading where it is usually the case that strength predictions underestimate measured laminate capability. The major effect responsible for these observations is that of nonlinearity in stress/strain behaviour, and this can result in significant redistribution of stress within a laminate. Nonlinear constitutive behaviour can be incorporated in laminate analysis[43] and the arising stress distributions calculated in an iterative manner.

Micromechanics of stress/strain behaviour

A message that arises from consideration of the lack of correlation of observed component behaviour to the results of simple elastic/failure analysis is that

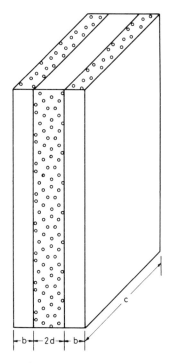

3.30   Cross-ply laminate.

composite materials can be considered to be more 'forgiving' than would be expected from the brittle attributes normally ascribed to them. Undertaking an analysis of stress/strain behaviour on a micromechanical level can provide a basis for rationalizing these observations. As a first step, consider the cross-ply laminate shown in Fig. 3.30. With the application of a uniaxial tensile load, transverse cracking of the inner lamina will result in a redistribution of the load and an increase in the stress on the outer longitudinal laminae (Fig. 3.31)[44,45] When the layers are elastically bonded there is an additional stress, $\Delta\sigma$, on the longitudinal layers which has a maximum value, $\Delta\sigma_0$, in the plane of the transverse crack:

$$\Delta\sigma = \Delta\sigma_0 \exp\left(-\phi^{1/2}y\right) \qquad\qquad [3.69]$$

where

$$\phi = \frac{E_c G_{12}}{E_1 E_2}\left(\frac{b+d}{bd^2}\right)$$

and E_c is the initial composite modulus.

The corresponding load, F, transferred back into the transverse layer at a distance, y, is given by:

$$F = 2bc\Delta\sigma_0\left[1 - \exp(-\phi^{1/2}y)\right] \qquad\qquad [3.70]$$

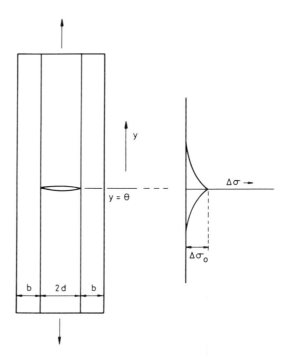

3.31   Transverse crack in a cross-ply laminate.

The first crack in the transverse lamina will occur when the applied stress on the composite is $E_c \, \varepsilon_2^*$ which is equivalent to a load of $2\sigma_2^* dc$. The next cracks will form when the ply is again loaded to a similar degree by shear transfer. This can be obtained by substituting the value of the required load $(2\sigma_2^* dc)$ into equation 3.69. The new value of $\Delta\sigma_0$ is given by:

$$\Delta\sigma_0 = \sigma_2^* \frac{d}{b}\left[1 - \exp(\phi^{1/2})\frac{s}{2}\right]^{-1} \qquad [3.71]$$

where $s/2$ is the crack spacing. Similar substitutions can be carried out for different values of crack spacing. In practice, the process will continue until the ultimate strength of the longitudinal plies is exceeded. Figure 3.32 shows a comparison of experimental and calculated crack spacing data for GRP.[46] Further consideration of the analysis can be conducted to yield minimum values of transverse ply cracking strain:

$$(\varepsilon_2^*)_{\min} = \frac{2bE_1\gamma_t\phi^{1/2}}{(b + d)E_2E_c} \qquad [3.72]$$

where γ_t is the fracture surface energy of the transverse ply.

Figure 3.33 shows the effect of ply thickness on the value of minimum

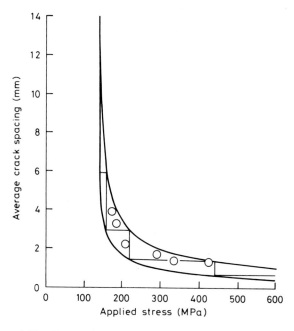

3.32   Comparison of experimental results with
theoretical curves of crack spacings as a function of
applied stress for GRP cross-ply laminates.

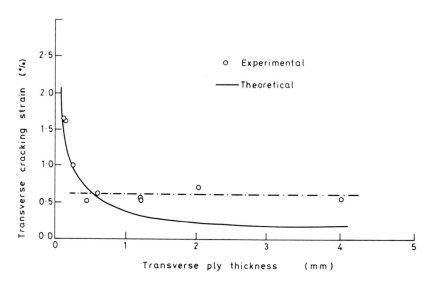

3.33   Transverse cracking strain for GRP cross-ply
laminates.

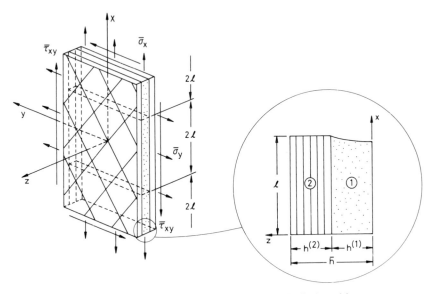

3.34   Composite laminate with a cracked ply subject
to in-plane loading.

cracking strain. At low values of transverse ply thickness the transverse failure strain tends to increase rapidly and this is borne out by the experimental results shown. This increase in apparent strength arises due to constraint of cracking through the presence of the outer stiff longitudinal plies. This is a good example where interaction between adjacent lamina gives rise to beneficial results which are not apparent with the classical laminate analysis.

The micromechanics of composite laminates has been the subject of considerable study and there are numerous models available.[47–54] Consider the cross-ply laminate shown in Fig. 3.34.[55] Uniformity of load, material and geometry are assumed in the y direction together with lamina orthotropy. The analysis leads to constitutive relationships for the cracked laminate:[55]

$$\bar{\sigma}_x = \bar{A}_{11}\bar{\varepsilon}_x + \bar{A}_{12}\bar{\varepsilon}_y$$

$$\bar{\sigma}_y = \bar{A}_{12}\bar{\varepsilon}_x + \bar{A}_{22}\bar{\varepsilon}_y$$

[3.73]

where $\bar{\sigma}$ and $\bar{\varepsilon}$ are average stresses and strains. The stiffness terms, $\bar{A}_{ij}$, for the damaged laminates are given by:

$$\bar{A}_{11} = \frac{h^{(1)}Q_{11}^{(1)} + h^{(2)}Q_{11}^{(2)}}{\beta_2 h}$$

$$\bar{A}_{12} = \frac{\beta_1 h^{(1)}Q_{12}^{(1)} + \beta_2 h^{(2)}Q_{11}^{(2)}}{\beta_2 h}$$

$$\bar{A}_{22} = \frac{h^{(1)}Q_{22}^{(1)} + h^{(2)}Q_{22}^{(2)}}{h} - \left(\frac{\beta_2 - \beta_1}{\beta_2}\right)\frac{h^{(1)}}{h}\frac{Q_{12}^{(1)2}}{Q_{11}^{(1)}}$$

and

$$\beta_1 = 1 - \frac{\tanh \alpha_1 l}{\alpha_1 l}$$

$$\beta_2 = 1 + \frac{h^{(1)}Q_{11}^{(1)}}{h^{(2)}Q_{11}^{(2)}} \cdot \frac{\tanh \alpha_1 l}{\alpha_1 l}$$

where Q_{ij} are the stiffness properties of the lamina, l is the crack spacing and α_1 is given by:

$$\alpha_1^2 = \frac{3Q_{55}^{(1)}Q_{55}^{(2)}}{h^{(1)}Q_{55}^{(2)} + h^{(2)}Q_{55}^{(1)}}\left(\frac{h^{(1)}Q_{11}^{(1)} + h^{(2)}Q_{11}^{(2)}}{h^{(1)}h^{(2)}Q_{11}^{(1)}Q_{11}^{(2)}}\right)$$

(Q_{55} is the out-of-plane shear stiffness term, $\tau_{yz} = Q_{55}\gamma_{yz}$.)
    Similarly for shear

$$\bar{\tau} = \bar{A}_{66}\bar{\gamma}_{xy} \qquad\qquad [3.74]$$

where

$$\bar{A}_{66} = \frac{h^{(1)}Q_{66}^{(1)} + h^{(2)}Q_{66}^{(2)}}{\beta_4 h}$$

and

$$\beta_4 = 1 + \frac{h^{(1)}Q_{66}^{(1)}}{h^{(2)}Q_{66}^{(2)}}\frac{\tanh \alpha_2 l}{\alpha_2 l}$$

$$\alpha_2^2 = \frac{3Q_{44}^{(1)}Q_{44}^{(2)}}{h^{(1)}Q_{44}^{(2)} + h^{(2)}Q_{44}^{(1)}}\left(\frac{h^{(1)}Q_{66}^{(1)} + h^{(1)}Q_{66}^{(2)}}{h^{(1)}h^{(2)}Q_{66}^{(1)}Q_{66}^{(2)}}\right)$$

(Q_{44} is the out-of-plane layer restrained by summing term $\tau_{xz} = Q_{44}\gamma_{xz}$.)
    Using the analysis it is possible to calculate the change in constitutive properties with crack density. Figure 3.35 shows the measured change in modulus values for the GRP flat plate specimens. Correlation between predicted values and experiment is good. It is not possible, however, to apply these elasticity calculations for the constitutive properties directly in a design-related situation. Measurements of crack density as a function of load are difficult to undertake and a means is required to be provided to link changes in stress/strain behaviour to externally applied stresses or strains. This is a fundamental difficulty with this type of analysis and it is likely that developments to achieve this link analytically will require the determination of materials-based parameters that are not amenable to easy evaluation. To a

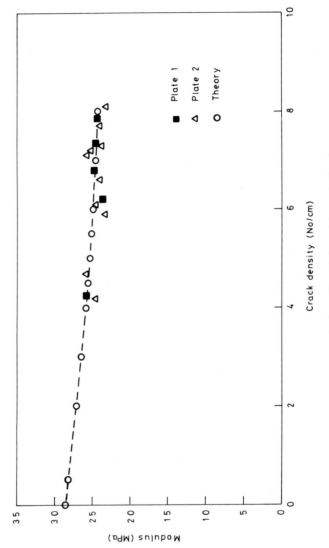

3.35 Variation of modulus with crack density.

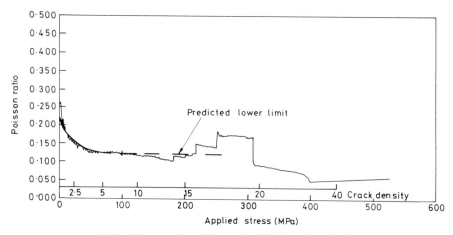

3.36   Measured and theoretical Poisson's ratio for
(0, 90)$_s$ GRP laminate.

greater or lesser extent a degree of empiricism or curve fitting to measured data
will be required. Figure 3.36 shows the measured variation of Poisson's ratio
with applied stress for a (0, 90)$_s$ GRP cross-ply laminate. Initially there is a
significant reduction where the value fails from its initial value of 0.26 to
approximately 0.12, after which it remains effectively unchanged. Also shown
in Fig. 3.36 are results of calculations using equation 3.75 for (0, 90)$_s$ cross-ply
laminates. The reduction of Poisson's ratio with crack density is of similar
form to the change with applied stress and the calculated and measured lower
limits are approximately equal. Because of the rapid change in value, the use of
Poisson's ratio measurements offers opportunities as a means of relating crack
density and stress. In Fig. 3.36 the crack density calculations and applied stress
plots are scaled with respect to the measured values of Poisson's ratio. Crack
density and applied stress can now be related for this material and then used to
explore the change of other properties with load. As an example, Fig. 3.37
shows the calculated stress/strain curve for the cross-ply GRP. Agreement
between theory and experiment is good.

Applying the micromechanical analysis to the problem of combined loading
yields good results (Fig. 3.38).[56] For a cylindrical specimen the envelope
shown would represent combinations of internal pressure and axial loading.
As can be seen, the relationship between the calculated envelope using a
conventional analysis (Tsai–Hill) and the experimental data is poor. The
analysis presented in equations 3.75 and 3.76 is not strictly applicable to
angle-ply laminates as the nature of cracking is more complex than with the
simple cross-ply systems. However, the ± 75° layup to which Fig. 3.38 refers is
sufficiently shallow so that some approximation to material behaviour can be
made. A further difference is that unlike a cross-ply laminate where the

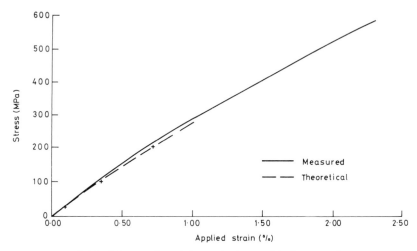

3.37   Measured and theoretical stress/strain curve for
(0, 90)ₛ GRP laminate.

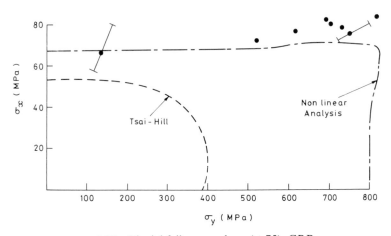

3.38   Biaxial failure envelope (±75)ₛ GRP.

restraining layers are of comparatively high modulus in the ± 75° system each
layer is of the same stiffness. However, the presence of the alternating + θ and
− θ layers will act as restraints to transverse cracks. Applying the micromechanics analysis gives rise to calculated nonlinear stress/strain behaviour and
this can be used in the failure analysis. The result of nonlinearities is to
redistribute the load within the material from the transverse direction on to the
reinforcement and this increases the apparent strength of the laminate. Figure
3.38 shows the effect of inelastic behaviour on the calculated failure envelope
and the degree of correlation with experimental data is much improved.

References

1 Ekvall J C, 'Elastic properties of orthotropic monofilament laminates', ASME Paper 61-AV-56, Los Angeles, 1961.

2 Adams D F and Tsai S W, 'The influence of random filament packing on the elastic properties of composite materials', *J Comp Mat*, **3**, 368–381, July 1969.

3 Whitney J M and Riley M B, 'Elastic properties of fiber reinforced composite materials', *AIAA Journal*, **5**, 1537–1542, September 1966.

4 Whitney J M, 'Elastic moduli of unidirectional composites with anisotropic filaments', *J Comp Mat*, **1**, 188–193, April 1967.

5 Hashin Z, 'On elastic behaviour of fibre reinforced materials of arbitrary transverse plane geometry', *J Mech Phys Solids*, **13**, 119–134, 1965.

6 Hashin Z and Rosen B W, 'The elastic moduli of fibre reinforced materials', Paper No 63-WA-175 *J Appl Mech*, ASME, 1964.

7 Tsai S W, 'Structural behaviour of composite materials', NASA CR-71, July 1964.

8 Halpin J C and Tsai S W, 'Effects of environmental factors on composite materials', AFML-TR 67–423, June 1969.

9 Tsai S W and Hahn H T, *Introduction to Composite Materials*, Technomic, Westport, Connecticut, 1980.

10 Hull D, *An Introduction to Composite Materials*, Cambridge University Press, Cambridge, 1981.

11 Noyes J V and Jones B H, 'Analytical design procedures for the strength and elastic properties of multilayer fiber composites', Proceedings of the AIAA/ASME 9th Structures, Dynamics and Materials Conference, Paper 68–336, 1968.

12 Darlington M W, McGinley P L and Smith G R, 'Creep anisotropy and structure in short-fibre-reinforced thermoplastics. 1 Prediction of 100–S creep modulus at small strains', *Plas Rub Mat App*, **2**, 51–58, 1977.

13 Cox H L, 'The elasticity and strength of paper and other fibrous materials', *Brit J App Phys*, **3**, 72–79, 1952.

14 Schapey R A, 'Thermal expansion coefficients of composite materials based on energy principles', *J Comp Mat*, **2**, 380–404, 1968.

15 Rosen B W, 'Thermomechanical properties of fibrous composites', *Proc R Soc*, **A319**, 79–94, 1970.

16 Jones R M, *Mechanics of Composite Materials*, Scripta, Washington, 1975.

17 Hancock P and Cuthbertson R C, 'Effect of fibre length and interfacial bond in glass fibre-epoxy resin composites', *J Mat Sci*, **5**, 762–768, 1970.

18 BS 3496, 'E-glass fibre chopped strand mat for the reinforcement of polyester resin systems', BSI, London, 1989.

19 Greszczuk L B, 'Consideration of failure modes in the design of composite materials', AGARD Conference Proceedings 163, Failure Modes of Composite Materials with Organic Matrices and their Consequences on Design, Paper 12, 1974.

20 Chamis C C and Sinclair J H, Proceedings of the 37th Annual Conference of the Society of Plastics Industry (SPI) Reinforced Plastics/Composites Institute, New York, 1982.

21 Chamis C C and Sinclair J H, Proceedings of the 37th Annual Conference of the Society of Plastics Industry (SPI) Reinforced Plastics/Composites Institute, New York, 1984.

22 Timoshenko S P, *Strength of Materials*, Van Nostrand Reinhold, London, 1958.

23 Hill R, 'A theory of the yielding and plastic flow of anisotropic metals', R Soc Proc, Series A, **193**, 281, 1948.

24 Tsai S W, 'Strength characterization of anisotropic materials', NASA-CR 224, 1965.

25 Azzi V D and Tsai S W, 'Anisotropic strength of composites', *Exp Mech*, **5**, 283, 1965.

26 Norris C B, 'Strength of orthotropic materials subjected to combined stresses', Forest Products Lab., FPL 1816, 1950.

27 Puppo A H and Evensen H A, 'Strength of anisotropic materials under combined stresses', *AIAA Journal*, **10**, 468, 1972.

28 Hotter H, Schelling H and Krauss H, 'An experimental study to determine the failure envelope of composite materials with tubular specimens under combined loads and comparison between several classical criteria', AGARD-CP-163, 1975.

29 Eckold G C, PhD Thesis, 'Strength and elastic properties of filament wound composites', UMIST, 1978.

30 Gol'denblat II and Kopnov V A, 'Strength of glass reinforced plastics in the complex stress state', *Mekhanika Polimerov*, **1**, 70, 1965.

31 Tsai S W and Wu E M, 'A general theory of strength for anisotropic materials', *Journal of Composite Materials*, **5**, 58, 1971.

32 Huang C L and Kimser P G, 'A criterion for strength for orthotropic materials', *Fibre Sci Tech*, **8**, 103, 1975.

33 Sinclair J H and Chamis C C, 'Fracture modes in off-axis fiber composites', Proceedings of the 34th SPI/RP Annual Technology Conference, Paper 22-A, Society of the Plastics Industry, New York, 1979.

34 Tsai S W and Wu E M, 'A general theory of strength for anisotropic materials', *J. Comp Mat*, **5**, 58–82, 1971.

35 Pipes R B and Cole B W, 'On the off axis strength test for anisotropic materials', *J. Comp Mat*, **7**, 246–256, 1973.

36 Johnson A F, 'Engineering Design Projects of GRP', BPF, London, 1975.

37 Knape W and Schneider W, 'The role of failure criteria in the fracture analysis of fibre-matrix composites', in *Deformation and Fracture of High Polymers*, Ed HH Kausch, J A Hassell and R I Jaffee, Plenum Press, New York, 1973, pp 543–556.

38 Eckold G C, 'A design method for filament wound GRP vessels and pipework', *Composites*, **16**, 41–47, 1985.

39 Greszczuk L B, 'Consideration of failure modes in the design of composite structures', AGARD CP-163, Munich, 1975.

40 Schewetz P T and Schwartz H S, *Fundamental Aspects of Fibre Reinforced Plastic Composites*, Wiley Interscience, New York, 1968.

41 Chai T T, 'Design for commercial filament winding', *SPE J*, **22**, 43, 1966.

42 Swanson S R and Trask B C, 'An examination of failure strength in $(0/\pm60)$ laminates under biaxial stress', *Composites*, **19**, 400–406, 1988.

43 Eckold G C, Leadbetter D, Soden P D and Griggs P R, 'Lamination theory in the prediction of failure envelopes for filament wound materials subjected to biaxial loading', *Composites*, **9**, 4, 243–246, 1978.

44 Aveston J and Kelly A, 'Theory of multiple fracture in fibrous composites', *J. Mat Sci*, **8**, 352–362, 1973.

45 Garrett K W and Bailey J E, 'Multiple transverse fracture in 90' cross ply laminates of a glass reinforced polyester', *J Mater Sci*, **12**, 157–168, 1977.

46 Parvizi A, Garrett K W and Bailey J E, 'Constrained cracking in glass fibre reinforced epoxy cross ply laminates', *J Mat Sci*, **13**, 195–201, 1978.

47 Hancox N L and McCartney L N, 'Critical appraisal of micro mechanical and continuum methods of predicting the effects of damage in and the properties of composite materials', Report of the Industrial Advisory Committee on Fracture Avoidance, 1989.

48 McCartney L N, 'New theoretical model of stress transfer between fibre and matrix in a uniaxially fibre reinforced composite', NPL Report DMA(A)168, November, 1988.

49 Ogin S L and Smith P A, 'A model for matrix cracking in cross ply laminates', *ESA Journal*, **11**, 45–60, 1987.

50 Kelly A and McCartney L N, 'Matrix cracking in fibre reinforced and laminated composites', *Proceedings of the International Conference on Composite Materials*, Vol. 3, Elsevier Applied Science, London, 1987, pp 210–222.

51 McCartney L N, 'Theory of stress transfer in a cross ply laminate containing a parallel array of transverse cracks', *J Mech Phys Solids*, **40**, 27–68, 1992.

52 Tan S C and Nuismer R J, 'A theory for progressive matrix cracking in composite laminates', *J Comp Mat*, **23**, 1029–1047, 1989.

53 Zhang J, Fan J and Soutis C, 'Analysis of multiple matrix cracking in $[\pm\theta_m/90_n]_s$ composite laminates, Part 1: In-plane stiffness properties', *Composites*, **23**, 291–298, 1992.

54 Joshi G P and Frantziskonis G, 'Damage evolution in laminated advanced composites', *Comp Struct*, **17**(1), 127–139, 1991.

55 Nuismer R J and Tan S C, 'Constitutive relations of a cracked composite lamina', *J. Comp Mat*, **22**, 306–321, 1988.

56 Highton J, Adeoye A B and Soden P D, 'Fracture stresses for $\pm$ 75° filament wound GRP tubes under biaxial loads', *J Strain Analysis*, **20**, 139–150, 1985.

4

BEAMS, PLATES AND SHELLS

In the previous chapter the basic principles for deriving a laminate construction to accommodate a variety of applied loadings were described. Essentially this provides the first stage in composite component design, that is tailoring the constituent materials to suit operating conditions. The next step is to assess the boundary conditions applied to the laminate in the form of forces, bending moments, displacements and curvatures. For very complex structures where these calculations would be intractable it is usual to employ a numerical technique such as finite elements to construct a model in which the component is subdivided into a number of small units. The response of each element is calculable and from these individual contributions and by applying conditions of compatibility across element boundaries the behaviour of the structure as a whole can be derived. There are, however, a number of technologically important structures whose behaviour can be derived from the principles of mechanics. Typical of these are beams, plates and shells. Many engineering systems consist of units or assemblies of such structures and the ability to quantify their behaviour is therefore a basic requirement. Even where a numerical analysis is proposed, the availability of expressions suitable for 'hand' calculations is necessary to allow validation of the computer model of the structure.

Beams

Beams are perhaps the most common type of structure found in engineering design and their behaviour under a host of applied loads and boundary conditions is well documented. For beams constructed from conventional isotropic materials the solutions for a variety of configurations of loading and edge constraint are well known and available from a series of standard texts.[1,2] For example, the solution for a simple cantilever of rectangular cross-section (Fig. 4.1), subject to an end load is given by:

$$M_{max} = PL$$

$$\sigma_{max} = 6PL/bt^2$$

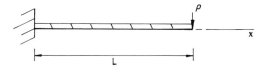

4.1   Bending deformation of a cantilever.

$$\delta_{max} = \frac{PL^3}{3EI}$$

$$\theta = \frac{PL^2}{2EI} \tag{4.1}$$

where M_{max} and σ_{max} are the maximum moment and stress due to the applied loading P located at the end of the beam $x = L$, δ_{max} is the deflection at the tip of the beam, θ the beam rotation, b and t are the width and thickness of beam respectively and EI is the flexural rigidity.

In the case of an anisotropic beam the solution to a first approximation may be obtained by introducing the stiffness along the axis of the beam into the above expressions. To understand the complete behaviour of such a beam, however, recourse must be made to the principles of elasticity.[3,4]

Consider the equilibrium of the small rectangular element of unit thickness shown in Fig. 4.2. In the absence of body forces the equation of equilibrium in the x direction is:

$$\left(N_x + \frac{\delta N_x}{\delta x} \cdot dx \right) - N_x + \left(N_{xy} + \frac{\delta N_{xy}}{\delta y} \cdot dy \right) - N_{xy} = 0 \tag{4.2}$$

At the limit, equation 4.2 becomes:

$$\frac{\partial N_x}{\partial x} + \frac{\partial N_{xy}}{\partial y} = 0 \tag{4.3}$$

Applying the same procedure in the y direction yields:

$$\frac{\partial N_{xy}}{\partial x} + \frac{\partial N_y}{\partial y} = 0 \tag{4.4}$$

Equations 4.2 and 4.3 are the governing equilibrium equations for the rectangular section shown in Fig. 4.2 under the conditions of plane stress. The next stage in the analysis is to drive the constitutive relations for the material. As has already been shown (Chapter 3) the compliance relationships for the material are:

$$\begin{bmatrix} \varepsilon_x \\ \varepsilon_y \\ \gamma_{xy} \end{bmatrix} = \begin{bmatrix} \bar{S}_{12} & \bar{S}_{12} & \bar{S}_{13} \\ \bar{S}_{21} & \bar{S}_{22} & \bar{S}_{13} \\ \bar{S}_{31} & \bar{S}_{32} & \bar{S}_{33} \end{bmatrix} \begin{bmatrix} N_x \\ N_y \\ N_{xy} \end{bmatrix} \tag{4.5}$$

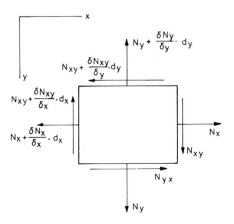

4.2   Equilibrium of a rectangular element.

The differential equations given above are usually used to determine the state of stress in a body subjected to the action of a given set of forces. Equations 4.3 and 4.4 are not sufficient in themselves and the elastic deformation of the body must also be considered. Consider a small element, $dx\,dy\,dz$ (Fig. 4.3). If the body undergoes a deformation of which u, v, and w are components of displacement of the point 0 in the x, y and z directions, the displacement in the x direction, δx_A, of an adjacent point A on the x axis is:

$$\delta x_A = u + \frac{\partial u}{\partial x} \cdot dx \qquad [4.6]$$

The increase in length of the element 0A due to the deformation is therefore $(\partial u/\partial x)\cdot dx$ and the unit elongation in the x direction is $\partial u/\partial x$. Similarly, in the y direction the unit elongation is $\partial v/\partial y$.

Shear deformations can be considered using a similar approach (Fig. 4.3). The displacements of point A in the y direction, δy_A, and that of point B in the x direction, δx_B, are:

$$\delta y_A = v + \left(\frac{\partial v}{\partial x}\right) dx \qquad [4.7]$$

$$\delta x_B = u + \left(\frac{\partial u}{\partial y}\right) dy \qquad [4.8]$$

Because of these displacements the new direction 0'A' of element 0A is inclined to the initial direction by $\partial v/\partial x$ and 0'B' is inclined to 0B by $\partial u/\partial y$. The shearing strain, i.e. the change of angle A0B, is therefore given by the sum $(\partial u/\partial y + \partial v/\partial x)$.

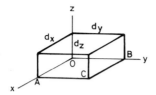

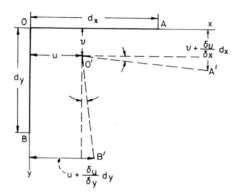

4.3 Deformation of an element of an elastic body.

The strain displacement equations can therefore be written as:

$$\varepsilon_x = \frac{\partial u}{\partial x}$$

$$\varepsilon_y = \frac{\partial v}{\partial y}$$

[4.9]

$$\gamma_{xy} = \frac{\partial u}{\partial y} + \frac{\partial v}{\partial x}$$

where ε_x, ε_y and γ_{xy} are the normal and shear strains in the xy plane.

Differentiating equations 4.9, the first twice with respect to y, the second twice with respect to x and the third once with respect to x and once with respect to y yields:

$$\frac{\partial^2 \varepsilon_x}{\partial y^2} + \frac{\partial^2 \varepsilon_y}{\partial x^2} = \frac{\partial^2 \gamma_{xy}}{\partial x \partial y}$$

[4.10]

This relation is known as the condition of compatibility.

The usual method of solving these equations is by use of stress function method.[3-5] Consider a function $F(x,y)$ where:

$$N_x = \frac{\partial^2 F}{\partial y^2}$$

$$N_y = \frac{\partial^2 F}{\partial x^2} \qquad [4.11]$$

$$N_{xy} = \frac{-\partial^2 F}{\partial x \partial y}$$

It can be easily shown by substitution that equations 4.3 and 4.4 are satisfied by this definition of $F(x,y)$.

The objective of the analysis is now to derive expressions for $F(x,y)$ which can be used to evaluate the stress components N_x, N_y and N_{xy}. Substituting expressions for the strain components ε_x, ε_y and γ_{xy} from the constitutive relation, equation 4.5, and expressing the stress components in terms of the stress function yields the following differential equation:

$$\bar{S}_{22}\frac{\partial^4 F}{\partial x^4} - 2\bar{S}_{23}\frac{\partial^4 F}{\partial x^3 \partial y} + (2\bar{S}_{12} + \bar{S}_{33})\frac{\partial^4 F}{\partial x^2 \partial y^2}$$

$$- 2\bar{S}_{13}\frac{\partial^4 F}{\partial x \partial y^3} + \bar{S}_{11}\frac{\partial^4 F}{\partial y^4} = 0 \qquad [4.12]$$

Considering the case of an isotropic material where the coupling terms in the compliance matrix become zero, equation 4.12 reduces to:

$$\frac{\partial^4 F}{\partial x^4} + \frac{2\partial^4 F}{\partial x^2 \partial y^2} + \frac{\partial^4 F}{\partial y^4} = 0 \qquad [4.13]$$

the form of which is found in many standard texts.[3,4]

For beams, particular solutions of equation 4.12 can be obtained for a range of loading conditions. These are often in the form of polynomials and, by taking solutions of different orders and applying suitable boundary conditions, a number of important problems can be considered.[6]

In the case of the beam shown in Fig. 4.4 it can be shown that the stress function has the following form:

$$F = \frac{P}{t^3}\left\{ -2xy^3 + \frac{3t^2 xy}{2} + \frac{\bar{S}_{13}}{2\bar{S}_{11}}(t^2 y^2 - 2y^4) \right\} \qquad [4.14]$$

where coordinates x and y are defined by reference to Fig. 4.4.

The stress resultants can now be determined using equation 4.11:

$$N_x = -12\frac{P}{t^3}xy + \frac{P}{t^3}\cdot\frac{\bar{S}_{13}}{\bar{S}_{11}}(t^2 - 12y^2)$$

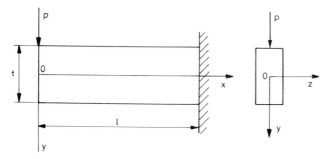

4.4   Rectangular cantilever subject to an end load.

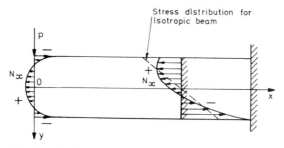

4.5   Parabolic stress distributions across the thickness
of an anisotropic beam.

$$N_y = 0 \qquad\qquad [4.15]$$

$$N_{xy} = \frac{-6P}{t^3}\left(\frac{t^2}{4} - y^2\right)$$

In an orthotropic beam in which the axis of the beam corresponds to the axis of orthotropy, the coupling coefficient $\bar{S}_{13}$ becomes zero and therefore equation 4.15 reduces to:

$$N_x = \frac{-12P}{t^3} xy$$

$$N_y = 0 \qquad\qquad [4.16]$$

$$N_{xy} = \frac{-6P}{t^3}\left(\frac{t^2}{4} - y^2\right)$$

Equations 4.16 are the same as those for an isotropic beam and as can be seen there is a linear distribution of stress across the thickness of the beam, i.e. maxima and minima at the outer surfaces and the neutral axis at the centroid.

For the case where $\bar{S}_{13} \neq 0$ the axial stress, N_x, is not distributed linearly across the thickness of the beam and a parabolic law results (Fig. 4.5). The

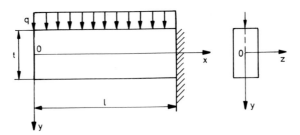

4.6   Rectangular cantilever subject to a distributed
load.

maximum value of the axial stress resultant which may no longer be located on
the outer surfaces $N_{x,max}$ is equal to:

$$N_{x,max} = \frac{6PL}{t^2}\left(1 + \frac{\bar{S}_{13}}{\bar{S}_{11}} \cdot \frac{t}{3L}\right) \qquad [4.17]$$

Uniformly loaded beams can be treated in a similar way (Fig. 4.6), except in
these cases a polynomial of higher order is required. For a cantilever subjected
to a uniformly distributed load the stress components are given by:

$$N_x = \frac{-6qx^2 y}{t^3} + q\left\{\frac{\bar{S}_{13}}{\bar{S}_{11}} \cdot \frac{x}{t}\left(1 - \frac{12y^2}{t^2}\right) + 2\left(\frac{\bar{S}_{12} + \bar{S}_{33}}{4\bar{S}_{11}} - \frac{\bar{S}_{13}^2}{\bar{S}_{11}^2}\right)\left(\frac{4y^3}{t^3} - \frac{3y}{5t}\right)\right\}$$

$$N_y = \frac{q}{2}\left(-1 + \frac{3y}{t} - \frac{4y^3}{t^3}\right) \qquad [4.18]$$

$$N_{xy} = \frac{-6qx}{t^3}\left(\frac{t^2}{4} - y^2\right) - q \cdot \frac{\bar{S}_{13}}{\bar{S}_{11}}\left(\frac{y}{t} - \frac{4y^3}{t^3}\right)$$

The first terms of equation 4.18 are those that would be determined by
elementary bending theory for an isotropic beam. The additional terms are
dependent on the relationships between elastic constants. For an orthotropic
laminate in which the beam axis corresponds to one of the principal material
directions:

$$N_x = \frac{-6qx^2 y}{t^3} + qm\left(\frac{4y^3}{t^3} - \frac{3y}{5t}\right)$$

$$N_y = \frac{q}{2}\left(-1 + \frac{3y}{t} - \frac{4y^3}{t^3}\right) \qquad [4.19]$$

$$N_{xy} = \frac{6qx}{t^3}\left(\frac{t^2}{4} - y^2\right)$$

where $m = (2\bar{S}_{12} + \bar{S}_{33})/2\bar{S}_{11}$.

Recalling the definition of the terms in the compliance matrix allows m to be

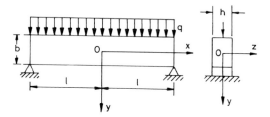

4.7 Supply supported beam subject to a distributed load.

expressed in terms of engineering constants, viz.:

$$m = \frac{1}{2}\left(\frac{E_{11}}{G_{12}} - 2v_{12}\right)$$ [4.20]

The greater the degree of anisotropy (or ratio of E_{11}/G_{12}), therefore the greater the value of m and the more significant the deviation from conventional beam theory.

Using equations 4.5 and 4.9 to determine the deflections, δ, at the end of the beam and the beam curvature, $1/\rho$, yields:

$$\delta = \frac{q\bar{S}_{11}L^4}{8I} - \frac{qt^2L^2}{8I}\left(3\bar{S}_{12} + 4\bar{S}_{33} - \frac{8\,\bar{S}_{13}^2}{3\,\bar{S}_{11}}\right)$$ [4.21]

$$\frac{1}{\rho} = \frac{M\bar{S}_{11}}{I} + \frac{qt^2}{40I}\left(3\bar{S}_{12} + 4\bar{S}_{33} - \frac{8\,\bar{S}_{13}^2}{3\,\bar{S}_{11}}\right)$$

where I is the moment of inertia of the beam cross-section and M is the bending moment at the cross-section at x and is given by:

$$M = \frac{-qx^2}{2}$$ [4.22]

Again the first terms in the above equation are those that would be determined from elementary bending theory. The deflection represented by the second term is primarily due to the contribution of shear deformation. Whilst this, to a certain extent, is present in all beams it is usually ignored in isotropic materials, except where the beam is short, as it is comparatively insignificant. This may not be the case for a composite beam, particularly if the reinforcement is aligned along the beam axis. This is due to the fact that in such a case $E_{11} >> G_{12}$ and therefore the contribution of the second terms in equation 4.21 can be considerable. To quantify this effect, consider the further case of a simply supported beam with a uniformly distributed load (Fig. 4.7).

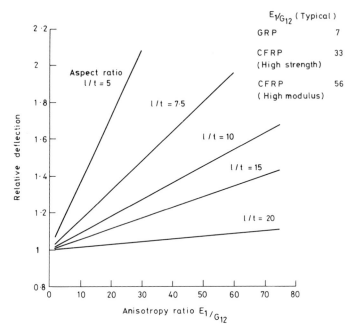

4.8   Effect of shear deformation on beam deflection.

Formulae for stress, deflection and curvature are derived as per the previous example:

$$N_x = \frac{6q}{t^2}(L^2 - x^2)y + q\left\{\frac{-\bar{S}_{13}}{\bar{S}_{11}} \cdot \frac{x}{t}\left(1 - \frac{12y^2}{t^2}\right)\right\}$$

$$+ 2q\left\{\left(\frac{2\bar{S}_{12} + \bar{S}_{33}}{4\bar{S}_{11}} - \frac{\bar{S}_{13}^2}{\bar{S}_{11}^2}\right)\left(\frac{4y^3}{t^3} - \frac{3y}{5t}\right)\right\} \qquad [4.23]$$

$$\delta = \frac{5q\bar{S}_{11}L^4}{24I} + \frac{qt^2L^2}{80I}\left(3\bar{S}_{12} + 4\bar{S}_{33} + \frac{32\,\bar{S}_{13}^2}{3\,\bar{S}_{11}}\right)$$

$$\frac{1}{\rho} = \frac{M\bar{S}_{11}}{I} + \frac{qt^2}{40I}\left(3\bar{S}_{12} + 4\bar{S}_{33} + \frac{32}{3} \cdot \frac{\bar{S}_{13}^2}{\bar{S}_{11}}\right)$$

Figure 4.8 shows the effect of shear deformation on the calculated deflection for a beam. As can be seen, the magnitude of the effect becomes significant as the degree of anisotropy increases.

An elementary derivation of the effect of shear forces on deflection and curvature can be carried out assuming the slope of the deflection curve of the

Table 4.1. Shear coefficient for geometrical forms

Section	α
Rectangle	3/2
Circle	4/3
Thin tube	2
Box / I	A/A_{web}

beam due to shear alone is approximately equal to the shear strain at the neutral axis.[2,3] The total curvature under the action of a distributed load can be shown to be:

$$\frac{d^2v}{dx^2} = \frac{M}{EI} - \frac{\alpha q}{GA}$$

[4.24]

Essentially q/A is the average shear stress obtained by dividing the shear force by the cross sectional area of the beam and α is a numerical factor by which the average shear stress must be multiplied to obtain the shear stress at the centroid of the cross-section. The value of α varies therefore with the geometry of the cross-section. Table 4.1 has typical values. Equation 4.24 can be integrated with the appropriate boundary conditions to provide solutions for a range of design cases. For the simply supported beam the total deflection is given by:

$$\delta = \frac{5q\bar{S}_{11}}{24I} L^4 \left(1 + \frac{12}{5} \cdot \frac{\alpha I \bar{S}_{33}}{\bar{S}_{11} bt L^2} \right)$$

[4.25]

Calculations based on this simple approach tend to exaggerate the effect of shear deformation and a judgement must be made as to which approach is most suitable for a given design problem.

Common engineering structures consist of laminated beams where individual layers are orthotropic. In this case the analysis is, to a certain extent, simplified with the proviso that the beam may not be symmetrical about its midplane.[5] The basic principles developed in Chapter 3 may be employed. Assuming that the total laminate thickness remains sufficiently thin that plane sections remain plane on deformation, the strain displacement relationships can be written as:

$$\varepsilon_x = \varepsilon_x^0 - z\kappa_x$$
$$\varepsilon_y = 0 \qquad\qquad\qquad [4.26]$$
$$\gamma_{xy} = 0$$

where ε_x, ε_y and γ_{xy} are the normal and shear strains in the x, y directions, ε_x^0 the midplane strain in the x direction and κ_x the curvature in the x direction.

The stress/strain relationships for the kth lamina are:

$$\sigma_x^{(k)} = \bar{Q}_{11}^{(k)}\,\varepsilon_x$$
$$\sigma_y^{(k)} = \bar{Q}_{12}^{(k)}\,\varepsilon_x \qquad\qquad\qquad [4.27]$$
$$\tau_{xy}^{(k)} = \bar{Q}_{13}^{(k)}\,\varepsilon_x$$

From the constitutive relationships for the laminate (Chapter 3) expressions for stress and moment resultants are therefore:

$$N_x = A_{11}\varepsilon_x^0 - B_{11}\kappa_x$$
$$N_y = A_{12}\varepsilon_x^0 - B_{12}\kappa_x \qquad\qquad\qquad [4.28]$$
$$N_{xy} = A_{13}\varepsilon_x^0 - B_{13}\kappa_x$$

$$M_x = B_{11}\varepsilon_x^0 - D_{11}\kappa_x$$
$$M_y = B_{12}\varepsilon_x^0 - D_{12}\kappa_x \qquad\qquad\qquad [4.29]$$
$$M_{xy} = B_{13}\varepsilon_x^0 - D_{13}\kappa_x$$

The A, B and D terms are the coefficients in the appropriate stiffness matrices.

Consider the beam shown in Fig. 4.9 subjected to a moment distribution $M_x(x)$ and a transverse force $q(x)$.[5] As the in-plane force, N_x, is equal to 0:

$$\varepsilon_x^0 = \frac{B_{11}}{A_{11}}\kappa_x \qquad\qquad\qquad [4.30]$$

which by substitution into equation 4.29 gives

$$\kappa_x = \left(\frac{A_{11}}{B_{11}^2 - A_{11}D_{11}}\right)M_x(x) \qquad\qquad\qquad [4.31]$$

Equations 4.30 and 4.31 are the governing equations for a laminated beam subjected to the boundary conditions which give rise to a bending moment distribution of $M_x(x)$. In form they are identical to those pertaining to an isotropic beam, i.e. $\kappa_x = M/EI$, except that the flexural rigidity EI is replaced

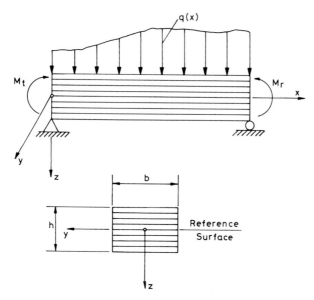

4.9 Laminated beam with applied moment and
transverse pressure.

by the term $(B_{11}^{2} - A_{11}D_{11})/A_{11}$. An important point to note is that unless $B_{11} = 0$, i.e. the laminate is symmetrical with respect to the centre-line, there is no unstrained reference surface or neutral axis. Also, the other components of force and moment resultant, N_y, N_{xy}, M_y and M_{xy} are non-zero and if deformations (warpage) are restrained they are developed at the support positions. The magnitude of these forces and moments can be determined by using the value of κ_x (equation 4.31) and equations 4.28 and 4.29.

Substitution of equation 4.30 and 4.31 into equation 4.26 yields solutions for the strain for each lamina.

$$\varepsilon_x = \frac{A_{11}}{B_{11}^{2} - A_{11}D_{11}}\left(\frac{B_{11}}{A_{11}} - z\right)M_x(x) \qquad [4.32]$$

which using the lamina constitutive equation gives rise to the following lamina stress systems:

$$\sigma_x^{(k)} = \bar{Q}_{11}^{(k)}\left(\frac{A_{11}}{B_{11}^{2} - A_{11}D_{11}}\right)\left(\frac{B_{11}}{A_{11}} - z\right)M_x(x)$$

$$\sigma_y^{(k)} = \bar{Q}_{12}^{(k)}\left(\frac{A_{11}}{B_{11}^{2} - A_{11}D_{11}}\right)\left(\frac{B_{11}}{A_{11}} - z\right)M_x(x) \qquad [4.33]$$

$$\tau_{xy}^{(k)} = \bar{Q}_{13}^{(k)}\left(\frac{A_{11}}{B_{11}^{2} - A_{11}D_{11}}\right)\left(\frac{B_{11}}{A_{11}} - z\right)M_x(x)$$

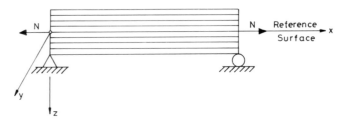

4.10   Laminated beam with axial load.

A further aspect related to the behaviour of beams, specifically those that are unsymmetrical with respect to the centre-line, occurs under simple tensile loading, i.e. $N_x = P$ and $M_x = 0$, (Fig. 4.10).[5] Due to the fact that $B_{11} \neq 0$, the beam will tend to bend as well as elongate. To restrain this induced curvature additional boundary forces and moments are necessary at support positions. From equation 4.29:

$$\varepsilon_x^0 = \frac{D_{11}}{B_{11}} \kappa_x \qquad\qquad [4.34]$$

which when substituted into equation 4.28 gives rise to:

$$\kappa_x = \left(\frac{B_{11}}{A_{11}D_{11} - B_{11}^2} \right) P \qquad\qquad [4.35]$$

By equating equations 4.31 and 4.35 the moment which must be applied to maintain flatness, i.e. $\kappa_x = 0$, can be derived, viz.:

$$M = \frac{B_{11} \cdot P}{A_{11}} \qquad\qquad [4.36]$$

Expressions can now be established for the stress distribution for each layer similar to those shown as equations 4.33.

The principles developed above can be used for any combination of load or support condition. The effect of laminate detail affects flexural rigidity only. The analysis of the beam itself is as per the corresponding isotropic case. One factor which has not been accounted for in the discussion of laminated beams is the effect of shear deformation. This must be considered in highly anisotropic systems and the approximate strength of materials approach will suffice for most purposes (see equations 4.24 and 4.25).

The ramifications of the beam analysis presented are important not only for component design purposes, but also in the interpretation of test data. To give an example, for certain applications such as chemically resistant structures, it is common to incorporate resin-rich layers on external surfaces for corrosion purposes and it can be shown that their presence can have a significant effect on measured properties.[6] Consider a laminate consisting of four layers of

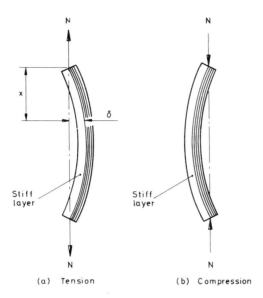

(a)   Tension   (b)   Compression

4.11   Effect of lateral deflection on an asymmetric
laminate (N = axial force).

chopped strand mat (CSM) and resin rich layers on both surfaces. Taking modulus and thickness values of 7.7 GPa and 4.2 mm, and 4.5 GPa and 0.5 mm/layers for the CSM and resin surfaces respectively and applying the derived theory yields a value of tensile modulus of 7.08 GPa and a flexural modulus of 6.19 GPa. Thus, despite the fact that the resin surfaces are very thin, the tensile modulus would be 8% lower than the CSM layers alone and the flexural modulus would be 20% lower. Similarly these expressions can be used to elucidate the behaviour of asymmetrical constructions. Whilst this analysis would be expected to give accurate results for low strain levels, care needs to be applied if deflections become significant. Under tensile conditions the effect of the non-zero components of the $[B]$ matrix can be overestimated. This is due to the fact that as the tension is applied, the less stiff layers tend to extend more, producing not only curvature, but also lateral deflection (Fig. 4.11). This, in turn, causes a correcting moment $P\delta$ to apply, which acts to reduce the curvature. The converse is true in compression where the induced bending moment will act to increase curvature. In both these cases the behaviour observed in practice will be dominated by the end constraints offered by the fixing details at the ends of the beam.

Plates

The analysis of flat plates is generally a complex problem. Although the generation of the basic equilibrium equation is straightforward, a solution in

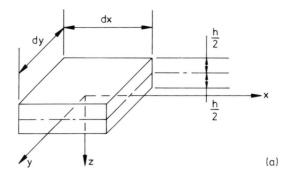

(a)

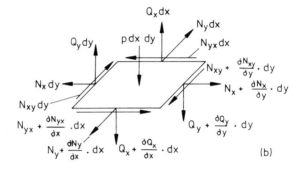

(b)

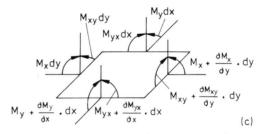

(c)

4.12   Forces and moments acting on a plate element.

closed form is rarely possible. For design purposes recourse is usually made to texts where the results of numerical calculations are represented in tabular form.[1] With composite structures the equations can become formidable and added complexity arises because of the coupling which arises between extensional and bending terms.

The governing equations for the out of plane bending of composite plates can be derived in the same way as for the element shown in Fig. 4.2. The forces and moments acting on a plate element are shown in Fig. 4.12. The equilibrium equations may be written as:[7]

$$\frac{\partial M_y}{\partial y} + \frac{\partial M_{xy}}{\partial x} = Q_x$$

$$\frac{\partial M_x}{\partial x} + \frac{\partial M_{xy}}{\partial y} = Q_y \qquad [4.37]$$

$$\frac{\partial Q_y}{\partial y} + \frac{\partial Q_x}{\partial y} = -q$$

Combining equations 4.37 using differentiation and substitution as before, yields:

$$\frac{\partial^2 M_x}{\partial x^2} + 2\frac{\partial^2 M_{xy}}{\partial x \partial y} + \frac{\partial^2 M_y}{\partial y^2} = -q \qquad [4.38]$$

Recalling that the curvatures are defined as:

$$\kappa_x = \frac{\partial^2 w}{\partial x^2}$$

$$\kappa_y = \frac{\partial^2 w}{\partial y^2} \qquad [4.39]$$

$$\kappa_{xy} = \frac{\partial^2 w}{\partial x \partial y}$$

and the stress and moment resultant/strain and curvature relationships derived in Chapter 3 give the following expressions for moment resultants:

$$M_x = -\left(D_{11}\frac{\partial^2 w}{\partial x^2} + D_{12}\frac{\partial^2 w}{\partial y^2} + 2D_{13}\frac{\partial^2 w}{\partial x \partial y} \right)$$

$$M_y = -\left(D_{12}\frac{\partial^2 w}{\partial x^2} + D_{22}\frac{\partial^2 w}{\partial y^2} + 2D_{23}\frac{\partial^2 w}{\partial x \partial y} \right) \qquad [4.40]$$

$$M_{xy} = -\left(D_{13}\frac{\partial^2 w}{\partial x^2} + D_{23}\frac{\partial^2 w}{\partial y^2} + 2D_{33}\frac{\partial^2 w}{\partial x \partial y} \right)$$

Substitution of these equations into equation 4.38 gives the governing differential equation for a unidirectional laminated plate, viz.:

$$D_{11}\frac{\partial^2 w}{\partial x^4} + 4D_{13}\frac{\partial^2 w}{\partial x^3 \partial y} + 2(D_{12} + 2D_{33})\frac{\partial^4 w}{\partial x^2 \partial y^2}$$

$$+ 4D_{23}\frac{\partial^2 w}{\partial x \partial y^3} + D_{22}\frac{\partial^4 w}{\partial y^4} = q \qquad [4.41]$$

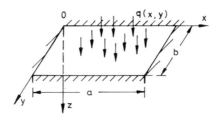

Figure 4.13   Bending of a rectangular plate with
simply supported edges.

Where the axes of the plate correspond to the principal material axes, the coupling terms in the stiffness matrix, D_{13} and D_{23}, become zero and equation 4.41 reduces to:

$$D_{11}\frac{\partial^4 w}{\partial x^4} + 2(D_{12} + 2D_{33})\frac{\partial^4 w}{\partial x^2 \partial y^2} + D_{22}\frac{\partial^4 w}{\partial y^4} = q \qquad [4.42]$$

For an isotropic plate $D_{11} = D_{22} = (D_{12} + 2D_{33}) = D$ and the equivalent expression for such a plate can now be easily obtained as:[6]

$$\frac{\partial^4 w}{\partial x^4} + 2\frac{\partial^4 w}{\partial x^2 \partial y^2} + \frac{\partial^4 w}{\partial y^2} = \frac{q}{D} \qquad [4.43]$$

The solution of these equations is usually obtained in series form. Consider the plate shown in Fig. 4.13. It is orthotropic with the principal material directions parallel to the plate axes. The edges are simply supported and the plate is subjected to a normal pressure loading $p(x,y)$. To obtain the solution for deflection equation 4.42 must be solved subject to the prevailing boundary conditions, viz.:

$$\text{at } x = 0 \text{ and } x = a \qquad w = M_x = 0$$

$$\text{at } y = 0 \text{ and } y = b \qquad w = M_y = 0$$

All of these conditions are satisfied by using a solution of the form:

$$w = \sum_{m=1}^{\infty} \sum_{n=1}^{\infty} A_{mn} \sin\frac{m\pi x}{a} \sin\frac{n\pi y}{b} \qquad [4.44]$$

where the coefficients A_{mn} are obtained from:

$$q = \sum_{m=1}^{\infty} \sum_{n=1}^{\infty} a_{mn} \sin\frac{m\pi x}{a} \sin\frac{n\pi y}{b} \qquad [4.45]$$

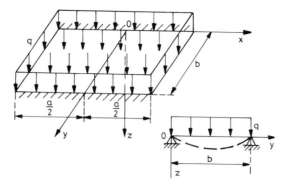

4.14 Rectangular plate subject to a uniform
transverse pressure load.

where

$$a_{mn} = \frac{4}{ab} \int_0^a \int_0^b q \sin \frac{m\pi x}{a} \sin \frac{n\pi y}{b} \, dx \, dy$$

This can be shown to be similar to the equivalent expression for an equivalent isotropic plate:[6]

$$w = \frac{1}{D\pi^4} \sum_{m=1}^{\infty} \sum_{n=1}^{\infty} \frac{a_{mn} \sin\left[(m\pi x)/a\right] \sin\left[(n\pi y)/b\right]}{\left[(m^2/a^2) + (n^2/b^2)\right]} \qquad [4.46]$$

Solution to these equations can be obtained through the application of appropriate numerical techniques after which values for moments and stresses may be obtained. For example, a plate with a uniformly distributed load (Fig. 4.14) can be shown to have maximum deflection and moment values as calculated from the following:[8]

$$w_{max} = \frac{qb^4}{D_{22}} \phi$$

$$M_{x,max} = \left(\mu_1 + \mu_2 \nu_{21} \sqrt{\frac{D_{11}}{D_{22}}} \right) \frac{qa^2}{\varepsilon^2}$$

$$M_{y,max} = \left(\mu_2 + \mu_1 \nu_{12} \sqrt{\frac{D_{22}}{D_{11}}} \right) qb^2 \qquad [4.47]$$

where

$$\varepsilon = \frac{a}{b} \left(\frac{D_{22}}{D_{11}} \right)^{1/4} \quad \text{for } a \geq b$$

and the coefficients ϕ, μ_1 and μ_2 are obtained from Table 4.2. Similar results

Table 4.2. Coefficients for plates with distributed pressure loading

ε	ϕ	μ_1	μ_2
1	0.004 07	0.0368	0.0368
1.5	0.007 72	0.0280	0.0728
2	0.010 13	0.0174	0.0964
2.5	0.011 50	0.0099	0.1100
3	0.012 23	0.0055	0.1172
5	0.012 97	0.0004	0.1245
∞	0.013 02	0.0000	0.1250

can be obtained for a variety of conditions and, in some cases, unexpected results may arise. For example, a plate with two simply supported and two fixed sides would be expected to have a maximum moment at the centre of the fixed sides. In an isotropic plate this is certainly the case. However, for certain orthotropic properties and plate conditions this may not prove to be a correct assessment and the maximum moment may occur at the centre. The point to note is that preconceived notions derived from experience with traditional materials need not apply – care needs to be taken.

For isotropic plates there are a number of texts with tabulated coefficients for deflection and bending moment.[1] Orthotropic plates can be treated in a similar way except there are more parameters which must be considered.[8] From equation 4.42 it can be seen that the flexural rigidity terms D_{11}, D_{22} and D'_{33}, defined as $(D_{12} + 2D_{33})$, are important and it has been shown that it is their ratios with respect to one another that can be used in developing a design procedure. Examining, for example, equation 4.47 it can be seen that deflection and bending moment for a plate under distributed load, q, can be expressed as:

$$w = \frac{\alpha q a^4}{D_{22}}$$

$$M_x = \beta_1 q a^2 \qquad\qquad [4.48]$$

$$M_y = \beta_2 q a^2$$

where α, β_1 and β_2 are constants which depend on elastic properties, plate aspect ratio, a/b, and edge conditions.

Similarly, for plates subjected to a transverse point load, P:

$$w = \frac{\alpha P a^2}{D_{22}}$$

$$M_x = \beta_1 P \qquad\qquad [4.49]$$

$$M_y = \beta_2 P$$

Table 4.3. Typical values of rigidity ratios for fibre reinforced plastics

Material	E_1 (GPa)	E_2 (GPa)	G (GPa)	v_{12}	D_{11}/D_{22}	D'_{33}/D_{22}
CSM/polyester	8	8	3	0.32	1	1
WR/polyester	15	15	4	0.15	1	0.67
Glass fabric/polyester	25	25	4	0.17	1	0.49
UD glass/polyester	40	10	4	0.30	4	1.04
Kevlar/epoxide	76	8	3	0.34	9.5	1.08
UD carbon/epoxide	148	10	4	0.31	14.8	1.09

For an isotropic material $D_{11}/D_{22} = 1$ and $D'_{33}/D_{22} = 0$.   WR = woven roving.

Table 4.4. Stiffness parameter, α, for simply supported plates under point loading

Unidirectional composite ($D'_{33}/D_{22} = 1$, $v_{12} = 0.3$)

D_{11}/D_{22}	a/b				
	4	2	1	0.5	0.25
1	0.001 07	0.004 17	0.011 6	0.0167	0.017 3
2	0.000 975	0.003 79	0.009 33	0.011 1	0.011 3
4	0.000 875	0.003 38	0.006 76	0.007 04	0.007 15
7	0.000 796	0.002 98	0.004 81	0.004 75	0.004 9
10	0.000 753	0.002 7	0.003 85	0.003 8	0.003 9
15	0.000 7	0.002 37	0.002 92	0.002 85	0.002 98

Woven roving ($D_{11}/D_{22} = 1$, $v_{12} = 0.15$)

D_{11}/D_{22}	a/b		
	1	2	3
1	0.0118	0.002 12	0.000 271
0.8	0.0131	0.002 26	0.000 286
0.6	0.0146	0.002 43	0.000 304
0.4	0.0165	0.002 63	0.000 325

Table 4.3 shows typical values of the ratios of flexural rigidities important for the determination of the plate coefficients. Data for unidirectional and woven roving material are given in Table 4.4. The case considered is that for centre point loading and simply supported edges.[8]

The results for the pressure loading conditions are shown in Figures 4.15–4.18. For high aspect ratio unidirectional plates it can be seen that there is little increase in stiffness with increasing anisotropy ratio (D_{11}/D_{22}) whereas

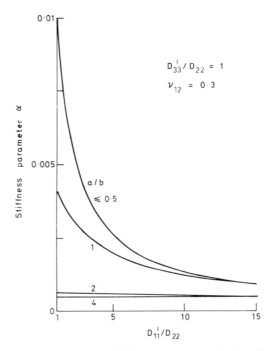

4.15   Variation of stiffness parameter for simply
supported rectangular plates under uniform pressure
(unidirectional reinforcement).

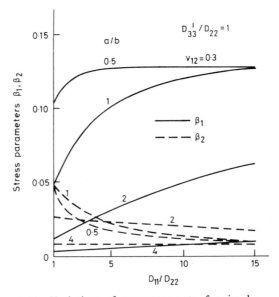

4.16   Variations of stress parameter for simply
supported rectangular plates under uniform pressure
(unidirectional reinforcement).

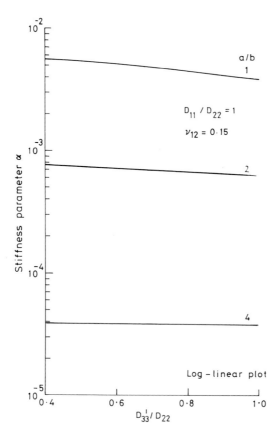

4.17 Variation of stiffness parameter for simply
supported rectangular plates under uniform pressure
(woven roving).

the converse is true where a/b is less than 1. As a result the stiffness of
rectangular plates can most effectively be increased by orientating the fibre
reinforcement along the shortest direction. Also, as D_{11}/D_{22} increases, β_1
increases and β_2 falls, indicating higher stresses in the fibre direction. A further
point to note is the fact that assuming isotropy, i.e. $D_{11} = D_{22}$, will result in
erroneous results. For example, using the lower rigidity will be conservative
for deflection but will underestimate fibre direction stresses. The same is true
for woven roving materials. Although $D_{11}/D_{22} = 1$, the second ratio is less
indicating a reduced stiffness, but more highly stressed panel.

An alternative to the tabular presentation of results is to assume an
expression in a form which satisfies the applied boundary conditions and then
to apply elastic analysis techniques to establish the details of the parameters in
the equations. One such approach results in the following.[8]

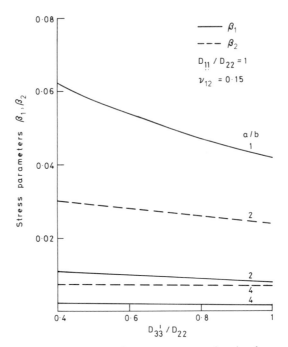

4.18   Variations of stress parameter for simply
supported rectangular plates under uniform pressure
(woven roving).

For distributed loading:

$$\alpha = n_1[D_{11}/D_{22} + 2(D'_{33}/D_{22})(a/b)^2 + (a/b)^4]^{-1}$$
$$\beta_1 = n_2\alpha[D_{11}/D_{22} + v_{12}(a/b)^2] \qquad [4.50]$$
$$\beta_2 = n_2\alpha[v_{12} + (a/b)^2]$$

and for centre-point loading:

$$\alpha = m_1(a/b)[D_{11}/D_{22} + 2(D'_{33}/D_{22})(a/b)^2 + (a/b)^4]^{-1} \qquad [4.51]$$

Values for the constants n_1, n_2 and m, are given in Table 4.5. Figure 4.19 shows a comparison between the tabulated results based on the more rigorous analysis and the more approximate expressions. There is good agreement over the whole normal range of materials with some divergences with very high anisotropy ratios.

For most cases, in-plane loading of plates can be catered for using the laminate analysis discussed in the previous chapter. As with beams, if the laminate is symmetrical with respect to its midplane, normal forces will give rise to normal strains only. Where there is asymmetry there will be a tendency

Table 4.5. Numerical coefficients for plate bending

| | Pressure load | | Point load |
a/b	n_1	n_2	m_1
1	0.0162	9.1	0.0464
0.5, 2	0.0157	8.9	0.0522
0.25, 4	0.0145	9.4	0.0773

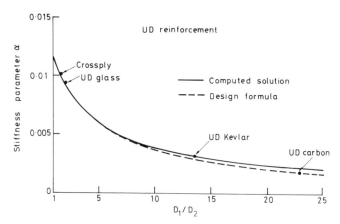

4.19 Accuracy of design formula for simply supported square plates under centre load.

for warpage and additional forces and moments will be generated at supports if the plate is restrained. A key technological problem which often features in design is that of a plate containing a circular hole. Expressions are given below for an orthotropic plate subject to tension, shear and bending.[9] In principle most design cases can then be rationalized by a summation of the contribution of each stress as appropriate. It must be emphasized, however, that these analyses give solutions for in-plane stresses only. The effect of interlaminar and through-thickness stresses which can cause delamination cannot be accounted for by a simple approach. The equations are given without derivation.

For an orthotropic plate subjected to a tensile force, q, applied at an angle θ to the principal material direction (Fig. 4.20), the stress distribution expressed in terms of rotation about the hole is given by:

$$\sigma_\theta = q \cdot \frac{E_\theta}{E_1} \left\{ [- \cos^2 \theta + (k + n) \sin^2 \theta] k \cos^2 \phi + [(1 + n) \cos^2 \theta \right.$$

$$\left. - k \sin^2 \theta] \sin^2 \phi - n (1 + k + n) \sin \theta \cos \theta \sin \phi \cos \phi \right\} \qquad [4.52]$$

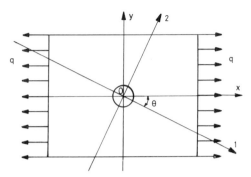

4.20   Plate with circular hole under tension.

where E_θ is the modulus in direction θ, E_1 and E_2 are the modulus values in the principal material directions, $k = (E_1/E_2)^{1/2}$ and:

$$n = \left\{ 2\left(\frac{E_1}{E_2} - v_{12} \right) + \frac{E_1}{G_{12}} \right\}^{1/2}$$

when $\phi = 0$, i.e. when the loading direction is parallel to the material axis, equation 4.52 reduces to:

$$\sigma_\theta = q \frac{E_\theta}{E_1} [- k \cos^2 \phi + (1 + n) \sin^2 \phi] \qquad [4.53]$$

This can be compared with the equivalent equation for an isotropic plate:

$$\sigma_\theta = q[1 - 2 \cos^2 (\phi - \theta)] \qquad [4.54]$$

Generally, the stress in an orthotropic plate will not be symmetric with respect to the line parallel or perpendicular to the applied force unless the condition $\theta = 0$ prevails. Nor will the maximum stress occur at the ends of the diameter normal to the applied force. Figure 4.21 shows a typical stress distribution for the case where the applied force is at 45° to the principal direction. The corresponding distribution where load is applied in the material direction is shown in Fig. 4.22. For the case $\theta = 0$ the maximum stress will be either $- q/k$ (at the ends of the diameter parallel to the forces) or $q(1 + n)$ (at the ends of the diameter perpendicular to the forces), depending on the value of the elastic constants.

For a plate subjected to a shear force, Q, applied at an angle θ to the principal material direction (Fig. 4.23), the stress distribution expressed in forms of rotation about the hole is given by:

$$\sigma_\theta = Q \frac{E_\phi}{2E_1} (1 + k + n)[- n \cos 2 \phi \sin 2 \theta + (1 + k) \cos 2 \theta + k - 1] \sin 2 \phi]$$

$$[4.55]$$

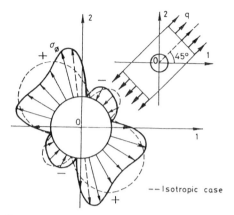

4.21   Stress distribution for tension applied at 45° to
the principal material direction.

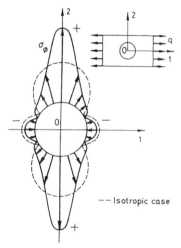

4.22   Stress distribution for tension applied in the
principal material direction.

When $\theta = 0$, this equation becomes:

$$\sigma_\theta = -Q \frac{E_\phi}{2E_1}(1 + k + n)n\sin^2\phi \qquad [4.56]$$

and for an isotropic plane

$$\sigma_\theta = 4Q\sin^2\phi \qquad [4.57]$$

An interesting consequence of the stress distribution described by equation 4.55 is that the case of $\theta = 45$, the optimum for a plate without a hole, is subject

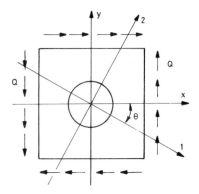

4.23   Plate with circular hole under shear.

to a high stress concentration (Fig. 4.24). The best laminate angles, in terms of stresses around the hole are 0° or 90° (Fig. 4.25).

For a plate subjected to an in-plane bending moment, M, applied at an angle θ to the principal material direction (Fig. 4.26), the stress distribution expressed in terms of rotation about the hole (radius a) is given by:

$$\sigma_\theta = \frac{Ma}{2I} \cdot \frac{E_\theta}{E_1} \Big\{ k[1 - k - (1 + k + n)\cos 2\theta]\sin^3 \phi \cos \theta + \{n^2 + k(k + 2n - 1)\}$$

$$+ [n(1 + n) + k(1 + k + 2n)]\cos 2\theta\} \, x \sin^2\phi \cos \phi \sin \theta \qquad [4.58]$$

$$- [(1 + n)^2 - k - (k + n + 1)(1 + n)\cos 2\theta] \, x \sin \phi \cos^2 \phi \cos \theta$$

$$+ [1 - k - (1 + k + n)\cos^2 \theta]\cos^3 \phi \sin \theta] \Big\}$$

When $\theta = 0$, this equation becomes

$$\sigma_\theta = \frac{Ma}{2I} \cdot \frac{E_\theta}{E_1}[1 - k - (1 + k + n)\cos 2\phi]\sin \phi \qquad [4.59]$$

Again, comparing with an isotropic plate

$$\sigma_\theta = \frac{Ma}{I} \cdot \sin \phi \cos 2\phi \qquad [4.60]$$

Figures 4.27 and 4.28 show stress distribution for the cases where the bending moments are applied at 45° to and in the principal material directions.

As has been stated, these analyses do not fully describe the total situation around a hole. For example, stress concentrations are independent of hole size, whereas measured strengths under, say, tension are relatively less for large holes than with those with a smaller radius. This effect can be rationalized through considerations of stress peak redistribution. Around a small hole stress contours are closer together and therefore there is more scope

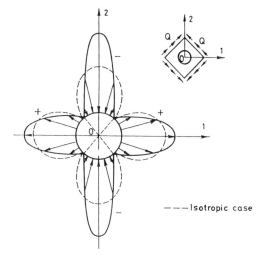

4.24   Stress distribution for shear applied at 45° to the
        principal material direction.

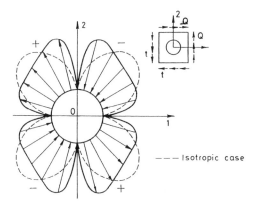

4.25   Stress distribution for shear applied in the
        principal material direction.

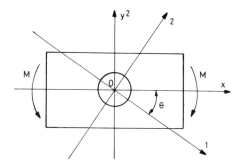

Figure 4.26   Plate with circular hole under bending.

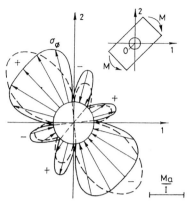

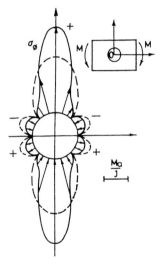

4.27   Stress distribution for bending applied at 45° to
the principal material direction.

4.28   Stress distribution for bending applied at the
principal material direction.

for this redistribution. Two criteria have been presented, both based on the stress system at some distance from the hole.[10-12]

For the plate shown in Fig. 4.29 the normal stress σ_y along the x axis in front of the hole can be approximated by:

$$\sigma_y(x,0) = \frac{\sigma}{2}\left\{ 2 + \left(\frac{r}{x}\right)^2 + 3\left(\frac{r}{x}\right)^4 - (k_t - 3)\left[5\left(\frac{r}{x}\right)^6 - 7\left(\frac{r}{x}\right)^8 \right]\right\} \quad [4.61]$$

where k_t is the stress concentration factor given by equation 4.52.

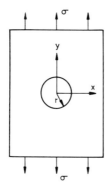

4.29   Plate with circular hole of radius r.

   The first criterion assumes failure occurs when σ_y at some distance, d_0, from the hole reaches the tensile strength of the parent material. The plate strength, σ_p, becomes:

$$\frac{\sigma_p}{\sigma_0} = \frac{2}{2 + p_1^2 + 3p_1^4 - (k_t - 3)(5p_1^6 - 7p_1^8)} \qquad [4.62]$$

where $p_1 = r/(r + d_0)$. For large holes, σ_p/σ_0 tends to $1/k_t$, the classical result, whereas as the hole becomes small σ_p/σ_0 approaches unity.

   The second criterion assumes failure to occur when the average value of σ_y over some fixed distance a_0, ahead of the hole reaches that of the parent material, i.e.:

$$\sigma_p = \frac{1}{a_0} \int_r^{r + a_0} \sigma_y(x,0)\, dx \qquad [4.63]$$

This yields

$$\frac{\sigma_p}{\sigma_0} = \frac{2(1 - p_2)}{2 - p_2^2 - p_2 + (k_t - 3)(p_2^6 - P_2^8)} \qquad [4.64]$$

where $p_2 = r/(r + a_0)$. The application of these criteria to the case of GRP gives the results in Fig. 4.30.

Shells

Shell type components of either single or double curvature provide efficient structural capability because of their ability to carry load as membranes, i.e. in the form of tensile, compressive or shear forces in the plane of the surface. Generally, this is much more effective than accommodating out of plane loads and bending moments. For example, in a curved surface with an internal pressure circumferential tensile stresses are generated, whereas to contain a

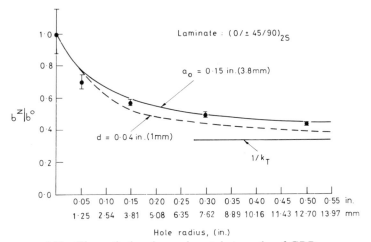

4.30 Theoretical and experimental strengths of GRP
plates with circular holes (σ_N/σ_0-ratio of notched to
un-notched strength).

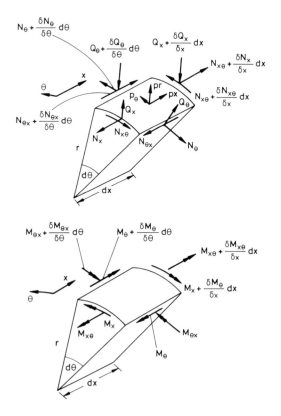

4.31 Stress and moment resultants for a cylindrical
shell.

similar pressure, a flat plate, where loads are reacted by bending, would need a much greater thickness. The general analysis of isotropic shell structures is widely described in the literature.[13-16] These can be rather complex and the following discussion is limited to cylindrical shells with an emphasis on utility in selected design situations. A further simplification is the neglect of shear deformation. Whilst this can be significant for very high levels of anisotropy the analysis presented has application for a wide range of commonly used material systems.

For the general case of asymmetric loading a shell element of length dx and width $r \cdot d\theta$ has the stress and moment resultants shown in Fig. 4.31, the directions shown being taken as positive. Proceeding as with beams and plates the resulting equilibrium equations can be written as:

$$\frac{r\partial N_x}{\partial x} + \frac{\partial N_{\theta x}}{\partial \theta} + rp_x = 0$$

$$\frac{\partial N_\theta}{\partial \theta} + \frac{r\partial N_{x\theta}}{\partial x} - Q_\theta + rp_\theta = 0 \qquad [4.65]$$

$$\frac{\partial Q_\theta}{\partial \theta} + \frac{r\partial Q_x}{\partial x} + N_\theta - rp_r = 0$$

$$\frac{\partial M_\theta}{\partial \theta} + \frac{r\partial M_{x\theta}}{\partial x} - rQ_\theta = 0$$

$$\frac{r\partial M_x}{\partial x} + \frac{\partial M_{\theta x}}{\partial \theta} - rQ_x = 0$$

$$rN_{x\theta} - rN_{\theta x} + M_{\theta x} = 0$$

Considering the case of symmetric loading, force and moment terms become independent with respect to θ. For $p_x = 0$ and $p_\theta = 0$, i.e. the only external load is a radial pressure p_r, the above equations reduce to:

$$r\frac{\partial N_{x\theta}}{\partial x} - Q_\theta = 0$$

$$r\frac{\partial Q_x}{\partial x} + N_\theta = rp_r \qquad [4.66]$$

$$\frac{\partial M_{x\theta}}{\partial x} - Q_\theta = 0$$

$$\frac{\partial M_x}{\partial x} - Q_x = 0$$

Substitution of the last expression into the second of equation 4.66 gives an expression in terms of moments and in-plane force, viz.:

$$\frac{\partial^2 M_x}{\partial x^2} + \frac{N_\theta}{r} = p_r \qquad [4.67]$$

Again strain displacement equations from the reference surface within the shell are required and these can be written as:

$$\varepsilon_x^0 = \frac{\partial u}{\partial x}$$

$$\varepsilon_\theta^0 = \frac{1}{r}\frac{\partial v}{\partial \theta} + \frac{1}{r}w_0$$

$$\gamma_{x\theta}^0 = \frac{\partial v}{\partial x} + \frac{1}{r}\frac{\partial u}{\partial \theta} \qquad\qquad [4.68]$$

$$\kappa_x = \frac{\partial^2 w}{\partial x^2}$$

$$\kappa_x = \frac{1}{r^2}\frac{\partial^2 w}{\partial \theta^2}$$

$$\kappa_{xy} = \frac{1}{r}\frac{\partial^2 w}{\partial x\partial \theta}$$

Manipulation of equations 4.68 gives rise to the compatibility equation:

$$-\frac{1}{r}\frac{\partial^2 w}{\partial x^2} + \frac{\partial^2 \varepsilon_y}{\partial x^2} + \frac{\partial^2 \varepsilon_x}{\partial y^2} - \frac{\partial^2 \gamma_{xy}}{\partial x\partial y} = 0 \qquad\qquad [4.69]$$

Recalling the relationships between stress and moment resultants and strains and curvatures derived in Chapter 3 allows reduction of the problem into one where the curvature, or through equation 4.68 transverse deflection w, and the stress resultants N_x, N_y and N_{xy} are the primary variables.

Recalling the laminate constitutive equations in matrix form:

$$[N] = [A][\varepsilon] + [B][\kappa]$$
$$\qquad\qquad [4.70]$$
$$[M] = [B][\varepsilon] + [D][\kappa]$$

Inversion of equation 4.70 followed by back-substitution gives:

$$[\varepsilon] = [A]^{-1}[N] - [A]^{-1}[B][\kappa]$$
$$\qquad\qquad [4.71]$$
$$[M] = [b][N] + [d][\kappa]$$

where the partially inverted stiffness matrices $[b]$ and $[d]$ are defined by:

$$[b] = [B][A]^{-1}$$

and

$$[d] = [D] - [B][A]^{-1}[B]$$

For axisymmetric loading the stress resultants are independent of the θ variable and under such circumstances N_x and $N_{x\theta}$ are constants. This means that the stress function (equation 4.11) can be written after changing to (x,θ) coordinates as:[17,18]

$$F(x,\theta) = F_0(x) + \tfrac{1}{2} N_x \theta^2 - N N_{x\theta} x\theta \qquad [4.72]$$

Substitution of equations 4.68 and 4.71 into both 4.69 and 4.67 and using equation 4.72 yields two coupled ordinary differential equations:

$$A_{22}^* \frac{d^4 F_0}{dx^4} - b_{12} \frac{d^4 w}{dx^4} - \frac{1}{r} \frac{d^2 w}{dx^2} = 0 \qquad [4.73]$$

$$b_{12} \frac{d^4 F_0}{dx^4} + d_{11} \frac{d^4 w}{dx^4} - \frac{1}{r} \frac{d^2 F_0}{dx^2} = p_r(x)$$

where A_{ij}^* are the coefficients of $[A]^{-1}$.

The presence of both dependent variables in each of the equations indicates coupling between membrane and bending effects.

Integrating the first of equation 4.73 twice gives:

$$\frac{d^2 F_0}{dx^2} = \frac{1}{A_{22}^*} \left\{ \frac{w}{r} + b_{12} \frac{d^2 w}{dx^2} - A_{12}^* N_x - A_{26}^* N_{x\theta} \right\} \qquad [4.74]$$

and substituting the results into the second gives:

$$(d_{11} A_{22}^* + b_{12}^2) \frac{d^4 w}{dx^4} + \frac{2b_{12}}{r} \cdot \frac{d^2 w}{dx^2} + \frac{w}{r^2} = \frac{A_{12}^*}{r} N_x + \frac{A_{26}^*}{r} N_{x\theta} + A_{22}^* p_r \quad [4.75]$$

This equation can now be solved with the input of appropriate design cases and laminate configurations. Generally, it is similar to that which may be derived for an isotropic shell and solutions are of the same form, differing only because of the presence of the additional terms containing the coupling constant b_{12}. If the laminate is designed such that all coupling effects are zero, the laminate problem becomes identical to that of an equivalent isotropic shell.

To explore the application of this analysis for specific cases three laminates types are considered: unbalanced symmetrical such as a single ply construction of constant angle θ, balanced asymmetric such as a $\pm\theta$ angle-ply, and a balanced symmetrical laminate such as a $(0, 90)_s$ cross-ply.[20]

Unbalanced symmetrical laminate. In this case the terms in the extentional/bending coupling matrix are all zero ($[B] = 0$) and therefore $b_{12} = 0$. Equation 4.75 becomes:

$$D_{11} \frac{d^4 w}{dx^4} + \frac{1}{A_{22}^* r^2} \cdot w = \frac{A_{22}^*}{r} N_x + \frac{A_{26}^*}{r} N_{x\theta} + A_{22}^* p_r \qquad [4.76]$$

The general solution of this expression can be shown to be

$$w = e^{-\alpha x}(k_1 \cos \beta x + k_2 \sin \beta x) + f(x) \qquad [4.77]$$

where $f(x)$ is the particular integral of equation 4.76 and is essentially a function of the main body forces N_x, $N_{x\theta}$ and p_r. For example, with internal pressure only $f(x) = r^2 A_{22}^* p_r$.

$$\beta = \left[\frac{b+d}{4D_{11}}\right]^{1/2} \qquad [4.78]$$

$$\alpha = \left[\frac{b-d}{4D_{11}}\right]^{1/2}$$

and

$$b = \frac{2}{r}(D_{11}/A_{22}^*)^{1/2}$$

$$d = -\frac{1}{r^2} \cdot \frac{D_{16}A_{26}^*}{A_{22}^*}$$

and the constants k_1 and k_2 are determined by the application of the appropriate boundary conditions. The presence of the shear/extension coupling term D_{16} is interesting to note. When a single ply cylinder of orientation θ is subject to normal loads, e.g. pressure and tension, it will tend to rotate. If restrained a torque will be established in the tube which should feature in the analysis.

Balanced, unsymmetrical laminate. For this construction the solution of equation 4.75 is of the same form as in the previous case, i.e.

$$w = e^{-\beta x}(k_1 \cos \beta x + k_2 \sin \beta x) + f(x) \qquad [4.79]$$

where it can be shown that

$$\beta^4 = \frac{1}{4r^2 d_{11} A_{22}^*} \qquad [4.80]$$

Balanced, symmetrical laminate. In this case the governing equation can be determined from equation 4.76 by putting the normal/shear coupling terms to zero. Again the solution is of the form

$$w = e^{-\beta x}(k_1 \cos \beta x + k_2 \sin \beta x) + f(x) \qquad [4.81]$$

Owing to the absence of coupling of any type $D_{11} = d_{11}$ and

$$\beta^4 = \frac{1}{4r^2 D_{11} A_{22}^*}$$

The value of β in this case can be easily defined in terms of engineering constants as:

$$\beta^4 = \frac{3}{r^2 t^2} \frac{E_2}{E_1}(1 - v_{12}v_{21}) \qquad [4.82]$$

Table 4.6. Edge solutions for long cylindrical shells

End condition	Deflected shape (w_x)	Slope (θ_x)	Longitudinal unit bending moment (M_x)	Longitudinal unit shear force (Q_x)
at $x = 0$ $Q = Q_0$	$\dfrac{Q_0}{2\beta^3 D_{11}} \cdot \gamma(\beta_x)$	$\dfrac{-Q_0}{2\beta^2 D_{11}} \cdot \phi(\beta x)$	$\dfrac{Q_0}{\beta} \cdot \zeta(\beta x)$	$Q_0 \cdot \psi(\beta x)$
at $x = 0$ $M = M_0$	$\dfrac{M_0}{2\beta^2 D_{11}} \cdot \psi(\beta x)$	$-\dfrac{M_0}{\beta D_{11}} \cdot \gamma(\beta x)$	$M_0 \cdot \phi(\beta x)$	$-2\beta M_0 \cdot \zeta(\beta x)$
at $x = 0$ $w = w_0$ $Q = Q$	$w_0 \cdot \phi(\beta x)$	$-2\beta w_0 \cdot \zeta(\beta x)$	$-2D_{11}\beta^2 w_0 \cdot \psi(\beta x)$	$4D_{11}\beta^3 w_0 \cdot \gamma(\beta x)$
at $x = 0$ $w = 0$ $\theta = \theta_0$	$\dfrac{\theta_0}{\beta} \cdot \zeta(\beta x)$	$\theta_0 \cdot \psi(\beta x)$	$-2D_{11}\beta\theta_0 \cdot \gamma(\beta x)$	$2D_{11}\beta^2\theta_0 \cdot \phi(\beta x)$

Continuing the simplification of the governing differential equation to the case of isotropy yields:

$$D\frac{d^4 w}{dx^4} + \frac{Etw}{r^2} = Z \qquad [4.83]$$

where Z are applied body forces. The solution of this is

$$w = e^{-\beta x}(k_1 \cos \beta x + k_2 \sin \beta x) + f(x) \qquad [4.84]$$

where

$$\beta^4 = \frac{3(1 - v^2)}{r^2 t^2}$$

Comparing the solution for the orthotropic and isotropic cases shows that the variation is simply related to the ratio of E_2/E_1, i.e. the degree of anisotropy, which appears in the definition of the constant β.

These equations can now be used to calculate stress distributions for many situations of important practical significance. The values for deflections w_x can be evaluated directly and those for rotation, θ_x, moment, M_x and shear force Q_x can be calculated using the conventional relationships:

$$\theta_x = \frac{dw}{dx}$$

$$M_x = D_{11}\frac{d^2 w}{dx^2} \qquad [4.85]$$

$$Q_x = D_{11}\frac{d^3 w}{dx^3}$$

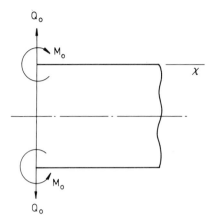

4.32   Cylindrical shell subject to bending moments
and shear forces.

Application of the analysis to a variety of end conditions yields the results given in Table 4.6. The functions used are defined as:

$$\phi(\beta x) = e^{-\beta x}(\cos \beta x + \sin \beta x)$$

$$\psi(\beta x) = e^{-\beta x}(\cos \beta x - \sin \beta x)$$

$$\gamma(\beta x) = e^{-\beta x}\cos \beta x \qquad\qquad\qquad [4.86]$$

$$\zeta(\beta x) = e^{-\beta x}\sin \beta x$$

The solutions given in Table 4.6 are similar in form but different in detail from those given in standard texts for isotropic shells.[13] This method of presentation allows direct comparison with the isotropic case.

For many practical situations applied boundary conditions can be reduced to bending moments, M_0, and shearing forces, Q_0 both uniformly distributed at the edge $x = 0$ (Fig. 4.32).

From Table 4.6 the expressions governing the response of the shell under these conditions are:

$$w_x = \frac{e^{-\beta x}}{2\beta^3 D_{11}}[Q_0 \cos \beta x + \beta M_0 (\cos \beta x - \sin \beta x)]$$

$$\theta_x = \frac{e^{-\beta x}}{2\beta^2 D_{11}}[Q_0 \cos \beta x + \sin \beta x) + 2\beta M_0 \cos \beta x] \qquad [4.87]$$

$$M_x = \frac{e^{-\beta x}}{\beta}[Q_0 \sin \beta x + \beta M_0 (\cos \beta x + \sin \beta x)]$$

$$Q_x = e^{-\beta x}[Q_0 (\cos \beta x - \sin \beta x) - 2\beta M_0 \sin \beta x]$$

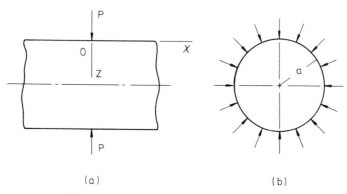

Figure 4.33 Cylindrical shell subject to load
uniformly distributed along a circular section.

Particular design cases can now be examined by evaluating the boundary
conditions relevant to each situation and establishing values for w, θ, M or Q
as appropriate. Consider a load uniformly distributed along a circular section
(Fig. 4.33). This would be representative of, for example, a stiffened cylinder
where deflections from pressure or thermal expansion are restrained.

From symmetry:

$$Q_0 = -\frac{P}{2}$$ [4.88]

Also because of symmetry, $\theta = 0$ at $x = 0$. Therefore from equation 4.85:

$$M_0 = \frac{p}{4\beta}$$ [4.89]

Therefore the maximum bending stress at the edge of the constraint is given
by:

$$\sigma_0 = \pm \frac{6p}{4\beta t^2}$$ [4.90]

Expressing the bending stress in terms of k, the ratio of principal moduli yields:

$$\sigma_0 = \pm \frac{1.14pr^{1/2}}{t^{3/2}k^{1/4}}$$ [4.91]

The distribution of stress as a function of distance from the restraint can be
found by inserting values for M_0 and Q_0 into equation 4.87. A similar case
arises if the restraint occurs at the edge of the shell, for example, a vessel or
tank fixed at its base (Fig. 4.34.)

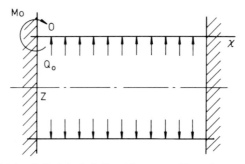

4.34   Cylindrical shell subject to uniform internal pressure.

Assuming the ends of the shell are rigid

$$w_0 = \frac{pr^2}{E_2 t} \qquad [4.92]$$

$$\theta_0 = 0$$

Therefore at $x = 0$, equation 4.85 gives:

$$-\frac{1}{2\beta^3 D_{11}}[Q_0 + \beta M_0] = \frac{pr^2}{E_2 t} \qquad [4.93]$$

Solving for M_0

$$\frac{1}{2\beta^2 D_{11}}[Q_0 + 2\beta M_0] = 0 \qquad [4.94]$$

$$M_0 = 2 \cdot \frac{\beta^2 D_{11}}{E_2} \cdot \frac{pr^2}{t}$$

The maximum stress at the edge is therefore given by:

$$\sigma_0 = \pm \left(\frac{3}{k}\right)^{1/2} \frac{pr}{t} \qquad [4.95]$$

Stress concentrations can also occur at the junction of shells of different curvature. A common example is the connection between a cylinder and hemispherical end cap under internal pressure (Fig. 4.35). With the application of pressure the two geometries will tend to expand radially by different amounts. Continuity of displacement and rotation occurs and therefore local bending is caused at the junction. As the bending is local the thickness and stiffness of the end cap and cylinder in the critical area can be assumed to be the same. On this basis the application of a shear force is sufficient to ensure compatibility of deflection and rotation at the junction. It can be shown that

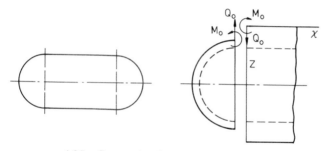

4.35 Connection between a cylinder and hemispherical end cap under internal pressure.

the relative deflection of the unrestrained cylinder and sphere is given by

$$w = \frac{pr^2}{2E_2 t} \qquad [4.96]$$

From equation 4.85:

$$Q_0 = \frac{\beta^3 D_{11} pr^2}{E_2 t} \qquad [4.97]$$

Substitution into the third expression of equation 4.87 yields:

$$M_x = D_{11}\beta^2 \frac{pr^2}{tE_2} e^{-\beta x} \sin \beta x \qquad [4.98]$$

i.e.

$$\sigma_x \approx \left(\frac{3}{k}\right)^{1/2} \frac{pr}{2t} \cdot \zeta(\beta x) \qquad [4.99]$$

It can be shown that $\zeta(\beta x)$ is a maximum at $x = \pi/4\beta$. Therefore

$$\sigma_{max} \approx \pm 0.134 \left(\frac{3}{k}\right)^{1/2} \cdot \frac{pr}{t} \qquad [4.100]$$

In the event of differences in thickness or elastic properties between cylinder and sphere, the edge moment, M_0, is not equal to 0 and equations 4.85 and the equivalents for a spherical shell must be equated. This approach can be adopted for the junction of any shell where there is a continuous change in tangent. In practice there are limitations due to the complexity that can arise in the equations governing the behaviour of shells of all but the most simple geometries. In this case recourse is normally made to numerical methods.

Temperature effects can also be important, particularly for GRP where significant through-thickness thermal gradients may be present due to the material's low thermal conductivity. It can be shown that the bending moments generated in a cylindrical shell with a linear through-thickness temperature gradient are given by:

$$M_x = (D_{11}\alpha_1 + D_{12}\alpha_2)\frac{\Delta T}{t}$$ [4.101]

$$M_y = (D_{12}\alpha_1 + D_{22}\alpha_2)\frac{\Delta T}{t}$$

At the free edge axial stresses must be zero, therefore

$$M_0 = (D_{11}\alpha_1 + D_{12}\alpha_2)\frac{\Delta T}{t}$$ [4.102]

The circumferential component of this moment M_θ is

$$M_x = v_{12}(D_{11}\alpha_1 + D_{12}\sigma_2)\frac{\Delta T}{t}$$ [4.103]

It can be shown that

$$N_\phi = -\frac{Eh}{a}w$$ [4.104]

Therefore from equations 4.85:

$$N_\phi = \frac{E_2 t}{r} \cdot \frac{M_0}{2\beta^2 D_{11}}$$ [4.105]

Adding the stresses arising from equations 4.102, 4.103 and 4.105 yields:

$$\sigma_\phi = \frac{\Delta T}{2}E_1\left\{\alpha_1(v_{21} - v_{12}) + k\alpha_2 + \frac{\sqrt{k}}{\sqrt{3}}(1 + v_{12})\right\}$$ [4.106]

As an example of the application of these expressions consider a pressure vessel of the following design:[21]

Radius	1000 mm
Thickness	10 mm
Design pressure	0.15 MPa

Table 4.7 shows typical elastic properties used in the design of GRP process equipment.[22] For the base of the vessel, assuming it is fully fixed, the maximum axial stress on the inside surface is shown in Figure 4.36 expressed as a function of winding angle. Figure 4.37 shows the distribution of stress for winding angles in the range $\pm 45°$ to hoop. Calculations to determine maximum stresses were performed using the data in Table 4.7 and equation 4.95. As can be seen stress levels fall with increasing winding angle and this is as a result of two factors:

Table 4.7. Elastic properties for filament wound GRP process equipment

Winding angle	Modulus	
	E_{hoop}	E_{axial}
Hoop	37 333	6133
75	31 733	6000
65	21 333	5867
55	13 067	6133
45	8 000	8000

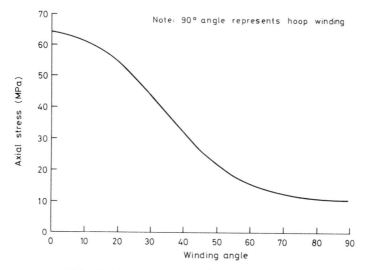

Note: 90° angle represents hoop winding

4.36 Axial stress concentrations as a function of winding angle.

- Reducing radial expansion due to increasing circumferential modulus.
- Reducing bending moment required to ensure compatibility due to reducing axial modulus.

The equivalent stress levels for the isotropic case corresponds to the ± 45° curve. It should be noted that this is only valid where the laminate is assumed to be symmetrical and balanced. The corresponding hoop stress distributions are shown in Fig. 4.38 and 4.39 and these are calculated from the summation of pressure membrane stress and the effect of the axial bending moment. Values of Poisson's ratio used in the latter case are shown in Figure 4.40. These are calculated using the elastic identity:

$$\frac{E_1}{E_2} = \frac{v_{12}}{v_{21}}$$ [4.107]

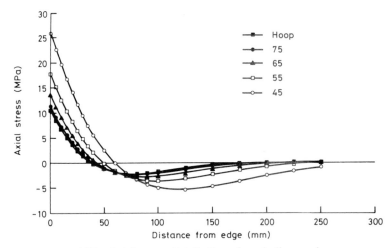

4.37   Axial stress distributions for winding angles
$\pm 45°$ to hoop.

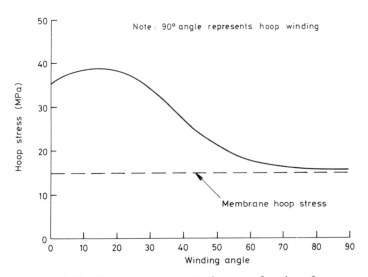

4.38   Hoop stress concentrations as a function of
winding angle.

Experimental measurements of Poisson's ratio, however, indicate that this expression does not hold for all angles and that the very low values indicated in Fig. 4.40 are not achieved. This means that the hoop stress values for high winding angles above will be somewhat higher, approximately $\pm 10\%$, than indicated in Fig. 4.38 and 4.39.

Comparing these results to the isotropic case, which is often the basis of

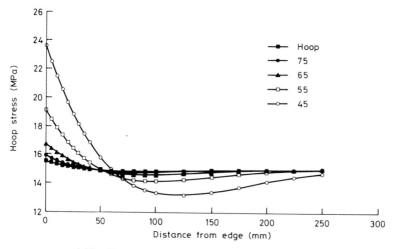

4.39   Hoop stress distributions for winding angles
±45° to hoop.

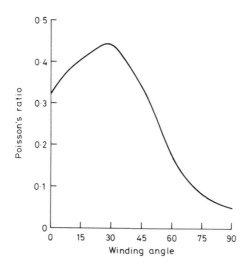

4.40   Poisson's ratio for GRP filament wound
laminates.

current practice for composite components, indicates that present methods lead to a degree of overdesign. Applying conditions of isotropy to equation 4.93 yields:

$$\sigma_0 = \pm \sqrt{3} \cdot \frac{pr}{t} \qquad [4.108]$$

i.e., the stress peak in the axial direction is $\sqrt{3}$ times the membrane hoop stress

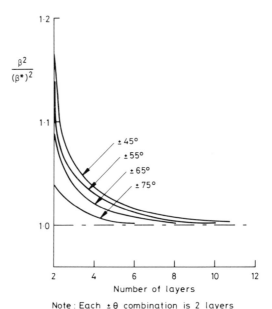

Note: Each ±θ combination is 2 layers

4.41   Effect of laminate asymmetry (β and β^* defined
by equations 4.80 and 4.81).

due to internal pressure. In, for example, common filament wound pressure
vessel designs of, say, $\pm 55°$ or $\pm 75°$, which have calculated stress peaks of
1.19 and 0.75 times the hoop stress respectively, the maximum stresses are
somewhat lower than would otherwise be indicated. For low winding angles
the stress calculation will be greater than $\sqrt{3}$ times the hoop stress, but for
such structures pressures are likely to be relatively low (otherwise the
reinforcement would be closer to the hoop direction).

A basic assumption in these calculations is that of laminate symmetry. The
significance of the effect of symmetry can be evaluated through consideration
of the value of β. Values for the different β's are given in equations 4.80 and
4.81. In Fig. 4.41 the influence of laminate asymmetry is shown and as can be
seen, the effect falls rapidly with increasing number of laminate layers; for layer
thickness of 0.5 mm the error in the calculation becomes small ($< 2\%$) for a
winding thicknesses at around 2–3 mm. For commercial process plant, where
thicknesses are considerably thicker than this, therefore the assumption of
symmetry is valid.

Using the example pressure vessel described above the effect of a uniform
radial load generated due to differential thermal expansion between the vessel
wall and a stiffening ring can be evaluated. Consider a steel ring of the
following characteristics:

Thickness   3 mm (Normally a rolled steel section is used
                             for stiffening to maximize I and Z properties)

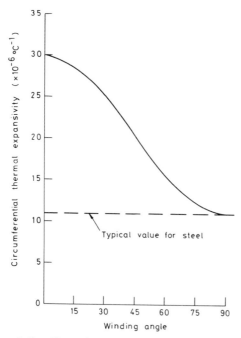

4.42 Circumferential thermal expansivity as a function of winding angle.

| Modulus | 210 GPa |
| Thermal expansivity | $11 \times 10^{-6} \, C^{-1}$ |

Typical values for the thermal expansion coefficients for filament wound laminates and their variations with angle are shown in Fig. 4.42. Current design standards prohibit the use of continuous steel reinforcement on GRP vessels over 60 °C due to perceived values of induced thermal strain. This design rule is based on thermal expansivity coefficients for chopped strand mat and woven roving laminates, typically $30 \times 10^{-6} \, °C^{-1}$ and $20 \times 10^{-6} \, °C^{-1}$ respectively. However, as can be seen, the use of such a temperature value as a limit is not appropriate since at high angles there is only a small mismatch of expansion rate.

Calculation of the stress concentrations arising due to the thermal expansion mismatch can be carried out using either the method employed to derive equation 4.91 or that for equation 4.95. Adopting the latter approach, the restrained thermal expansion of the shell due to the ring is given by:

$$w_0 \, \alpha_2 \Delta T - \left(\frac{E_2 \sigma_2 t + E_s \alpha_s t_s}{E_2 t + E_s t_s} \right) \Delta T \qquad [4.109]$$

where E_2 and σ_2 are the circumferential modulus and expansivity of the shell, t

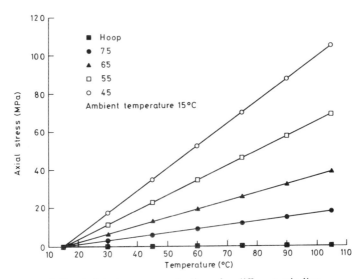

4.43   Axial stress concentrations for different winding
angles as a function of temperature.

the shell thickness and E_s, α_s, t_s are the corresponding properties of the ring. ΔT is the temperature change between fabrication and operating levels.

Using this definition of w_0 and following the analysis, equations 4.92–4.95 yield:

$$\sigma_0 = \pm (3E_1 \cdot E_2)^{1/2} \Delta T \alpha \qquad [4.110]$$

where

$$\alpha = \alpha_2 - \left(\frac{E_2 \alpha_2 t + E_s \alpha_s t_s}{E_2 t + E_s t_s} \right)$$

Figure 4.43 shows calculated maximum axial stresses and their variation with temperature. The effect of winding angle is clear and shows there is potential to control stresses due to thermal expansion mismatch to ensure that they are within allowable limits.

Another laminate configuration which often arises in design is one where the lack of symmetry in stiffness, for example, a (0, 90) laminate or an overwind on to an isotropic material. Here normal loads will give rise to bending strains due to the coupling coefficients in the stiffness matrix.

The solution of equation 4.73 becomes somewhat more complex because b_{12} terms are now non-zero. Following the procedure as with previous cases the solution can be shown to be:

$$w = \lambda^{\mu_1 \beta x} (k_2 \cos \mu_2 \beta x + k_2 \sin \mu_2 \beta x) + f(x) \qquad [4.111]$$

where

$$\beta^2 = \frac{1}{r} \cdot \frac{1}{(d_{11}A_{22}^* + b_{12}^2)^{1/2}}$$

and

$$(\mu_2 + i\mu_1) = (\lambda_2 + i\lambda_1)^{1/2}$$

$$\lambda_2 = - b_{12}/(d_{11}A_{22}^* + b_{12}^2)^{1/2}$$

$$\lambda_1 = (d_{11}a_{23}^*)^{1/2}/(d_{11}A_{22}^* + b_{12}^2)^{1/2}$$

Equations for parameters such as bending and bending moment must also be modified to take into account the influence of coupling.

Using equation 4.69, an expression for the axial bending moment can be derived:

$$M_x = b_{11}N_x + b_{12}N_y + d_{11}\frac{d^2w}{dx^2} \qquad [4.112]$$

Such equations can be used to develop solutions for shells of almost any laminate layup type. The extent by which asymmetry changes the bending moment and resulting stresses is clearly dependent on the particular case and the applicability of the more simplified results derived for balanced and/or symmetrical laminates will be determined by the value of elastic constants and the disposition of layers.

Buckling

The problem of structural instability, or buckling, is a common design issue for almost all types of geometry and loading condition. Whilst it is most commonly considered for cases where compression is present, torsion, pressure and even tension can potentially lead to an unstable structure. Buckling is a phenomenon which is governed by, primarily, stiffness and as a consequence modulus and geometry are the important design parameters. Strength values may also feature but only in that instability may follow an initial failure. This could be important in, for example, an impact event when the structure may be sufficiently weakened to allow buckling provided the geometry was flexible enough. If the arrangement could be provided with more rigidity, the initial damage could perhaps be sustained without collapse.

The simplest configuration where buckling can occur is a simple beam under uniaxial compressive load. Consider such a column subjected to a small lateral force which produces a small deflection. If the deflection disappears, i.e. the column returns to its straight form when the lateral force is removed, the column is said to be stable. On the other hand if the deflection remains, it is considered to be unstable and as such it will collapse on the application of further axial load. The failure load (critical load) for this mode of deformation is that which is just sufficient to keep the bar in a slightly bent form.

This mode of failure most commonly occurs in structures with a large aspect ratio. For an isotropic material, the critical load, p_{cr}, in a pin-jointed strut is given by the well-known Euler equation:[2]

$$p_{cr} = \frac{\pi^2 EI}{4L^2} \qquad [4.113]$$

where E is the modulus of the material, I is the second moment of area of the cross-section of the component and L is the component length.

For a simple laminated column the governing differential equation for the deflected surface can be written as:

$$\frac{\partial^2 w}{\partial x^2} + \frac{B_{11}}{B_{11}^2 A_{11} D_{11}} \cdot N_x = 0 \qquad [4.114]$$

The solution can be expressed in the form:

$$w = A \sin \alpha x + B \cos \alpha x \qquad [4.115]$$

where

$$\alpha = \left[\frac{A_{11} N_x}{B_{11}^2 - A_{11} D_{11}} \right]$$

Equation 4.115 can now be solved with the application of suitable boundary conditions. For a column pinned at each end, the critical buckling load is given by:

$$N_{cr} = \frac{B_{11}^2 - A_{11} D_{11}}{A_{11}} \cdot \left(\frac{\pi}{L} \right)^2 \qquad [4.116]$$

When $[B] = 0$, i.e. there is no coupling between normal forces and bending effects, equation 4.116 reduces to that for an isotropic material.

The derivation of the above equations only takes into account forces due to applied compressive stresses. When buckling occurs, however, there will be shearing forces acting on the cross-section of the bar. This is similar to the shear deformation that is seen with bending. To account for these additional forces it is possible to include a correction factor by which the Euler critical load, p_{cr}, should be multiplied, i.e.

$$\frac{1}{1 + (np_{cr}/AG)} \qquad [4.117]$$

where n is a numerical factor depending on the shape of the cross-section (for circular cross-section $n = 1.11$), A is the cross-section area of the cylinder and G is the transverse shear modulus of the material.

When considering column buckling it is worth noting that this mode of deformation can also occur with other types of loading, for example a cylinder under internal pressure (without end closures). Under such loading it is the fluid which is under pressure inside the cylinder that acts as a strut, in fact one of zero modulus. The only factor which prevents the fluid buckling is the shell wall, and thus the shell, depending on its geometry and elastic constants, may fail by column buckling. To quantify this effect, a Euler equation similar to equation 4.116 may be used to calculate the critical buckling pressure, p_{cr}:

$$(\pi r^2)p_{cr} = \frac{\pi^2 A_{11}I}{4hL^2} \qquad [4.118]$$

where r is the radius of the cylindrical vessel. Composites can be particularly prone to this effect as there is a tendency to reinforce such components in the circumferential direction. Axial modulus and therefore buckling load can be comparatively low.

When shells have small thickness/diameter ratios the wall may collapse under uniaxial compressive radial loading. This is known as shell buckling. For an isotropic cylindrical shell the critical stress at which this occurs, σ_{cr}, is given by:

$$\sigma_{cr} = \frac{kEh}{r\sqrt{3(1 - v^2)}} \qquad [4.119]$$

where E is the modulus of the material, v its Poisson's ratio, h is the thickness of the shell and r is its radius, k is an empirical constant which has the range $0 < k < 1$.

The presence of an empirical factor is a common occurrence in buckling studies as most buckling theories predict critical stresses in excess of the experimentally observed values. To investigate the discrepancies between theory and experiment reference can be made to the post-buckling behaviour of compressed cylindrical shells. This work showed that, although to originate buckling in an ideal cylinder the compressive stress given by equation 4.119 (with $k = 1$) is required, the load necessary to keep the cylinder in a buckled condition rapidly diminishes with increasing deflection (to about a third of the maximum). This demonstrates the reasons for the discrepancy, since in practical structures there are always initial deflections and imperfections which mean that bending starts when the load is small and the deflections reach the values at which further continuation of buckling requires a much smaller load than that for the ideal case.

For a composite cylinder the differential equations of equilibrium, which are applicable during buckling can be written as:[23,24]

$$r\frac{\partial N}{\partial x} + \frac{\partial N_{\theta x}}{\partial \theta} - P\left(\frac{\partial u^2}{\partial \theta^2} - r\frac{\partial w}{\partial x}\right) - rC\frac{\partial u^2}{\partial x^2} - 2T\frac{\partial u^2}{\partial x \partial \theta} = 0$$

$$r\frac{\partial N_\theta}{\partial \theta} + r^2\frac{\partial N_{x\theta}}{\partial x} + \frac{\partial M_\theta}{\partial \theta} + r\frac{\partial M_{x\theta}}{\partial x} - Pr\left(\frac{\partial v^2}{\partial \theta^2} + \frac{\partial^2 w}{\partial \theta}\right)$$
$$- r^2 C\frac{\partial^2 v}{\partial x^2} - 2rT\left(\frac{\partial v^2}{\partial x \partial \theta} + \frac{\partial w}{\partial x}\right) = 0 \qquad\qquad \text{[4.120]}$$

$$\frac{\partial^2 M_\theta}{\partial \theta^2} + r\left(\frac{\partial^2 M_{x\theta}}{\partial x \partial \theta} + \frac{\partial^2 M_{\theta x}}{\partial x \partial \theta}\right) + r^2\frac{\partial^2 M_x}{\partial x^2} - rN_\theta$$
$$- Pr\left(r\frac{\partial u}{\partial x} - \frac{\partial v}{\partial \theta} + \frac{\partial^2 w}{\partial \theta^2}\right) + r^2 C\frac{\partial^2 w}{\partial x^2} + 2rT\left(\frac{\partial v}{\partial x} - \frac{\partial^2 w}{\partial x \partial \theta}\right) = 0$$

where x, θ and z refer to the axial, circumferential and radial directions respectively, u, v and w are the displacements in those directions, N and M are stress and moment resultants, P, C and T are the applied external radial pressure, axial compression per unit length and the tangential force per unit length respectively and r is the radius of the cylinder.

Equations 4.120 can be expressed in terms of displacements by using definitions for stress and moment resultants and strains and by remembering the orthotropic constitutive equations (see Chapter 3). One possible solution of the resulting equations is to assume displacements of the following form:

$$u = U\sin\left(\frac{\lambda x}{r} + n\theta\right)$$

$$v = V\sin\left(\frac{\lambda x}{r} + n\theta\right)$$

$$w = W\sin\left(\frac{\lambda x}{r} + n\theta\right) \qquad\qquad \text{[4.121]}$$

$$\lambda = \frac{m\pi r}{L}$$

where substitution of equation 4.121 into equation 4.120 yields:

$$\begin{cases} (F_{11} - \lambda^2 q_2 - 2n\lambda q_3 + n^2 q_1)(F_{12})(F_{13} + \lambda q_1) \\ (F_{12})(F_{22} - \lambda^2 q_2 - 2n\lambda q_3 - n^2 q_1)(F_{23} - 2\lambda q_5 - nq_1) \\ (F_{13} + \lambda q_1)(F_{23} - 2\lambda q_3 - nq_1)(F_{33} - \lambda^2 q_2 - 2n\lambda q_3 - n^2 q_1) \end{cases} \begin{Bmatrix} U \\ V \\ W \end{Bmatrix} = 0$$
$$\text{[4.122]}$$

where U, V and W are constants, L is the length of the cylinder, m and n are defined as the number of half waves in the axial direction and the number of waves in the circumferential direction respectively and:

$$F_{11} = (\bar{A}_{11} + \bar{B}_{11})\lambda^2 + 2n\bar{A}_{16}\lambda + n^2(\bar{A}_{66} - B_{66} + \bar{D}_{66})$$

$$F_{12} = (\bar{A}_{16} + 2\bar{B}_{16} + \bar{D}_{16})\lambda^2 + n(\bar{A}_{12} + \bar{A}_{66} + \bar{B}_{12} + \bar{B}_{66})\lambda + n^2\bar{A}_{26}$$

$$F_{13} = (\bar{B}_{11} + \bar{D}_{11})\lambda^3 + n(3\bar{B}_{16} + \bar{D}_{16})\lambda^2 + [n^2(\bar{B}_{12} + \bar{B}_{66} - \bar{D}_{66})$$
$$+ \bar{A}_{12}]\lambda + n^3(\bar{B}_{66} - \bar{D}_{26}) + n(\bar{A}_{26} - \bar{B}_{26} + \bar{D}_{26})$$

$$F_{22} = (\bar{A}_{66} + 3\bar{B}_{66} + 3\bar{D}_{66})\lambda^2 + 2n(\bar{A}_{26} + 2\bar{B}_{26} + \bar{D}_{26}) + n^2(1 + \bar{B}_{22})$$

$$F_{23} = (\bar{B}_{16} + 2\bar{D}_{16})\lambda^3 + n(\bar{B}_{12} + 2\bar{B}_{66} + \bar{D}_{12} + 3\bar{D}_{66})\lambda^2 + [n^2(3\bar{B}_{26}$$
$$+ 2\bar{D}_{26}) + \bar{A}_{26} + \bar{B}_{26}]\lambda + n^3\bar{B}_{22} + n$$

$$F_{33} = \bar{D}_{11}\lambda^4 + 4n\bar{D}_{16}\lambda^3 + 2[n^2(\bar{D}_{12} + 2\bar{D}_{66}) + \bar{B}_{12}]\lambda^2 + 2n(2n^2\bar{D}_{26}$$
$$+ 2\bar{B}_{26} - \bar{D}_{26})\lambda + (n^2 - 1)^2\bar{D}_{22} + (2n^2 - 1)\bar{B}_{22} + 1$$

$$q_1 = \frac{Pr}{A_{22}}; q_2 = \frac{C}{A_{22}}; q_3 = \frac{T}{A_{22}}$$

and

$$\bar{A}_{ij} = \frac{A_{ij}}{A_{22}}; \bar{B}_{ij} = \frac{B_{ij}}{aA_{22}}; \bar{D}_{ij} = \frac{D_{ij}}{a^2 A_{22}}$$

where $i, j = 1, 2, 6$.

In order to have a non-trivial solution to equation 4.122 the determinant of the load matrix in equation 4.122 must be zero. Employing this condition results in the following expression:

$$H_1 P^2 + H_2 PT + H_3 T^2 + H_4 P + H_5 T + H_6 + H_7 C^2 + H_8 PC \quad [4.123]$$
$$+ H_9 CT + H_{10} C = 0$$

where

$$\frac{H_1 \cdot r^2}{A_{22}^2} = n^4(F_{11} + F_{22} + F_{33}) - 2n\lambda F_{12} + 2n^2\lambda F_{13}$$
$$- \lambda^2 F_{22} - 2n^3 F_{23} - n^2 F_{11}$$

$$\frac{H_2 \cdot r}{A_{22}^2} = 4\lambda[n^3(F_{11} + F_{22} + F_{33}) - \lambda F_{12} + n\lambda F_{13} - nF_{11} - 2n^2 F_{23}]$$

$$\frac{H_3}{A_{22}^2} = 4\lambda^2[n^2(F_{11} + F_{22} + F_{33}) - F_{11} - 2nF_{23}]$$

$$\frac{H_4 r}{A_{22}} = n^2(F_{12}^2 + F_{13}^2 + F_{23}^2 - F_{11}F_{22} - F_{22}F_{33} - F_{11}F_{33}) + 2n(F_{11}F_{23}$$

$$- F_{12}F_{13}) + 2\lambda(F_{12}F_{23} - F_{13}F_{22})$$

$$\frac{H_5}{A_{22}} = 2n\lambda(F_{12}^2 + F_{13}^2 + F_{23}^2 - F_{11}F_{22} - F_{22}F_{33}) + 4\lambda(F_{11}F_{23} - F_{12}F_{13})$$

$$H_6 = F_{11}F_{22}F_{33} + 2F_{12}F_{13}F_{23} - F_{11}F_{23}^2 - F_{22}F_{13}^2 - F_{33}F_{12}^2$$

$$\frac{H_7}{A_{22}^2} = \lambda^4(F_{11} + F_{22} + F_{33})$$

$$\frac{H_8 \cdot r}{A_{22}^2} = 2n^2\lambda^2(F_{11} + F_{22} + F_{33}) + 2\lambda^3 F_{13} - 2n\lambda^2 F_{23}$$

$$\frac{H_9}{A_{22}^2} = 4\lambda^3[n(F_{11} + F_{22} + F_{33}) - F_{23}]$$

$$\frac{H_{10}}{A_{22}} = \lambda^2(F_{12}^2 + F_{23}^2 + F_{13}^2 - F_{11}F_{22} - F_{22}F_{33} - F_{11}F_{33})$$

From equation 4.123 critical buckling loads can now be calculated under combinations of axial load, pressure and torque.

These equations provide a means by which the critical buckling loads can be calculated, but owing to the lengthy arithmetical procedures required for their solution they cannot be regarded as simple design equations. To overcome this it is often assumed that shell buckling occurs by a simpler mechanism and then these results are compared with the more rigorous theory. Consider the deformation mode shown in Fig. 4.44, where the shell buckles in a series of rings symmetrical with respect to the axis of the cylinder.[25] Since in any one cross-section all the points on the middle surface of the shell have the same displacements, only a small rectangular element of the shell need be examined. Consider such an element, dx, of unit width under the action of stress resultants N_x and N_y in the x and y directions respectively (Fig. 4.45).

For an anisotropic material the stress resultants can be calculated using the constitutive equations derived in Chapter 3, i.e.

$$N_x = A_{11}\varepsilon_x + A_{12}\varepsilon_y$$

$$N_y = A_{12}\varepsilon_x + A_{22}\varepsilon_y$$

[4.124]

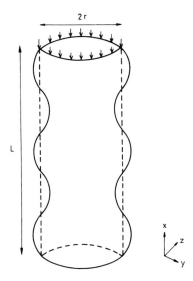

4.44 Deformation mode for axisymmetric buckling.

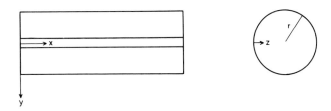

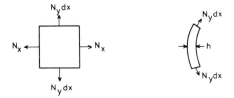

4.45 Shell element.

where A_{11}, A_{12} and A_{22} are stiffness terms and ε_x and ε_y are the elastic strains in the x and y directions respectively. For the case of axial loading, N_x is constant and N_y is dependent on the radial displacement of points on the strip during deformation. The strain of the middle surface of the shell in the circumferential direction, ε_y, is given by:

$$\varepsilon_y = -w/r \qquad [4.125]$$

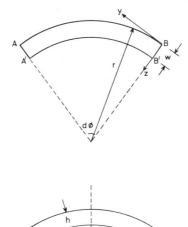

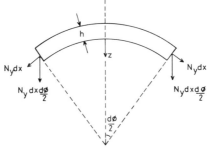

4.46   Displacement of middle surface.

where w is the displacement of the middle surface in the z direction (Fig. 4.46).
   Combining equations 4.124 and 4.125 results in the following equation for N_y:

$$N_y = \frac{1}{A_{11}}[A_{12}N_x - \frac{w}{r}(A_{11}A_{22} - A_{12}^2)] \qquad [4.126]$$

Because of the curved surface N_y has a component in the radial direction and its magnitude, per unit length, is:

$$\frac{1}{rA_{11}}[A_{12}N_x - \frac{w}{a}(A_{11}A_{22} - A_{12}^2)] \qquad [4.127]$$

Owing to the curvature of the strip which occurs in the xy plane due to bending, N_x also has a component in the radial direction, i.e. $N_x\kappa_x$, where κ_x is the change in curvature in the x direction. For a cylindrical shell this becomes:

$$N_x \cdot \frac{d^2w}{dx^2} \qquad [4.128]$$

The equation of equilibrium for a closed cylindrical shell in the radial direction can be written as:

$$-D_{11}\frac{d^4w}{dx^4} + \{\text{forces acting in the radial direction}\} = 0 \quad [4.129]$$

where D_{11} is the flexural rigidity of the material in the axial direction.

Combining these equations, the differential equation for the bending of the strip is:

$$D_{11}\frac{d^4w}{dx^4} = \frac{1}{rA_{11}}\left\{A_{12}N_x - \frac{w}{r}(A_{11}A_{22} - A_{12}^2)\right\} + N_x\frac{d^2w}{dx^2} \quad [4.130]$$

In applying equation 4.130 to a buckling situation, the displacement, w, should be measured not from the unstrained middle surface but from the middle surface after uniform compression is applied. Therefore, w should be replaced by:

$$w + A_{12}^*N_x r \quad [4.131]$$

where A_{12}^* is the corresponding term to A_{12} in the inverse of the constitutive equation. Using equation 4.131, the differential equation for symmetrical buckling of a cylindrical shell becomes:

$$D_{11}\frac{d^4w}{dx^4} - N_x\frac{d^2w}{dx^2} + C_1w - C_2N_x = 0 \quad [4.132]$$

where

$$C_1 = \frac{1}{r^2}\left(A_{22} - \frac{A_{12}^2}{A_{11}}\right)$$

$$C_2 = \frac{1}{rA_{11}}\{A_{12} - A_{12}^*(A_{11}A_{22} - A_{12}^2)\}$$

By assuming that the radial deformation is sinusoidal, the complementary function of equation 4.132 can be written as:

$$w = -A\sin\lambda x \quad [4.133]$$

where $\lambda = m\pi/L$, A is an arbitrary constant, L is the length of the cylinder and m is the number of half waves formed in the axial direction. A particular integral of equation 4.132 is:

$$w = \left(\frac{C_2}{C_1}\right)N_x \quad [4.134]$$

Hence the general solution can be written as

$$w = -A\sin\lambda_x + \left(\frac{C_2}{C_1}\right)N_x \quad [4.135]$$

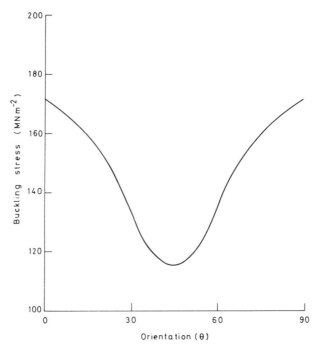

4.47   Effect of fibre orientation on axisymmetric
buckling mode.

Solving equation 4.132 for N_x, and then minimizing with respect to λ, results in the following expression for the minimum critical buckling load:

$$N_x = \frac{2}{r}\sqrt{\left(A_{22} - \frac{A_{12}^2}{A_{11}}\right)D_{11}} \qquad [4.136]$$

By the insertion of conditions of isotropy, equation 4.136 reduces to 4.119.

Figure 4.47 shows the variation of critical load with fibre direction for GRP angle-ply laminates as calculated from equation 4.136.[25] A rather unexpected result is that the maximum critical load value occurs at both the 0° and 90° orientations. In this buckling mode stretching of the shell occurs in both the axial and circumferential directions and the corresponding stiffnesses make an equal contribution. Critical loads as calculated by the more general analysis (equation 4.123) are shown in Fig. 4.48, again for GRP angle-ply laminates. There are differences in predicted response and this is due to the influence of the circumferential mode shapes which are not addressed in the derivation of equation 4.136. Interestingly, critical load maxima still occur at two orientations, one close to the axial direction and the second close to the hoop direction. Absolute values of predicted load also tend to be different and this is

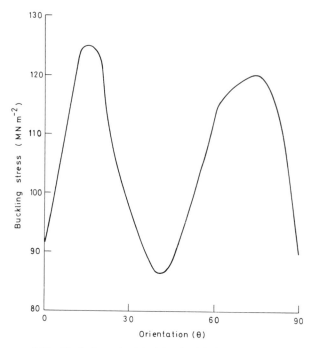

4.48   Variation of critical load with fibre orientation.

caused by the added circumferential waves providing a lower energy solution. The choice of analysis method must be carefully made, either can be used for ranking materials or geometrical configurations, and in a sense the variations in critical load predictions are not necessarily of prime importance, in that the influence of imperfections can be the more significant. Typical factors used to scale analytical results are of the order of 0.3–0.4.

The buckling of plates can be treated in a similar manner.[5] As with other types of plate loading, closed-form solutions for the general case are not available. However, with assumptions regarding laminate orientation some simplifications are possible. As indicated in the section on plates the deflection of an orthotropic plate can be represented by a differential equation of the form:

$$D_{11}\frac{\partial^4 w}{\partial x^4} + 2(D_{12} + D_{33})\frac{\partial^4 w}{\partial x^2 \partial y^2} + D_{22}\frac{\partial^4 w}{\partial y^4} = q$$
$$+ N_x\frac{\partial^2 w}{\partial x^2} + N_y\frac{\partial^2 w}{\partial y^2} + 2N_{xy}\frac{\partial^2 w}{\partial x \partial y}$$

[4.137]

Equation 4.119 has been solved for a number of specific cases. For example, consider a simply supported plate with an axial compressive force, p, acting, i.e.:

$$N_x = -p$$

$$N_y = N_{xy} = q = 0$$

Equation 4.137 becomes:

$$D_{11}\frac{\partial^2 w}{\partial x^4} + 2(D_{12} + 2D_{33})\frac{\partial^4 w}{\partial x^2 \partial y^2} + D_{22}\frac{\partial^2 w}{\partial x^2} + p\frac{\partial^2 w}{\partial x^2} = 0 \quad [4.138]$$

The boundary conditions for simply supported edges are as follows:

$$w = 0 \text{ at } x = 0, \, x = a, \, y = 0, \, y = b$$

$$M_x = 0 \text{ at } x = 0, \, x = a$$

$$M_y = 0 \text{ at } y = 0, \, y = b$$

Using the definitions of moment resultants

$$M_x = -\left(D_{11}\frac{\partial^2 w}{\partial x^2} + D_{12}\frac{\partial^2 w}{\partial y^2}\right)$$

$$[4.139]$$

$$M_y = -\left(D_{12}\frac{\partial^2 w}{\partial x^2} + D_{22}\frac{\partial^2 w}{\partial y^2}\right)$$

Considering these expressions in conjunction with the boundary conditions gives the following:

$$\frac{\partial^2 w}{\partial x^2} + v_{21}\cdot\frac{\partial^2 w}{\partial y^2} = 0 \qquad \text{at } x = 0, \, x = a$$

$$[4.140]$$

$$\frac{\partial^2 w}{\partial y^2} + v_{12}\cdot\frac{\partial^2 w}{\partial x^2} = 0 \qquad \text{at } y = 0, \, y = b$$

The solution of equation 4.140 which satisfies all of the above conditions is of the form

$$w = A_{mn} \sin\frac{m\pi x}{a} \sin\frac{n\pi y}{b} \qquad [4.141]$$

Substitution of the solution gives

$$A_{mn}\left\{\pi^4\left[D_{11}\left(\frac{m}{a}\right)^4 + 2(D_{12} + 2D_{33})\left(\frac{mn}{ab}\right)^2 + D_{22}\left(\frac{n}{b}\right)^4\right]\right.$$

$$\left. - p\pi^2\left(\frac{m}{a}\right)^2\right\} = 0$$

$$[4.142]$$

Taking the non-zero solution and noting that n must equal 1 for the minimum

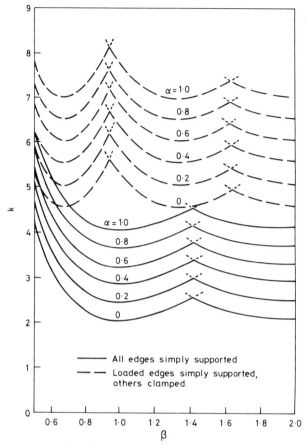

4.49   Coefficients for plate buckling under
compressive loading (simply supported edges).

value of p gives the following expression for critical buckling load:

$$p_{cr} = \frac{\pi^2}{b^2}\left[D_{11}\left(\frac{mb}{a}\right)^2 + 2(D_{12} + 2D_{33}) + D_{22}\left(\frac{a}{mb}\right)^2\right] \quad [4.143]$$

Solving for the minimum value of P_{cr} as a function of m will give the appropriate value of critical load. Further analysis shows that the simply supported case, together with other boundary conditions, can be represented in the form:

$$P_{cr} = k\frac{\pi}{tb^2}(D_{11}D_{22})^{1/2} \quad [4.144]$$

where k is a constant depending on boundary conditions, the values of which are plotted in Fig. 4.49 and 4.50 as a function of dimensionless parameters α

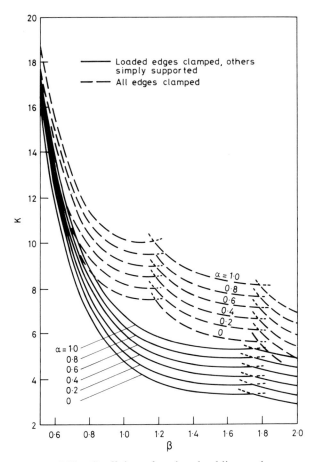

4.50 Coefficients for plate buckling under
compressive loading (clamped edges).

and β for two sets of edge restraint.[26] The parameters α and β are defined as:

$$\alpha = \frac{D_{12} + 2D_{33}}{(D_{11}D_{22})^{1/2}}$$

$$\beta = \left(\frac{a}{b}\right) \cdot \left(\frac{D_{22}}{D_{11}}\right)^{1/2}$$

The stability of plates under combined loading can be treated in a similar way, giving rise to the following equation defining instability:

$$P_x\left(\frac{m}{a}\right)^2 + P_y\left(\frac{n}{b}\right)^2 = \pi^2\left[D_{11}\left(\frac{m}{a}\right)^4 + 2(D_{12} + 2D_{33})\left(\frac{mn}{ab}\right)^2 \right.$$

$$\left. + D_{22}\left(\frac{n}{b}\right)^4\right] \qquad [4.145]$$

For combinations of tension and compression the tensile load should be taken with a minus sign.

As with beams and shells the buckling analysis of plates cannot be undertaken without caution. Analytical results should be treated with some circumspection and where possible correlation with data should be undertaken to establish a firm basis for design. Instability, because of its nature, can be a severe event – there is often little or no warning and its consequences can be dramatic. This is the case for any material, be it metallic and homogeneous or reinforced. For composites the analysis may be more complex owing to anisotropy but, on the other hand, the ability to enhance stiffness in a highly preferential manner offers many advantages in structural design.

References

1 Roark R J and Young W C, *Formulas for Stress and Strain*, McGraw-Hill, New York, 1977.

2 Timoshenko S P and Gere J M, *Mechanics of Materials*, Von Nostrand Reinhold, London, 1972.

3 Timoshenko S P and Gooder J N, *Theory of Elasticity*, McGraw-Hill, New York, 1951.

4 Love A E H, *Treatise on the Mathematical Theory of Elasticity*, Dover, New York, 1927.

5 Calcote L R, *The Analysis of Laminated Composite Structures*, Van Nostrand Reinhold, New York, 1969.

6 Soden P D and Eckold G C, Design of GRP pressure vessels, in *Developments in GRP Technology 1*, Ed B Harris, Applied Science, London, 1983.

7 Timoshenko S and Woinowsky-Kreiger S, *Theory of plates and shells*, McGraw-Hill, New York, 1959.

8 Johnson A F and Sims G D, *Composite Structures*, Vol 2, Ed I H Marshall, Applied Science, London, 1983.

9 Lekhnitskii S G, *Anisotropic Plates*, Gordon and Breach, New York, 1968.

10 Waddoups M E, Eisenmann J R and Kaminskii B E, 'Macroscopic fracture mechanics of advanced composite materials', *J. Comp. Mat*, **4**, 446, 1971.

11 Whitney J W and Nuismer R J, 'Stress fracture criteria for laminated composites containing stress concentrations', *J Comp Mat*, **8**, 253, 1974.

12 Nuismer R J and Whitney J M, *Fracture Mechanics of Composites*, ASTM STP 593, 1975.

13 Gill S S, *Stress Analysis of Pressure Vessels and Pressure Vessel Components*, Pergamon, Oxford, 1970.

14 Flugge W, *Stress in Shells*, Springer-Verlag, New York, 1960.

15 Novozhilov V V, *Thin Shell Theory*, Noordhoff, Gronigen, Netherlands, 1964.

16 Kraus H, *Thin Elastic Shells*, Wiley, Chichester, UK, 1967.

17 Dong S B, Pister K S and Taylor R L, 'On the theory of laminated amisotropic shells and plates', *J Aerospace Sciences*, **29**, 969, 1962.

18 Reissner E and Stavsky Y, 'Bending and stretching of certain types of heterogeneous aeolotropic elastic plates', *J App Mech*, Paper No 61–APM–21, 1961.

19 Gulati S T and Essenbury F, 'Effects on anisotropy in axisymmetric cylindrical shells', *J App Mech*, **34**, 659, 1967.
20 Pagano N J and Whitney J M, 'Geometric design of composite cylindrical characterisation specimens', *J Comp Mat*, **4**, 360, 1970.
21 Eckold G C, *Composite Structures*, Submitted for publication.
22 BS 4994, 'Design and construction of vessels and tanks in reinforced plastics', 1987.
23 Becker H and Gerard G, 'Elastic stability of orthotropic shells', *J Aerospace Sci*, **29**, 505, 1962.
24 Cheng S and Ho B P C, 'Stability of heterogeneous anisotropic cylindrical shells under combined loading, *AIAA Journal*, **1**, 892, 1962.
25 Eckold G C, 'Strength and elastic properties of filament wound composites, University of Manchester, 1978.
26 Stausky Y and Hoff N J, 'Mechanics of composite structures', in AGH Dietz (Ed.), *Composite Engineering Laminates*, MIT Press, Cambridge, Massachusetts, pp. 5–59, 1969.

5

ASPECTS OF DESIGN

When reviewing initial component feasibility or undertaking a scoping exercise on a design it is often the case that considerations can be limited to the major elements of the system – typically beams, plates or shells of various forms and sizes. Applied loadings and environmental conditions are also commonly considered differently in the stages of design. Initially the major loads such as tension, torques and pressures are employed to assess an overall laminate construction. Global estimates of weight, stiffness and strength can usually be made of sufficient accuracy to allow judgements to be made as to whether or not to proceed to the next phase of the product development programme.

In the next design step an altogether different level of detail must be accommodated. For example, individual elements must be joined together not only to achieve physical connection, but also to ensure efficient load transfer from one part to another, perturbations in applied stress fields due to the effects of free edges and discontinuities need to be minimized and account must be taken of fabrication-induced influences. At this point more rigour is also applied to evaluation of the operating environment. Interactions between mechanical loads, cyclic stresses and transients such as impact events or temperature excursions may all feature in the assessment. Additionally, permutations in materials of construction, e.g. hybrid combinations and sandwich structures, may need to be considered to achieve a higher level of structural optimization.

Inevitably the boundaries between the different levels of design will not be well defined and certainly the conclusions made in the early stages will need to be revisited in the context of the results of the more detailed study. This is especially the case with composites as the most efficient engineering solutions are usually those where all requirements are fully integrated within the design. Of course, the ideal situation would be one where all of these issues are addressed at the outset, but the economics of component development usually dictate a staged approach. It falls on the experience of the engineer, therefore, to attempt to ensure that the initial design basis is sufficiently robust to take into account aspects of performance which will arise should the programme continue to proceed.

Jointing

Ideally, load-bearing structures would be designed without joints or connections, eliminating a source of added weight, complexity and weakness. In reality this is seldom possible for a number of technical, commercial or practical reasons such as size restrictions during moulding, requirements for disassembly for transportation, inspection or repair, and the inclusion of structural fittings or bearings, all of which may be called for in the component design. The main purpose of a structural jointing configuration is to transfer load from one component to another. As a consequence there is likely to be a complex stress distribution in the joint region as well as in the joining feature itself and an objective of a given design will be to minimize stress concentrations arising in order to enhance structural efficiency. This is especially true with composites as they are often associated with a weight-saving concept, and also rapid variations in stress tend to cause significant through-thickness tensile stresses which can cause failure due to low strengths in these directions. Primary methods of attachment for composite materials are mechanical fastening, adhesive bonding or some combination of these.

Adhesive bonding

Adhesive bonding offers potential advantages over other fastening methods such as riveting or bolting as it does not reduce the adherend strength and is efficient in terms of low weight and increased stiffness. This is particularly true in the case of thin structural systems. However, the need for careful design and surface preparation, and the effects of low composite transverse strength and moisture degradation of the substrate/adhesive interface can limit their use in specific areas. Typical advantages of bonding include:

- Ability to join thin sheets and dissimilar materials.
- Improved structural efficiency, often with fewer pieces.
- Superior fatigue performance compared with bolted construction.
- Smooth external appearance.
- Sealant behaviour between dissimilar adherends thus reducing electrochemical corrosion and moisture ingress.
- Good damping characteristics.
- Avoidance of bolt holes which act as severe stress concentrators.

Figure 5.1 shows a number of basic joint configurations.

The purpose of an adhesive is to join surfaces together, using a mixture of chemical and physical bonds, and to transmit structural loads without separation. Modern structural adhesives are generally composed of mixtures of several different polymers, each of which are added to satisfy certain

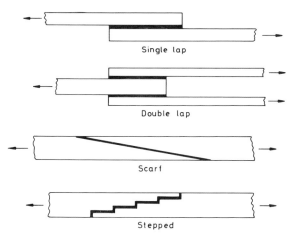

Figure 5.1   Basic joint configurations.

fabrication conditions or to improve properties required in the final joint. Toughness is one property where the design of the adhesive system, often on the molecular scale, can be effective in achieving enhanced behaviour. The base resin is commonly an epoxide, an acrylic or a polyurethane, but they can be blended with a variety of constituents to improve processing or in-service performance. Generally, the stronger adhesives solidify by chemical reaction whereas weaker adhesives rely on some physical change or natural surface tackiness. New formulations and applications are constantly in demand, and since there is not yet a universal adhesive, it is difficult to be fully acquainted with new developments and the detailed aspects of adhesive selection and joint design.[1-5] Recently, information on adhesives and selection procedures has been compiled either in a desktop database[6] or as a computer-based expert system.[7]

Most surfaces need to be cleaned before adhesive application; structural bonds and metals usually require chemical pretreatments. Metals, for example, require surface preparation to ensure that the metal oxide is firmly attached and its morphology is suitable for bonding. Joint durability in the long term is critically dependent on surface condition even though initial joint strengths with untreated surfaces may be satisfactory. This is due to the damaging effects of moisture attacking the interface and is exacerbated by combinations of high stress levels and elevated temperature. The uncured adhesive should be capable of spreading freely over the surface to ensure good wetting and to displace entrapped air and any residual traces of contaminants. Pretreatments for composite materials usually involve manual abrasion or dry alumina grit blasting to remove traces of release agent.[1]

The distribution of stresses within an adhesive bond subjected to tensile or shear loads is uneven along the bond length and has a marked stress

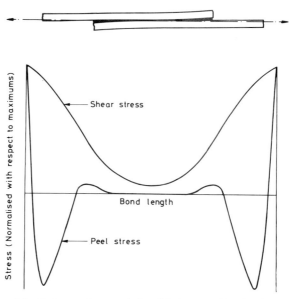

5.2   Deformations and simplified stress distributions
for a single lap joint.

concentration at the ends of the joint overlap. Figure 5.2 shows the
deformations and stress distributions for a single lap joint subjected to a
simple tensile load. (Note that calculated stress distributions are approximate
and are derived from a simple analysis – for example, shear stresses at the free
edge must be zero, but this is not indicated by the calculation method.) Both
the shear and peel stresses are important in the context of design. As the joint
length reduces, so does the length of bond area in the centre which is at zero
stress. For very short joints the stress distribution can be considered to be
effectively constant along the length. The effect of stress peaks at joint edges
can be made more severe for dissimilar adherends where the stiffness change
can influence the distribution of load. Figure 5.3 shows a typical design case, in
this case the shear strain distribution in a coaxial CFRP/ steel joint subjected
to a torsional load.[8] Calculations of these stresses – either by finite element
analysis or by classical analysis based on continuum mechanics – is essential in
design to ensure that all of the key stress values are identified. Generally,
analysis by either of these techniques should be capable of taking into account
factors such as nonlinearity of the adhesive, orthotropic adherends, shear
deflection in the composite, thermal stresses and changing geometrical
parameters.

   In order to modify the nature of the stress or strain distribution to suit
design criteria, a number of simple guidelines can be applied. For example,
increasing the adhesive thickness reduces the peak strains and flattens the

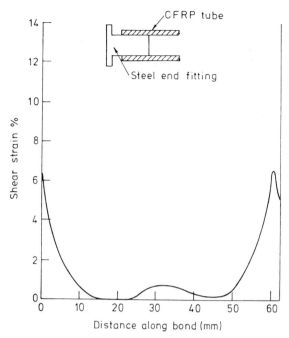

5.3   Strain distribution for a CFRP/steel joint subject
to a torsional load.

profile of the strain distribution. Conversely, decreasing the adhesive thickness increases maximum strains. Increasing the length of the joint will reduce the strains due to mechanical loads; however, this reduction will only occur until the strains in the centre of the joint reach a minimum, after which there is little worthwhile reduction. In some cases strains due to thermal mismatch between the adherends can be an important factor and these tend to become more significant with longer bond lengths. To reduce the very high stress concentrations at the edges of the joint a more compliant adhesive can be used. If the adherend properties, applied loads and joint geometry remain constant then peak stresses or strains within the joint can be controlled to a certain extent through careful selection of adhesive modulus.

Validation of predictive models can be performed by experimental stress analysis. Laser Moiré interferometry is a powerful technique which can be applied to joints as a means of producing a full field strain map of the adhesive layer and adherends.[9] Figure 5.4 shows Moiré displacement fringes for a thick lap shear test joint under increasing loads. Average peel strains across the adhesive thickness and along the bond length can be readily calculated from fringe displacement. The peel strains are compressive in the central region and become tensile near the ends of the overlap. At 3000 N the fringes are continuous over the whole bonded region. The peel strain concentration at the

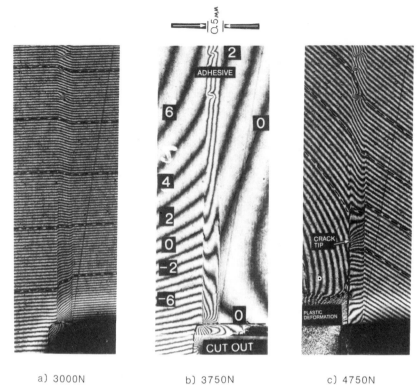

a) 3000N   b) 3750N   c) 4750N

5.4   Moiré displacement fringes close to the cut out of
a joint as a function of load.

cut-out can be clearly seen. At loads greater than 4000 N small interfacial
cracks are observed growing from the cut-outs and extend with increasing
load. Plastic deformation is also observed in the steel adherend adjacent to the
cut-out.

The geometry of the adherends can have a dramatic effect on joint
performance, not only affecting the stresses within the parent substrates but
also the method of load transfer. Figures 5.5 – 5.7 show different types of joint,
together with the corresponding stress distributions.[10] These joints are of
tubular form with different geometries of end fitting. Each type of joint has its
own characteristics and the selection of one configuration in preference to
another depends on the requirements of each particular design case.

With the wide number of options available it is important to specify what
type of joint would represent an optimum design. The aim of an optimization
procedure is to arrive at a design which is most effective in terms of weight,
material utilization, ease of manufacture and cost, whilst fully satisfying
operating requirements. It is governed by the load case under consideration
and the way in which the stress distribution varies with design variables. In an

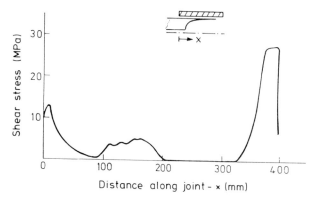

5.5   Adhesive shear stress distribution – profiled joint.

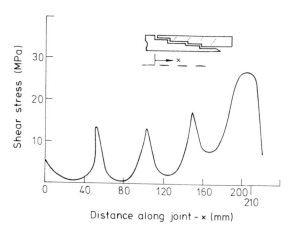

5.6   Adhesive shear stress distribution – stepped joint.

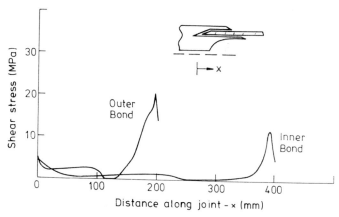

5.7   Adhesive shear stress distribution – double joint.

adhesive joint, different criteria may apply, depending on whether the loading is static, short term, creep or fatigue.

Possible design criteria for a short-term static load could be:

- The maximum adhesive strain must not exceed the strain to failure for the adhesive.
- The overlap length should be as small as possible for structural efficiency.
- The strain should ideally be constant across the whole joint length so as to make best use of the adhesive area.
- The maximum strain in the joint should not be sensitive to small changes in joint length.

In the application of such criteria the starting point would be to assume the adhesive layer thickness is initially set to the maximum allowable value to obtain a constant strain distribution. A minimum value of joint length to support the applied load would then be calculated by assuming that the average stress in the joint is at the elastic limit for the adhesive.

For a fatigue load a different approach should be adopted. Here a region of low stress within the joint would be desirable to allow some scope for redistribution of peak stresses during cyclic loading. At the joint edges the maximum adhesive strain must not exceed the allowable strain for the adhesive. What constitutes a maximum adhesive strain would need to be derived by experience but, for example, an exponential law based upon the failure strain, elastic strain limit and the number of cycles could be used. Again, quantification of the extent of the desired low stress area would need to be the subject of some consideration, but a basis along the lines that more than 50% of the joint length should be stressed below 10% of the elastic limit for the adhesive could be established. In the calculation the first step would be to set the thickness of the adhesive to a minimum value so as to maximize the unstressed region in the joint. A minimum value of joint length to support the applied load would then be calculated by assuming that the average stress in the joint was at the elastic limit for the adhesive. An initial stress analysis would be carried out with a joint length several times the minimum value. If the resultant maximum shear strain was greater than the allowable strain, then the adhesive thickness would be increased progressively until the maximum strain falls below the allowable strain, or until the maximum adhesive thickness is reached. If the strain falls below the allowable value, then the adhesive thickness is fixed and the length of the joint is altered to ensure that the required unstressed length of joint is obtained. Figure 5.8 compares the strain distributions for the static and fatigue criteria.

The procedures described provide a means of determining optimum values for bond line thickness and overlap length. Clearly, to define the engineering details of the joint completely there are many more aspects of the arrangement that need to be considered. Figure 5.9 shows a joint between a composite (glass

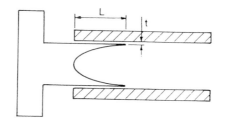

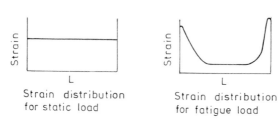

5.8 Possible criteria for design optimization.

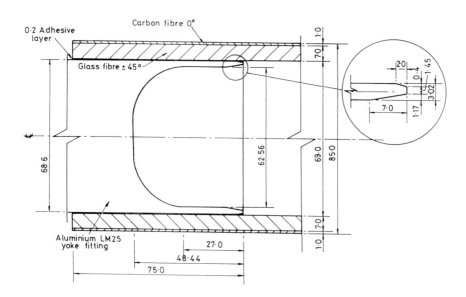

5.9 Profiled end fitting design for hybrid composite/aluminium shaft.

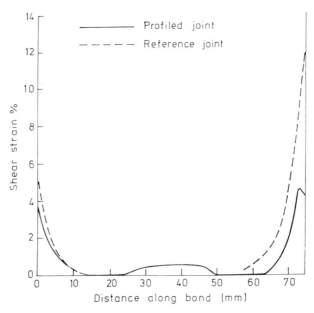

5.10   Calculated strain distribution for hybrid shaft.

fibre/carbon fibre hybrid) driveshaft and a metal end fitting.[11] Figure 5.10 shows the associated strain distribution compared with a reference design that consists of a simple plain ended plug fitting. Profiling the end fitting reduces strain concentrations significantly. In this case the analysis is nonlinear, and this reduction in strain at working load would not proportionally result in the same increase in maximum strength. In fact, an increase in strength of approximately 50% would be predicted for these designs. However, reduction in strain levels at working load would give a significant improvement on fatigue life. The effect of increasing bondline thickness locally at the edge is beneficial, but it is interesting to note that an optimum amount of thickening can be observed from the analysis. Figure 5.11 shows the effect of an increase in bondline thickness. Attention to details of the design such as adherend geometry can pay dividends in terms of component performance in both the short and long term.

Mechanical fastening

Mechanically fastened joints have the advantages of ease of assembly and disassembly, and of little or no preparation of surfaces; they also often exhibit good properties in thermal cycling or high humidity conditions. Their main disadvantages – increased weight, stress concentrations and low joint stiffness – often render them unfavourable for highly stressed, thin composite skins

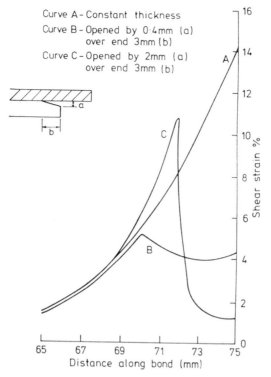

Curve A – Constant thickness
Curve B – Opened by 0·4mm (a)
over end 3mm (b)
Curve C – Opened by 2mm (a)
over end 3mm (b)

5.11   Effect of variation of bondline opening on peak shear strain.

where bonding tends to be preferred. This is usually more effective at joining relatively thick sections of materials as loads can be transferred through the laminate thickness.

Failure of mechanically fastened joints can occur in a number of ways, including tensile, shear-out or bearing modes, and in varying degrees of magnitude from resin cracking to complete failure. Bolted connections are most commonly used, although in some designs which are comparatively lightly loaded, other types of fasteners, such as screws, rivets and bolts can be used. Because of high stress concentrations associated with thread forms, etc., this type of connection can significantly weaken the surrounding composite.

Figure 5.12 shows the stress distributions around a pin-loaded hole.[12] The maximum bearing stresses for pin-loaded holes vary around the circumference, reaching a maximum value in the direction of load transfer. This maximum stress tends to become larger as the clearance in the pin fit is increased and therefore a good fit can improve the bearing strength considerably. Drilled and reamed holes perform better than moulded holes, probably because in the latter case the fibres are not evenly distributed and

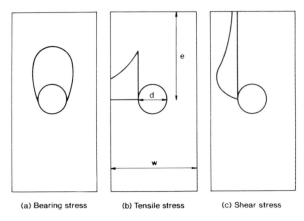

(a) Bearing stress   (b) Tensile stress   (c) Shear stress

5.12   Typical in-plane stress distributions around a
pin-loaded hole.

leave resin-rich regions adjacent to the opening. The maximum bearing stress
for a pin-loaded hole is reduced if the diameter to thickness ratio is greater
than 1, but this does not appear to be true when clamping pressure is present,
as is the case with bolted connections. Generally, mechanical fastening is
efficient in joining relatively thick composites as the load can be distributed
through the laminate thickness. This can be dependent on the laminate
construction, for example, the inclusion of $\pm$ 45° plies has been shown to be
beneficial. Interference fit fasteners give the best results and where multi-bolt
arrays are employed accurate alignment is paramount to ensure load transfer
within the bolt configuration is as per expectations. For joints between
adherends of uniform thickness there is little advantage in having more than
two fasteners in a line as those at the end of a sequence would carry most of the
load.[13] The load can be made more uniform if laminate thickness is varied
between rows of fasteners. The optimum geometry in terms of pitch number
and diameter depends primarily on the properties of the material as the
interactions between stress systems around adjacent holes can have adverse
effects.

   In terms of joint performance there are three major stress components
around a pin-loaded hole; bearing on the loaded side of the pin, tensile on the
net cross-section at the pin position and shear (Fig. 5.12).

   Simplistically, average values for these stresses are given by

$$\sigma_t = P/(w - d)t$$

$$\sigma_b = P/dt \qquad\qquad [5.1]$$

$$\sigma_s = P/2et$$

where σ_t, σ_b and σ_s are tensile, bearing and shear stresses respectively, w and t

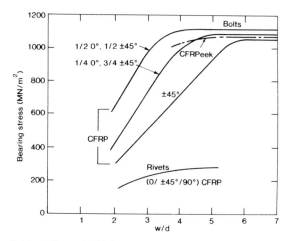

5.13 Effect of width on bearing failure stress of single hole joints in CFRP.

are the width and thickness of the plate, e is the length from the bolt hole to the end of the plate, and P is the applied load.

From the distributions shown in Fig. 5.12 and the above equations it can be seen that joint performance is strongly dependent on geometry. Relatively wide joints will fail in bearing and as width is reduced, the failure mode changes to that of tension. Variations in the value of plate edge/hole distance, e, has an analogous effect. To achieve adequate bearing strength the joint must have sufficient end distance. Figures 5.13 and 5.14 show the effect of geometry on calculated bearing stress.[14] The plateau regions correspond to those configurations where failure is dominated by bearing. The effect of laminate configuration can also be seen. For very simple laminate orientations the probable failure mode can be deduced intuitively. For example, a unidirectional material loaded parallel to the fibres would be expected to fail in shear whereas for the converse case, where fibres are perpendicular to the load, failure would be in tension. Complex laminates, on the other hand, vary considerably. Figure 5.15 shows the influence of fibre orientation on the failure mode of bolted joints in $0/ \pm 45°$ type laminates.[15] As the percentage content of the $\pm 45°$ plies is increased, the shear strength of the laminate increases and bearing becomes the observed failure mode. An additional increase will cause further change to a tensile mode because of the low laminate strength in tension.

Welding

Welding processes are not normally associated with the joining of composite materials and it is only with the advent of thermoplastics that the methods

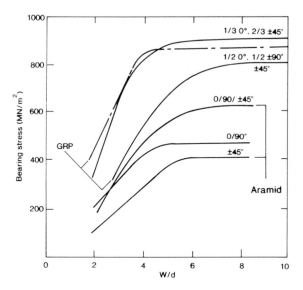

5.14   Effect of width on bearing failure stress of single
hole joints in GRP and aramid composites.

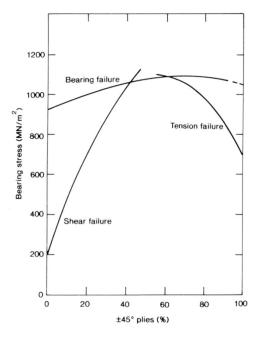

5.15   Influence of fibre orientation on failure mode of
bolted joints in $0/\pm45°$ CFRP.

have become of interest. Welding has attractions as it is potentially faster and more easily automated than adhesive bonding, while mechanical fasteners do not provide a continuous joint and drilling can be difficult and costly.

The welding processes which are potentially available can be divided into two groups:

- Processes involving mechanical movement – these include ultrasonic welding, friction welding (spin, angular and orbital) and vibration (linear friction) welding.
- Processes involving external heating – these include hot plate welding, hot gas welding and resistive and inductive implant welding.

Studies on welded structures indicate that for short fibre reinforced systems the techniques produce satisfactory results, but for continuous systems there are significant difficulties.[15] Discontinuity of load transfer and disruption of the fibre arrangement close to the weld are key issues that must be considered.

Impact

Damage of composite structures through impact events is perhaps one of the most important aspects of behaviour inhibiting widespread application. For ductile systems the materials are able to dissipate the incident kinetic energy through elastic and plastic deformation. Although this may result in some cases to permanent deformation, its effect is often localized. In composites, however, the scope for plastic deformation is limited and the consequences of an impact event can lead to a substantial amount of damage, the influence of which on residual properties is difficult to predict. The variables controlling the events during an impact include material properties, boundary conditions, deformation/failure mechanisms, environmental factors, imposed constraints and the parameters defining the impact event itself.

During an impact, a stress field is established on contact. A series of stress waves is then propagated through the thickness of the material which may or may not cause damage. In order to provide an understanding of the phenomena concerned, consider an elementary one-dimensional model.[16] For an incident compressive wave of magnitude σ_m with pulse length λ (Fig. 5.16), the wave will be reflected from the free surface with a net tensile stress σ_t defined by:

$$\sigma_t = \sigma_m - \sigma_i \qquad [5.2]$$

where σ_i is the compressional incident stress at the same point as the leading edge of the reflected wave. If $\sigma_m > \sigma_f$ where σ_f is the tensile failure stress there will be some position where failure occurs. At this instant:

$$\sigma_f = \sigma_m = \sigma_i \qquad [5.3]$$

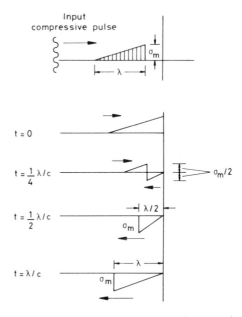

5.16   Net stress for reflection of load pulse at various
times.

It can be shown from geometry that the position of failure, t_1, measured from
the free surface is given by:

$$t_1 = \frac{\sigma_f}{\sigma_m} \frac{\lambda}{2}$$

[5.4]

Therefore, if $\sigma_m = \sigma_f$ failure occurs at $\lambda/2$ from the free surface and $\sigma_f < \sigma_f$ no
fracture will occur and if $\sigma_m > \sigma_f$ multiple fractures may occur. In this latter
case with the aftermath of each failure there will be a new wave and a new free
surface. Further analysis shows:

$$t_n = \frac{\sigma_f}{\sigma_{m,n}} \cdot \frac{\lambda_n}{2}$$

[5.5]

where $\lambda_n = \lambda_{n-1} - 2t_{n-1}$ and $\sigma_{m,n} = \sigma_i$ at the instant of the $(n-1)^{th}$ failure and
n denotes the number of the failures which for a given wave will be:

$$n = \frac{\sigma_m}{\sigma_f}$$

[5.6]

Such an analysis can only be considered as qualitative but it does allow an
appreciation of the phenomena involved during impact. For example, it can
readily be seen how the result of an impact can manifest itself. With the
application of load the dynamic stress system which is established which may

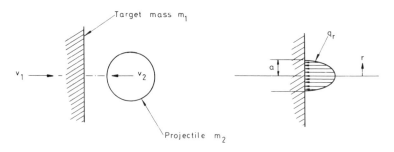

5.17   Pressure distributions due to impact.

result in damage propagation at a number of sites within the thickness of the material. Composites with their low transverse tensile strength can be prone to this type of effect. In the case of carbon composites which are opaque the damage may not be apparent without the application of a sophisticated inspection method. Such 'barely visible damage' is a major design issue.

The first stage in an attempt to provide a quantitative assessment is to derive relationships for the contacting force and resulting stress distribution for an impact event. Proceeding, using a simple analysis which ignores system vibrations, to consider the contact between a stationary semi-infinite target and an impactor gives expressions for the rates of change of velocity during impact:

$$- P = m_1 \frac{dv_1}{dt}$$
$$- P = m_2 \frac{dv_2}{dt}$$

[5.7]

where m_1 and v_1 are the mass and velocity of the target and m_2 and v_2 those for the projectile (Fig. 5.17). The assumption regarding system vibrations must only be regarded as approximate, but it provides a useful starting point which can be justified if the contact times are long in comparison with vibration periods.

The velocity of approach at the point of contact is:

$$\frac{d\alpha}{dt} = v_1 + v_2$$

[5.8]

where α is the distance the two components approach one another due to the local compression at the point of contact. Combining these equations yields:

$$\frac{d\alpha^2}{dt^2} = - P \frac{(m_1 + m_2)}{m_1 m_2}$$

[5.9]

The assumptions regarding system vibrations allows the static equations for two bodies in Hertzian contact to be used:[17]

$$P = n\alpha^{3/2} \qquad [5.10]$$

where

$$n = \frac{4r^{1/2}}{3\pi(k_1 + k_2)}$$

where r is the radius of the impactor and the constants k_1 and k_2 are given by:

$$k_1 = \frac{1 - v_1^2}{\pi E_1}$$

$$k_2 = \frac{1 - v_2^2}{\pi E_2}$$

where E_1, v_1 and E_2, v_2 are modulus values and Poisson's ratio for target and impactor respectively. Further manipulation using equations 5.8 and 5.10 allows an expression to be derived for the maximum deformation α_1, viz.:

$$\alpha_1 = \left(\frac{5v^2}{4Mn}\right)^{2/5} \qquad [5.11]$$

where v is the approach velocity of the two bodies at the beginning of impact and

$$M = \frac{1}{m_1} = \frac{1}{m_2}$$

For Hertzian contact (Fig. 5.17) between a spherical body pressed into a flat surface, the radius of the contact area, a, is given by:[17]

$$a = \left[\frac{3\pi P}{4}(k_1 + k_2)r_1\right]^{1/3} \qquad [5.12]$$

Combining equations 5.10 – 5.12 gives the maximum radius of area of contact during impact:

$$a_{max} = r^{1/2}\left(\frac{5v^2}{4Mn}\right)^{1/5} \qquad [5.13]$$

and

$$P = n^{2/5}\left(\frac{5v^2}{4M}\right)^{3/5} \qquad [5.14]$$

Employing the relationship for pressure distribution, $q_{x,y}$, for a Herztian contact,[17] i.e.:

$$q_{x,y} = q_0\left[1 - \frac{x^2}{a^2} - \frac{y^2}{a^2}\right]^{1/2} \qquad [5.15]$$

where q_0 is the surface pressure at the centre of contact ($x = 0$, $y = 0$). At the boundary of the contact area $q_{x,y} = 0$.

Summing the pressures over the contact area yields:

$$q_0 = \frac{3P}{2\pi a^2} \quad [5.16]$$

Combining equations 5.13–5.16 gives the following expression for the distribution of surface pressure as a function of impact velocity:

$$q_r = \left(\frac{3n}{2\pi R}\right)\left(\frac{5v^2}{4Mn}\right)^{1/5}\left[1 - \left(\frac{r}{a}\right)^2\right]^{1/2} \quad [5.17]$$

To obtain the duration of a given impact recourse can be made to the derivation of equation 5.11 and it can be shown[17] that:

$$t = \frac{2\alpha_1}{v}\int_0^x \frac{dx}{(1 - x^{5/2})^{1/2}} \quad [5.18]$$

where $x = \alpha/\alpha_1$.

The total duration, t_0, is therefore

$$t_0 = 2.94\frac{\alpha_1}{v}$$

$$t_0 = 2.94\left[\frac{5}{4Mnv^{1/2}}\right]^{2/5} \quad [5.19]$$

Knowledge of the magnitude and distribution of surface pressures and the contact area as a function of velocity and time allows calculation of the distribution of internal stresses within the impacted plate.[18] For isotropic materials the maximum tensile, σ_t, compressive, σ_c and shear, σ_s, stresses are given by:[17]

$$\sigma_t = \frac{(1 - 2v_1)}{3}q_t$$

$$\sigma_c = q_t \quad [5.20]$$

$$\sigma_s = \left[\frac{1 - 2v_1}{4} + \sqrt{2}\frac{(1 + v_1)^{3/2}}{9}\right]q_t$$

where q_t is the maximum surface pressure at time t. Furthermore the internal stresses distributions can be calculated using the derived values of q_t. Figure 5.18 shows the solution for an isotropic material and Fig. 5.19 shows the positions of maximum stress.

For composite materials the situation is somewhat more complicated as there are no closed-form solutions for the Hertzian contact problem or for the

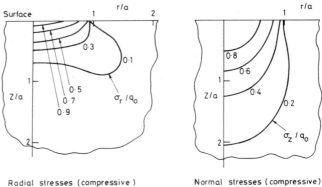

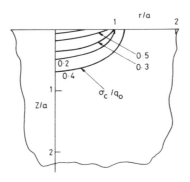

5.18   Internal stresses for a solid subjected to a
       surface pressure caused by impact.

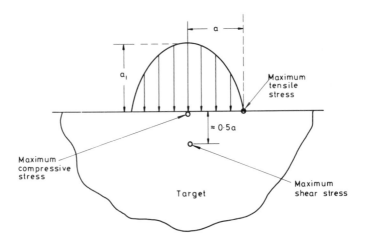

5.19   Position of maximum stress during impact.

Table 5.1. Values of stiffness term k_1 for GRP and CFRP laminates

Material	$k_1 \times 10^6$ (Equation 5.21)	$k_1 \times 10^6$ (Finite element)
Unidirectional GRP	0.068	0.060
Quasi-isotropic GRP	0.114	0.112
Quasi-isotropic CFRP	0.218	0.209

stress distribution. However, for quasi-isotropic materials an approximation may be made where the value of k_1 in equation 5.10 is replaced by

$$k_1 = \left(\frac{1}{\pi E_3}\right) \left\{ \frac{\left[\left(\frac{E_3}{(1 - v_{12}^2)E_1}\right)^{1/2} + \frac{G_{23}}{E_1}\right]^2 - \left[\frac{v_{23}}{1 - v_{12}} + \frac{G_{23}}{E_1}\right]^2}{\frac{4G_{23}}{(1 - v_{12}^2)E_1}} \right\} \quad [5.21]$$

where E_1 and v_{12} are the in-plane modulus and Poisson's ratio and E_3, v_{23} and G_{23} are the transverse modulus, Poisson's ratio and shear modulus respectively.

Using this expression the surface pressure distribution, may be determined. Table 5.1 shows the values of k_1 as calculated approximately from equation 5.21 and those derived from a finite element analysis.[17] Given the pressure distributions, internal stress levels can now be obtained using numerical methods, e.g. finite elements. Evidence from indentation experiments suggests that k_1 values derived from equation 5.21 can be used with orthotropic materials as, to a first approximation, it can be argued that stiffness values in the through-thickness direction will be similar for different laminate types and that these properties will dominate. Figure 5.20 gives distributions of stresses for quasi-isotropic layups. The values shown are normal and radial compressive stresses which arise as a result of an impact-induced surface pressure.

The next step in the process is to examine stress distributions with respect to a failure criterion. The simplest approach would be to use the maximum stress criterion and then to calculate the threshold velocity necessary to initiate tensile, compressive or shear failure. This could then be used to map the extent of failure as a function of time. The benefits of static design criteria are uncertain for impact conditions, but they can be used to achieve a ranking of materials. For example, the analysis can be extended to provide expressions for the calculation of damage threshold velocities for different modes of failure.[17,19]. For tensile failure:

$$v_t = k \left[\frac{k_1^2}{(\sigma_t/q_0)^{5/2}} \right] (\sigma_t^*)^{5/2} \quad [5.22]$$

where $k = r^3/m_2$, v_t is the threshold velocity, σ_t the maximum tensile stress, σ_t^*

Table 5.2. Threshold damage velocities for GRP and CFRP laminates

Material	Tensile failure $(v_T/k \times 10^2)$	Compressive failure $(v_C/k \times 10^2)$	Shear failure $(v_S/k \times 10^2)$
Glass epoxy	60.5	0.94	0.571
Thornel 75 epoxy	6.2	0.06	0.014

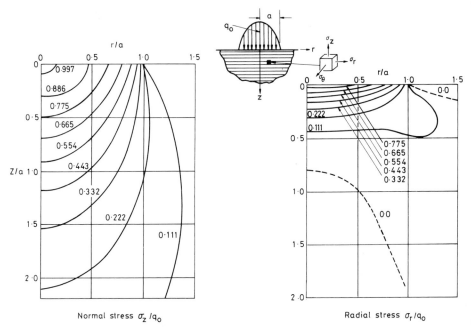

5.20   Stress distributions in quasi-isotropic GRP plate
subjected to impact surface pressures.

the tensile strength and r the impactor radius. Similar expressions can be obtained for shear and compressive failure. Typical values for GRP and CFRP relative to the impactor constant k are shown in Table 5.2. As can be seen, shear failure is indicated to be the dominant mode of failure, followed by compression. In the latter case damage would take the form of local crushing. A point to note is the relative superiority of glass reinforced systems over carbon reinforced materials. The extent of the damage zone can also be assessed as a function of surface pressure. Figure 5.21 shows damage area for a CFRP material with increasing surface pressure. Failure is initiated as the point of contact and thereafter grows with increasing pressure.

A somewhat different response occurs if the composite target is relatively flexible. In addition to contact forces the material will undergo bending deformation, the significance of which will be related to structural parameters

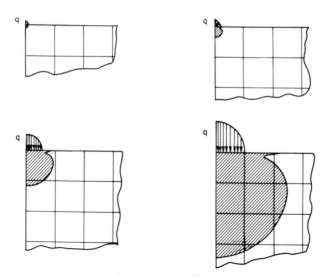

5.21 Propagation of damage zone with increasing
surface pressure.

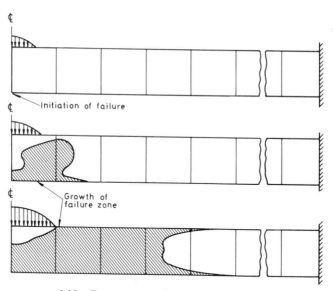

5.22 Damage zone for a plate structure.

such as stiffness and edge boundary conditions. The effect of the bending stress
will be additive to the stress due to surface pressure and this could change both
the position of damage initiation and its subsequent progression. Figure 5.22
shows an example of a relatively thin plate where the bending is sufficiently
great to cause tensile failure on the opposite surface to where pressure is

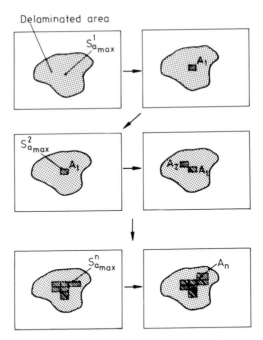

5.23   Steps in calculating extent of delamination.

applied. Further damage develops from this area and eventually propagates through the full thickness.

Conventional failure criteria for composites (see Chapter 3) do not address all of the aspects of behaviour encountered in an impact. Both the different types of mode of failure and the dynamics of the situation need to be accommodated in the analysis.[20-22] Figure 5.23 shows the steps employed to be used in one method to calculate delamination shapes and sizes.[23] The model is initially implemented through the calculation of available strain energy per unit area at every point of a laminate interface. The region of highest strain energy is identified, at which point a small element A_1 is considered and the available energy is compared to that for delamination. If there is an excess of energy, the area is considered to have delaminated. The calculation is repeated for the remaining area in a step-wise manner until the total extent of delamination is calculated. Figure 5.24 shows a comparison between measured and calculated results. Correlation is good, but the general applicability of the model beyond the range of the test specimens has yet to be demonstrated.

The response of material to local contact stress is a partial solution to the problem of impact. As with all dynamic load cases the behaviour of the structure as a whole and propagation of waves within the structure are vitally important, particularly with composites having low transverse properties.

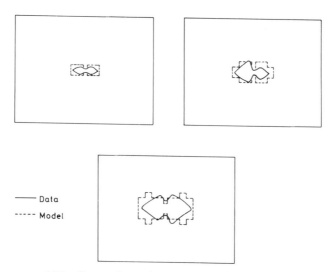

5.24   Comparison of measured and calculated
delaminated areas (cross-ply CFRP).

Computer codes which can accommodate the complexities of the calculations
are becoming available and these need to be linked to the type of failure
criterion which is appropriate to impact. Figure 5.25 shows the results of such
a dynamic calculation where the interactions of the projectile and plate are
simulated explicitly, together with the dynamic response of the specimen. In
this case penetration of the plate is predicted.[24]

During an impact event there are a large number of possible damage
mechanisms. Each has a range of characteristics and affects the properties of
the composite in different ways. Damage modes include delamination, matrix
cracking, splitting, debonding, fibre pull-out and fibre breakage.[25,26] Energy
absorption is a parameter which is often used to describe impact-type events
and Table 5.3 shows fracture energies for CFRP laminates with different resin
systems.[27] As a rule those modes that involve matrix or interphase failure
absorb much less energy than those concerned with fibre fracture or pull-out.

Fibres are the dominant constituent when considering most property
characteristics and the same is true when assessing impact performance. For
low velocity impact the stored energy capability of the fibre is of key
importance. As a result materials such as aramids, and to a lesser extent glass,
which have large areas under their stress/strain curves, offer relatively good
performance. Composites made from these materials tend to fail in a
progressive manner through delamination. Carbon fibre systems, on the other
hand, can be brittle and fail catastrophically at the maximum load. Figure 5.26
shows impact performance of composites as a function of fibre properties
measured in terms of the area under the stress/strain curve.[28] Whilst there is

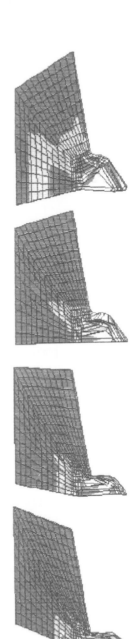

5.25 Dynamic calculations for impact response.

Table 5.3. Energy absorption characteristics of different failure modes for CFRP

Composite failure mode	Resin	Typical fracture energy (kJ/m)
Splitting	Epoxy PEEK	0.1–1 3.8
Delamination	Epoxy PEEK	0.1 2.2
Transverse fibre fracture	Epoxy PEEK	20 128
Fibre pull-out	Polyester Bismaleimide	26 800
Debonding	Epoxy	6

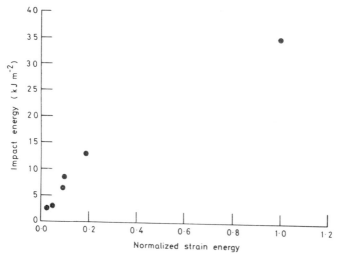

5.26   Variation of impact energy with strain energy of fibres.

no formal relationship governed by the mechanics of the situation the two parameters tend to increase or decrease in concert. Clearly, this is only of value for guidance as any analysis which is intended to evaluate impact performance must consider energy dissipation in all failure processes.

Of interest to designers of composite structures is the concept of residual strength, the load-carrying capability after an impact event. Such a value is often used as a means of quantifying material behaviour. Care must be taken in using this criterion in isolation as, although impact resistance of certain materials may be good and therefore relative reductions in strength low, this may not be too helpful in design as high extensibility fibres tend to have low mechanical properties in the first instance. This can be overcome by the use of

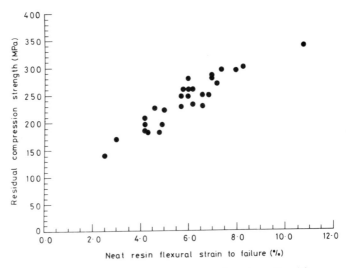

5.27   Variation of residual compression strength with
resin failure strain.

hybrid materials where the attributes of each constituent are used. A further
consequence of impact can be the initiation of modes of failure which would
not occur in an undamaged laminate. The effects of compressive load, for
example, can be particularly damaging as any delamination could seriously
affect elastic stability.

Matrix properties are, of course, a key consideration for impact performance. Not only do they provide the mechanism of load transfer into the fibres,
but if damaged during the impact the resulting cracks could allow ingress of
moisture, etc., which could then cause degradation. Figure 5.27 shows the
relationship between the impact resistance of a composite, measured as
residual compression strength, as a function of resin strain to failure.[29] Again,
as with Fig. 5.26 it is possible to derive a 'rule of thumb' as to the effect of
constituent properties. Because of the importance of impact there have been a
number of attempts to improve the energy-absorbing characteristics of matrix
materials. These have included the use of plasticizers, the addition of rubber or
thermoplastic particles, control of cross-link density (the lower the density of
cross-links, the more flexible the resin), the use of thermoplastic matrices and
the use of interlayers within plies. Although improvements can be achieved
(Fig. 5.28), in most cases the enhanced resin toughness is not wholly
transferred to the composite.[30] The properties of the interface are also
important, but perhaps more difficult to control given a preselected matrix/
fibre combination. For weak interfaces failure is generally through large areas
of delamination. This can be used to advantage if containment of projectiles or
debris is important. If residual strength is the more important criterion, an

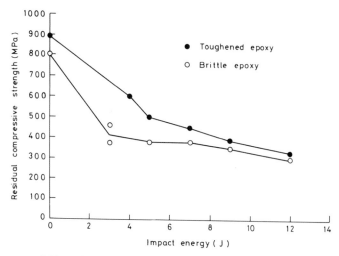

5.28   Effect of resin toughness on residual compressive strength.

interface region of greater strength would be preferred where damage is more localized in nature.

Laminate construction and orientation can be an important factor in the design for impact performance. Simple unidirectional materials do not perform particularly well, primarily because of the high stresses generated transverse to the fibre direction. Also the tendency for delamination is increased where the disposition of individual plies leads to large discontinuities in stiffness. Adopting woven materials or three-dimensional stitched fabrics which tend to promote increased through-thickness tensile strength can be used to good effect (Fig. 5.29).[31]

For a number of applications, particularly in the transport industry, the control of energy absorption under impact conditions is an important design feature. In metal structures this is often done by making use of the work done during plastic deformation. The simplest structural form where use is made of this effect is the inverted cylinder. Here a thin walled tube is punched on to a radiused die to achieve either internal or external inversion (Fig. 5.30), and an approximately linear energy absorption characteristic is obtained (Fig. 5.31).[32] Such devices are used for collapsible steering wheels, seat anchors and landing dampers. In composites there is little or no scope for gross plastic deformation of this nature, although there are considerable opportunities for energy absorption. This arises through the promotion of fibre/matrix debonding over the large surface area of interface. The key to achieving a high level of impact absorption is to design the component such that as great a volume of material as possible becomes involved in the failure process. For example, in a simple composite tubular structure under compression failure by

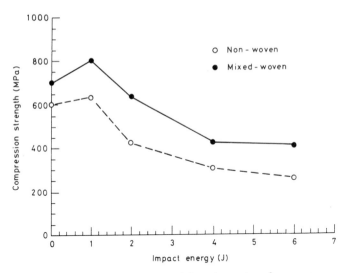

5.29 Effect of woven material on impact performance.

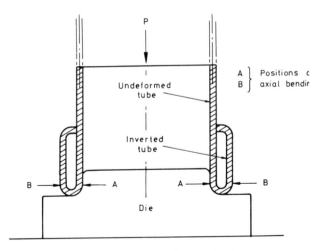

5.30 External inversion of a metal tube.

simple fracture in the central region of the cylinder is likely. Figure 5.32 shows the force displacement plot for such an event and, as can be seen, although the initial peak force is high, the overall area absorbed is small.[33] By providing a chamfer to the edges of the cylinder, however, a different mode of failure can be initiated.[34] The high stress levels in the chamfered regions result in local crushing and this can then propagate through the tube as a crush zone. The force displacement curve for this mechanism is also shown in Fig. 5.32. Although the initial failure load is lower than that for the plane tube,

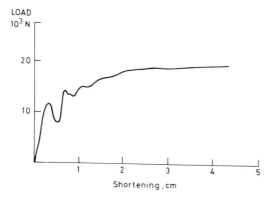

5.31   Load-shortening curves for external inversion of
an aluminium tube.

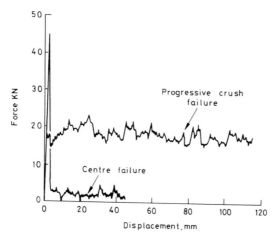

5.32   Force displacement curves for composite
cylinders.

energy-absorbing characteristics are much more attractive. Figure 5.33 shows
a schematic representation of the types of crush mechanism which can be
obtained.[33] The details of these vary according to material properties and
geometrical arrangements. Expressing energy absorption in a specific sense,
i.e. area under the force displacement curve divided by material density, is a
useful means of ranking materials especially if weight is an important feature of
the design. Table 5.4 shows specific energy absorption for tubular structures
for a number of materials. On this basis, the composite options compare well;
however, it should be noted that these values are only realized if crushing
mechanisms such as those shown in Figure 5.33 can be achieved.

Table 5.4. Specific energy absorption for tubular structures

Material	Specific energy $(J/kg^{-1} \times 10^{-3})$
Mild steel	25–29
Aluminium	11–16
GRP (filament wound)	38–41
GRP (woven cloth)	58–65
CFRP (0/90)	56

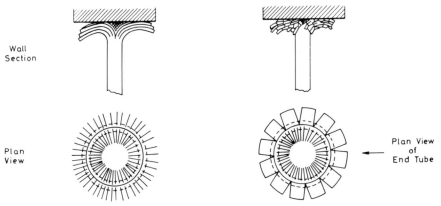

Wall
Section

Plan
View

Plan View
of
End Tube

5.33   Schematic representation of possible crush
mechanisms.

Free edge effects

The laminate analysis described in Chapter 3 provides a simple means of evaluating the response of a composite under a variety of loading conditions. However, the analysis only considers in-plane stresses and so does not rigorously accommodate certain boundary conditions. This is most significant at free edges where interlaminar shear and normal stresses are not considered and also when the condition that the in-plane shear stress must be zero is violated. These stress systems can be of great significance as laminated composites generally have low strength values in the through-thickness direction.

To provide an understanding of how these stresses are generated, consider the case of a symmetrical laminate. On application of a normal force there will be laminate extentional strains, but no laminate shear strains. There will, however, be shear strains in individual lamina due to coupling terms in the lamina stiffness matrix (see Chapter 3). The shearing stresses, τ_{xy}, associated with these strains cannot exist at the free edges of the laminate and the only

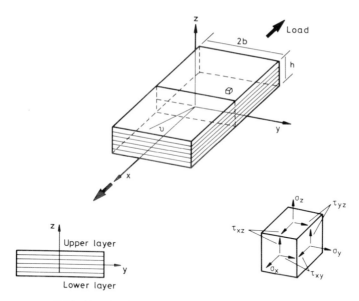

5.34 Interlaminar shear stresses in a composite
laminate.

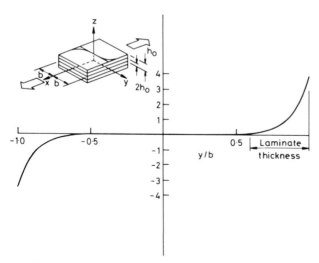

5.35 Axial displacement distribution at the laminate
surface.

possible means of satisfying equilibrium therefore is for shear strains acting on
the face of the lamina, τ_{xz}, to be present. Figure 5.34 shows the physical
situation. If unrestrained, individual laminae under a normal load will have
shear deformation. The action of the interlaminar shear stress which prevents
this strain causes an axial displacement distribution as shown in Fig. 5.35.[35]

The equilibrium equations for a plate under these stress components are:

$$\frac{\partial \tau_{xy}}{\partial y} + \frac{\partial \tau_{zx}}{\partial z} = 0$$

$$\frac{\partial \sigma_y}{\partial y} + \frac{\partial \tau_{yz}}{\partial z} = 0 \qquad [5.23]$$

$$\frac{\partial \tau_{yz}}{\partial y} + \frac{\partial \sigma_z}{\partial z} = 0$$

Use of these expressions together with displacement and constitutive equations allow the problem of interlaminar stress to be formulated. However, solutions to the resulting equations are not available in closed form and numerical techniques must be used. Figure 5.36 shows the results for a $45/-45/-45/45$ laminate under uniaxial tension. Away from the edge the stresses are equivalent to those of laminate theory. As the free edge is approached there are significant perturbations in the stress field. The in-plane normal stress, σ_x, rapidly reduces, the in-plane shear stress, τ_{xy}, approaches zero and the interlaminar shear stress τ_{xz} increases rapidly to its maximum value. Experience from a range of calculations indicates that the critical distance within which these stress gradients occur is of the order of the thickness of the laminate. The variation of interlaminar shear stress with orientation within an angle-ply laminate is shown in Figure 5.37. As would be expected, the shear stress is zero for 0 and 90° angles.[36]

To provide a rapid means of calculating these critical stress components a number of approximate methods have been developed.[37,38] These essentially entail assuming a form for the stress distribution and then deriving a set of expressions which fit the desired shape. For an angle-ply laminate, a third-order polynominal is appropriate. Assuming that $\tau_{xy} = 0$ at the free edge and equal to the laminate theory value at a distance h from the edge, where h is the laminate thickness, the value of τ_{xy} for the ith layer is:

$$(\tau_{xy})_i = \tau_i \left(1 - \frac{y^3}{h^3} \right) \qquad [5.24]$$

where h is the thickness of the laminate and τ_i is the value of τ_{xy} given by laminate theory for the ith layer.

Using equation 5.23 and the boundary condition which applies in the centre of the laminate:

$$\tau_{xz} = \frac{\partial \tau_{xz}}{\partial y} = 0 \qquad [5.25]$$

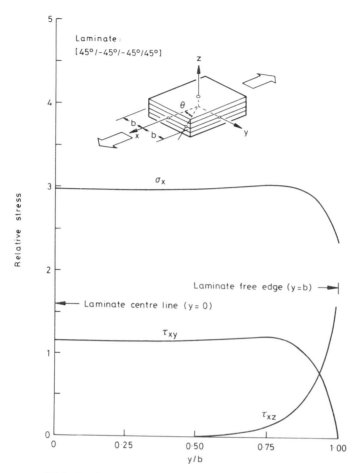

5.36   Variation of interlaminar stress for a symmetric
$\pm 45°$ laminate.

yields:

$$(\tau_{xz})_i = \frac{3h_i}{h^3}\tau_i\left(1 + \frac{h_{i-1}}{h_i}\cdot\frac{\tau_{i-1}}{\tau_i}\right)y^2 \qquad [5.26]$$

where h_i is the thickness of the ith layer.

The maximum value of interlaminar shear stress occurs at the edge and is given by:

$$(\tau_{xz})_{i,\max} = \frac{3}{h}(\tau_i h_i + \tau_{i-1}h_{i-1}) \qquad [5.27]$$

For the case of a balanced laminate where all layers are of equal thickness

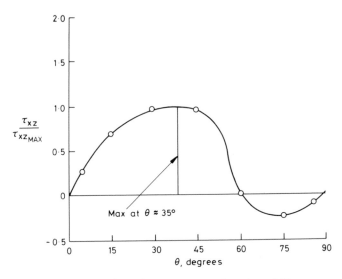

5.37 Interlaminar shear stress as a function of fibre
orientation for angle-ply laminate.

$$(\tau_{xy})_{max} = 3\frac{h_i}{h}\tau_i \text{ for odd interfaces}$$ [5.28]

$$(\tau_{xy})_{max} = 0 \text{ for even interfaces}$$

Figure 5.38 shows the shape of the resulting distribution and Figure 5.39 compares calculated results for a $\pm45°$ laminate with those obtained from a numerical approach.[36,38] The approximate calculation provides reasonable agreement with the more rigorous analyses. An observation which may be made from equation 5.27 is that the interlaminar shear stresses become lower as the number of layers within a given laminate thickness increases, i.e. as the value of the thickness ratio h_i/h reduces. This is easily rationalized as the greater the relative thickness of the layer, the greater the force needed to maintain compatibility with adjacent lamina. As the shear area over which this force acts is constant, the stress is increased by a corresponding amount.

Interlaminar normal stresses can be treated in a similar manner. These arise where there is a Poisson's ratio mismatch in the laminate. Therefore, for an angle-ply laminate where there is no such mismatch the normal stress components are zero. Cross-ply laminates, on the other hand, can have significant variations in Poisson's ratio between layers. The moment due to the value of σ_y which must be zero at the free edge is balanced by the equivalent moment from σ_z; the interlaminar normal force. The distribution of σ_z is often assumed to be of the form shown in Fig. 5.40. Again, approximate methods are

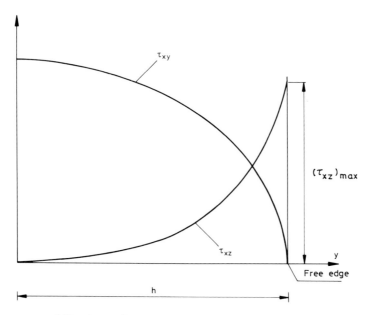

5.38 Approximate stress distributions for angle-ply laminates.

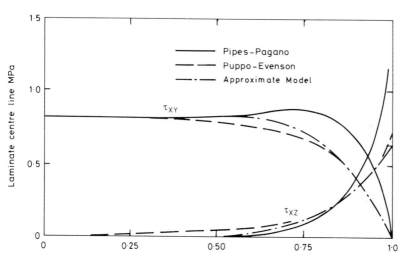

5.39 Interlaminar shear stresses in an angle-ply laminate.

available.[37] For a cross-ply laminate:

$$(\sigma_z)_i = \frac{-9}{7} \frac{(\sigma_y)_i}{h} z^2 - \frac{18}{7} \frac{h_{i-1}}{h^2} (\sigma_y)_{i-1} \cdot z - \frac{(\sigma_z)_{i-1,\max}}{5} \cdot h_{i-1} \quad [5.29]$$

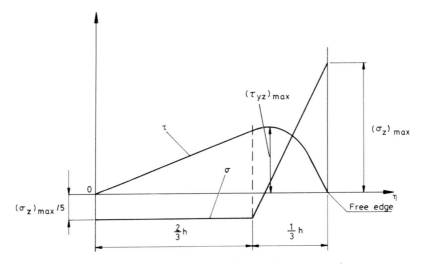

5.40   Approximate stress distributions for a cross-ply
laminate.

for $0 \leq y \leq \frac{2}{3}h$ and

$$(\sigma_z)_i = \left[\frac{9}{7} \frac{(\sigma_y)_i}{h^2} z^2 - \frac{18}{7} \frac{h_{i-1}}{h^2} \cdot (\sigma_y)_{i-1} \cdot z + \frac{(\sigma_z)_{i-1,max}}{5} \cdot h_{i-1} \right] \qquad [5.30]$$
$$x \left[18 \frac{(y - \frac{2}{3}h)}{h} - 1 \right]$$

for $\frac{2}{3}h \leq y \leq h$ where $(\sigma_y)_i$ is the value of σ_y for the ith layer calculated using laminate theory.

The maximum value for the normal interlaminar stresses can be written as:

$$(\sigma_z)_{i,max} = \frac{45}{7h} \sum_{k=0}^{i} \left[h_k (\sigma_y)_k + 2h_{k-1} (\sigma_y)_{k-1} \right] \qquad [5.31]$$

Using equation 5.28 and the boundary condition that $\tau_{xz} = 0$ at the edge and at a distance h from the edge gives:

$$(\tau_{yz})_i = \frac{18}{7} \frac{h_i}{h^2} \left[(\sigma_y)_i + \frac{h_{i-1}}{h_i} (\sigma_y)_{i-1} \right] y \qquad 5.32]$$

for $0 \leq y \leq \frac{2}{3}h$ and

$$(\tau_{yz})_i = \frac{18}{7} \frac{h_i}{h^2} \left[(\sigma_y)_i + \frac{h_{i-1}}{h_i} (\sigma_y)_{i-1} \right] \left[\frac{-9}{h} y^2 + 13_z - 4h \right] \qquad [5.33]$$

for $\frac{2}{3}h \leq y \leq h$.

Figure 5.41 shows the results for the interlaminar normal stresses in a

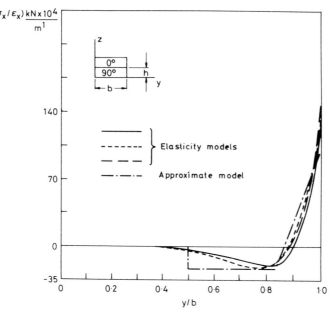

5.41 Interlaminar normal stresses in a cross-ply laminate.

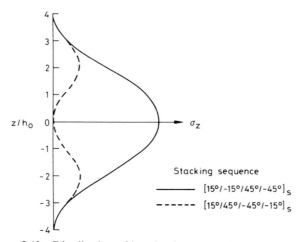

5.42 Distribution of interlaminar normal stress for different laminate configurations.

cross-ply laminate as calculated by different methods.[39,40] From examining equation 5.31 it can be seen that the distribution of σ_z is strongly influenced by the laminate stacking sequence and number of layers. Figure 5.42 shows relative σ_z distributions for different layups containing the same number and type of lamina.[41] As the interlaminar normal stress is a key parameter governing the tendency for delamination, the choice of which lamina in Fig. 5.42 would result in the stronger laminate is clear. Reconfiguring the laminate with the $\pm 45°$ on the outside will generate compressive σ_y stresses which will, in turn, cause the σ_z component to become negative and as a result delaminations would be expected to be suppressed.

The availability of approximate methods such as these is a valuable aid to the designer as solution of the elasticity problem is not straightforward. Even many of the finite element codes which are in use do not have element formulations which allow for the calculation of the interlaminar stress components. Other approximations have been proposed all of which are based on matching numerical results; for example:[42]

$$\sigma_y = \frac{\sigma_y(z)}{c}[1 - 2e^{-2\pi\bar{y}}\sin \pi y + \cos \pi \bar{y}] \qquad [5.34]$$

$$\tau_{xy} = \frac{\tau_{xy}(z)}{c}[1 - e^{-2\pi\bar{y}}\cos \pi \bar{y}] \qquad [5.35]$$

for $0 \le y \le h$ where $c = (1 + e^{-2\pi})$ and $\bar{y} = y/h$, and where $\sigma_y(z)$ and $\tau_{xy}(z)$ are determined from laminate theory. Substitution into the equilibrium equations 5.23 yields:

$$\tau_{yz} = -4\pi\frac{\tau_{yz}(z)}{c}e^{-2\pi\bar{y}}\sin 2\pi\bar{y} \qquad [5.36]$$

$$\sigma_z = 4\pi^2\frac{\sigma_z(z)}{c}e^{-2\bar{y}}(\cos \bar{y} - 2\sin \bar{y}) \qquad [5.37]$$

$$\tau_{xz} = -\pi\frac{\tau_{xz}(z)}{c}e^{-2\bar{y}}(\sin \bar{y} + 2\cos \bar{y}) \qquad [5.38]$$

where

$$\tau_{yz}(z),\ \sigma_z(z),\ \tau_{xz}(z) = -\int_{-h/2}^{z}\left(\frac{\sigma_y(z)}{dy},\ \frac{\tau_{yz}(z)}{dy},\ \frac{\tau_{xy}(z)}{dy}\right)dz$$

Figure 5.43 shows a comparison between calculations using the approximate method and numerical results.

It is clear from the preceding discussion that these through-thickness and interlaminar effects can be of vital importance in design. Their occurrence is not restricted to free laminate edges, but can feature at holes, cutouts, changes of section and discontinuities. Although they are very local in nature, being restricted to approximately one thickness from the edge, their influence is such that in order to control stress levels to within acceptable units, changes in the overall laminate sequence may prove to be necessary.

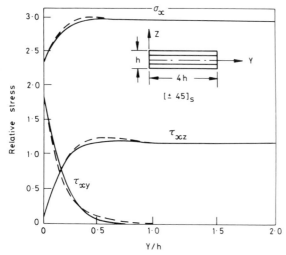

5.43   Comparison of solutions for interlaminar
stresses.

Hybrid composites

In the context of structural materials the term composite is normally used to
describe a material in which two phases are combined on a microscopic level
to produce a system where the properties of both constituents are brought to
the fore. Hybrid composites are a simple extension of this concept where a
number of material systems are used to optimize performance. In principle,
any combination of dissimilar materials can be considered as a hybrid. As an
example, for structural applications it can be shown that there can be
advantages by reducing the cost of expensive carbon fibre laminates by
incorporating quantities of glass reinforced material. A beam with the carbon
layers disposed on the external surfaces can still retain a high stiffness value
despite a large percentage of glass content. Figures 5.44 and 5.45 show the
flexural properties for such a beam and, as can be seen, a specimen of 50%
CFRP/50% GRP (by volume) possesses 90% of the properties of a bar made
entirely from CFRP.[43] The presence of the glass in this case should not just be
considered as a low cost filler as it does impart useful properties to the
composite as a whole. Figure 5.46 shows the impact energy (Izod) for the
hybrid beam.[44] The impact strength for the 50/50 combination is improved by
a factor of two and tests indicate that under impact conditions the failure
properties are also improved. Figure 5.47 shows the load deflection characteristics of the beam. As the load increases, the structure no longer breaks
catastrophically but can support around 25% of the peak load after the
carbon has failed. Another example is shown in Fig. 5.48. This is a tubular

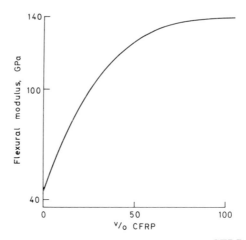

5.44   Flexural modulus versus percentage CFRP
content for a hybrid beam.

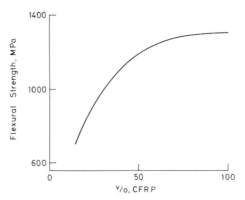

5.45   Variation of flexural strength with CFRP
content for a hybrid beam.

structure design to transmit torsional loading. It consists of an inner layer of
GRP at ± 45° to provide torsional stiffness, a thin layer of CFRP at 0°
disposed externally to the GRP to provide maximum contribution to axial
stiffness and finally an outer layer of aramid fibres. As a structure the
component is very effective, the glass makes up the bulk of its volume, and
therefore is instrumental in minimizing cost, the effect of carbon layers is to
increase whirling frequencies, thereby reducing support arrangements (compared with its steel counterpart), and the aramid provides abrasion resistance.

In terms of design, the hybrid concept gives the engineer added flexibility to
tailor properties for specific requirements. For convenience hybrids are often
split into four general types:[45]

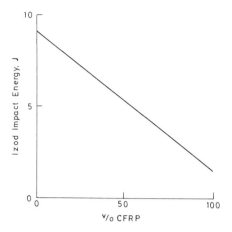

5.46   Variation of Izod impact energy with CFRP
content for a hybrid beam.

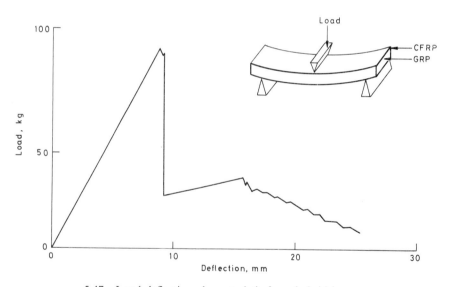

5.47   Load deflection characteristic for a hybrid beam.

- Dispersed fibre. This is the finest level of hybridization where two or more
  types of fibre are intimately mixed and dispersed in a common matrix. In
  practice, this mixing is done at the tow level as opposed to individual fibres.
- Dispersed fibre ply. This consists of an array of two or more types of fibre
  lamina to give a hybrid laminate. The thickness and dispersion of plies is
  determined by the requirements of the design.

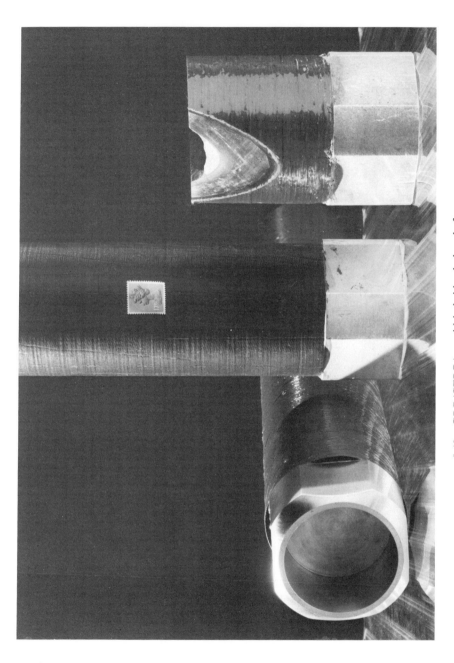

5.48 GRP/CFRP/aramid hybrid tubular shaft.

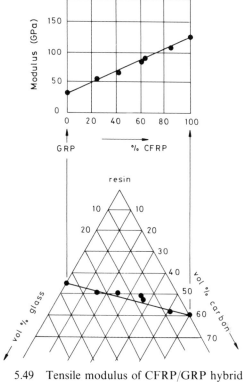

5.49 Tensile modulus of CFRP/GRP hybrid composites.

- Fibre skin and core. This type of laminate comprises fibre reinforced outer skins separated by an inner core also of fibre reinforced material. The beam considered in Fig. 5.44 and 5.45 is an example of such a design.
- Fibre skin, non-fibre core. Materials of this type include the sandwich type constructions.

The stiffness characteristics of hybrids can be evaluated using the rule of mixtures approach. Care must be applied for flexural properties where due account must be taken of the disposition of plies within the laminate. Conventional methods to calculate the position of the neutral axis and the contributions of each layer can be utilized.[46] For accurate prediction of properties, regard should be given to the detailed composition of the system. For example, glass and carbon laminates tend inherently to have different resin contents and this fact may be apparent in the hybrid. Figure 5.49 shows the effect. Laminates of 100% GRP or CFRP have different reinforcement fractions; a simple rule of mixture calculation without recourse to this type of diagram will give only approximate results.[47,48]

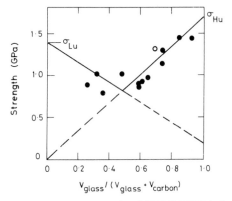

5.50   Tensile strength of CFRP/GRP hybrid
composites.

Because of the differences in failure behaviour in the various fibre systems
which may comprise a hybrid, the prediction of strength is a less straightforward task. For the case of a small amount of high strain to failure fibre in a
composite, where the balance of reinforcement is low strain to failure, the
composite strength will be reduced in proportion to the fraction of material
added. The hybrid composite strength, σ_c^*, is given by:

$$\sigma_c^* = \sigma_L^* v_L + \varepsilon_L^* E_h v_h \qquad [5.39]$$

where σ_L^*, ε_L^* and v_L are the strength, failure strain and volume fraction of the
fibres of low extensibility and E_h and v_h are the modulus and volume fraction of
high extensibility material. Failure occurs catastrophically since the high
extensibility fibres cannot carry the load when the low extensibility fibres fail.
As the proportion of the higher extensibility component increases, the point is
reached where there is sufficient material present to carry load after initial
failure. The strength of the hybrid is now:

$$\sigma_c^* = \sigma_h^* v_h \qquad [5.40]$$

where σ_h^* is the strength of the high strain to failure component.
    The critical composition at which this changeover occurs is given by:

$$v_{crit} = \frac{\sigma_L^*}{\sigma_L^* + \sigma_h^* - \varepsilon_L^* E_h} \qquad [5.41]$$

Figure 5.50 shows the results of these calculations for a series of glass/carbon
hybrids.[49] The details of the failure process in hybrids at a micromechanics
level is quite complex and will depend on the nature of the layup. On a
microscopic scale an intimately mixed material will behave differently from
one where the hybrid comprises discrete plies. Statistical methods have been
used with some success where, for example, it has been shown that the failure

Table 5.5. Impact resistance of hybrid composites

Composite	Unnotched Izod impact strength (J/m)
CFRP	1495
Aramid	2562
GRP	3843
75% graphite–25% aramid	1815
50% graphite–50% aramid	2349
75% graphite–25% glass	2349
50% graphite–50% glass	2989

process is strongly influenced by the statistical spread of failure strain of the two components and this can be used to demonstrate that the high extensibility fibres in a hybrid can behave like crack arrestors. The net effect is somewhat higher strength levels than indicated in Fig. 5.50.[50]

Toughness is another property where the use of the hybrid concept is attractive, particularly in the reinforcement of brittle CFRP with GRP or aramid fibres. Prediction of toughness characteristics is not straightforward, but there is some evidence supporting the use of a rule of mixtures type approach with some consideration given to the details of the composition.[45] Table 5.5 shows values of Izod impact strength of different fibre composition.

Cyclic loading is also a key area of performance where different fibre systems have contrasting behaviour. For example, consider glass and carbon composites. GRP behaves classically, with short fatigue lifetimes at high stress amplitudes and an increasing lifetime as the stress amplitude is lowered, whereas carbon appears unaffected by fatigue until the cycling stress approaches static strength values. For unidirectional laminates of CFRP and GRP a rule of mixtures approach appears to be adequate. Figure 5.51 shows typical S–N curves for hybrids of different construction both under ambient conditions and after exposure to temperature and moisture.[51] For both the unidirectional and quasi-isotropic material the benefits of the carbon content are evident. Taking the fatigue strengths at 10^6 cycles and plotting the values obtained as a function of composition gives the results in Fig. 5.52. Also plotted is the expected rule of mixtures behaviour. Dividing the fatigue strengths by the static failure stress would provide indications of the degree of property enhancement but it should be remembered that hybrid strengths are given by equations of the form of 5.39 and 5.40, and do not follow the rule of mixtures approach. In another example, Fig. 5.53 shows results from glass/carbon quasi-isotropic laminates, and those for the hybrids are equivalent to those of the all-carbon material and significantly above those of glass. The precise mechanism for this effect is not fully understood – other results even indicate an increased fatigue performance of hybrids over all-carbon.

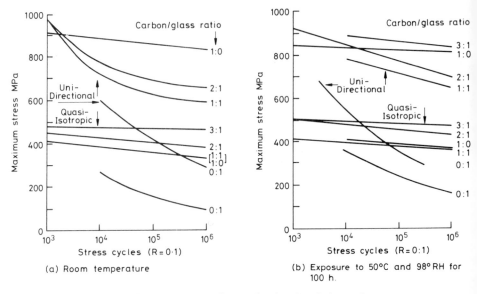

5.51   S–N curves for tension/tension fatigue of
CFRP/GRP hybrids.

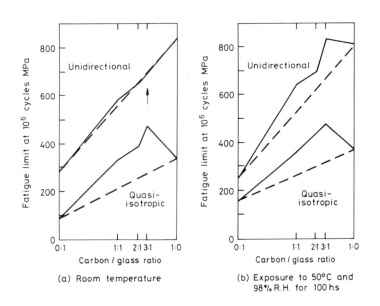

5.52   Variation of fatigue behaviour with composites
of CFRP/GRP hybrids (solid line – measured, dotted
line – rule of mixtures).

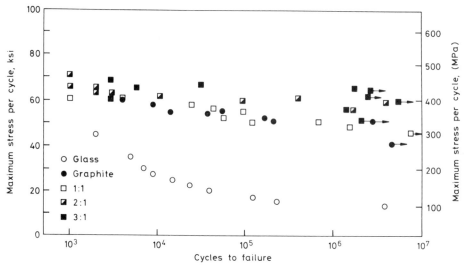

5.53 Comparison of fatigue stress/strain behaviour of
quasi-isotropic CFRP/GRP hybrid composites.

The most common structural hybrids are in the form of sandwich materials. These consist of two thin outer layers of strong, stiff material which are separated by a relatively thick layer of a core material. They are most commonly used where transverse forces and pressures must be accommodated by bending. Essentially the spacing apart of the load-bearing components creates a structure of high flexural rigidity. Although of comparatively low strength and stiffness, the core should not be considered as wholly non-structural as it must be capable of transferring load to the outer skins through shear and must resist crushing and buckling on the application of concentrated forces.[52]

The stiffness of a sandwich beam can be easily determined using conventional expressions for flexural rigidity.[46] For simple geometries where the skin materials are of common thickness and separated by a core of a different material, the bending stiffness, EI, is given by:

$$EI = \tfrac{1}{12}[E_s b(h^3 - c^3) + E_c bc^3] \qquad [5.42]$$

where E_s and E_c are moduli of skin and core respectively, c is the core thickness and h the overall thickness of the structure and b is the width. Assuming that the core is of low stiffness ($E_s > > E_c$) equation 5.42 reduces to:

$$EI = \frac{E_s b}{12}\left(1 - \frac{c^3}{h^3}\right) \qquad [5.43]$$

In order to design a sandwich panel it is often useful first to calculate the thickness of the structure as if it were to be constructed as a monolithic material. Simple strength of materials expressions can be used at this stage.[53] The overall thickness of an equivalent sandwich panel can then be determined from:

$$t_T = (h^3 + c^3)^{0.33} \qquad [5.44]$$

where t_T and h are thicknesses of the sandwich panel and of the equivalent monolithic panel respectively.

The skin thickness, t_s, is then:

$$t_s = \tfrac{1}{2}(t_T - c) \qquad [5.45]$$

Stresses in the skin layers can be calculated in a straightforward way from simple bending theory.[46]

A key factor in the performance of a sandwich is the capability of the core material to withstand shear forces, both in terms of stiffness and strength, as it is through the core that load is transmitted to the faces. Shear stress and shear deformation therefore need to be carefully considered. Depending on the application, and more particularly the details of the loading mechanism, there are numerous other considerations which may apply. Typical amongst these are local crushing of the core, buckling, and face wrinkling and dimpling. For the latter two conditions approximate stresses, σ_{cr}, are given by:

$$\sigma_{cr} = \frac{2E_s}{(1 - v_s^2)}\left(\frac{t_s}{5}\right)^2 \qquad [5.46]$$

where v_s is the Poisson's ratio of the skin.

$$\sigma_{cr} = 0.82E_s\left(\frac{E_c t_s}{E_s t_c}\right)^{1/2} \qquad [5.47]$$

Figure 5.54 summarizes the range of loading which may need to be addressed when considering sandwich panel design. Fittings and joints can also be critical features of a good sandwich panel design and schematic details of possible configurations are shown in Fig. 5.55. Again, the main issue here is in the mechanism of load transfer into the facings without imposing excessive stress levels into the core.

Flexible composites

The majority of thermosetting matrix materials in common use are regarded to have low strain to failure. Opportunities are available to enhance this characteristic with thermoplastic materials, but even here values are relatively low. Using elastomers as matrices allows materials with a much larger usable range of deformation to be manufactured.[54] Such systems are already in wide use in a number of circumstances where the material is not immediately

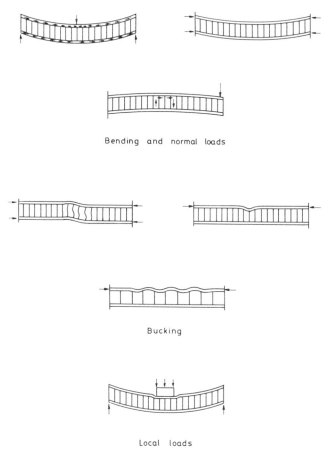

Bending and normal loads

Bucking

Local loads

5.54 Design issues for sandwich panels.

recognizable as a composite. Examples include pneumatic tyres, conveyor belts, flexible hoses and coated fabrics. The analysis of flexible composites can proceed, at least in the first instance, with the use of the classical lamination theory described in Chapter 3. However, as deformation becomes large and as a result of the peculiar elastic characteristics of elastomers, constitutive behaviour becomes very nonlinear and this must be considered in any attempt to predict properties.

Cord/rubber composites for pneumatic types are perhaps the most common application of flexible composites. These are complex elastomeric systems containing a rubber matrix of low modulus and high extensibility, a reinforcing cord of much higher stiffness and lower extensibility, and an adhesive film which effects a bond between the two. The combination is subjected to fluctuating tensile and compressive loads, high temperatures and moisture. Figure 5.56 shows the two common rubber tyre designs.[55]

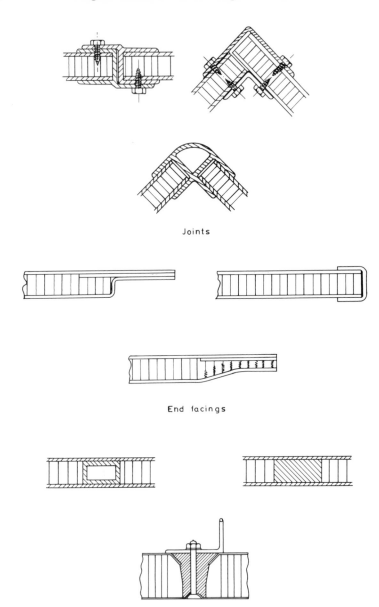

Joints

End facings

Local attachments

5.55   Typical fitting and joint details for sandwich
panels.

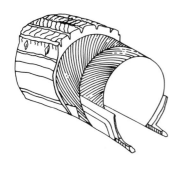
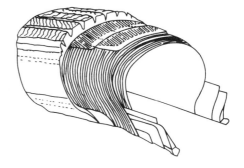

(a) Bias ply tyre   (b) Radial ply tyre

5.56   Rubber tyre configurations.

Assuming the matrix to be isotropic and incompressible and the cords to be inextensible, certain deductions can be made with respect to the linear elastic behaviour of a unidirectional laminate.[56] Taking $E_f >> E_m$ and $v_m = 0.5$ can be shown to give:

$$E_1 \simeq E_f V_f \gg E_2$$

$$E_2 \simeq \frac{4\,E_m}{3\,v_m}$$

$$v_{21} \simeq 0$$

$$G_{12} \simeq \frac{G_m}{v_m} \simeq \frac{E_2}{4}$$

[5.48]

Based on these equations the terms of the reduced stiffness matrix $[Q]$ for an off-axis laminate are given by:[57]

$$\bar{Q}_{11} \simeq E_2 + E_1 \cos^4 \theta$$

$$\bar{Q}_{22} \simeq E_2 + E_1 \sin^4 \theta$$

$$\bar{Q}_{66} \simeq 0.25\, E_2 + E_1 \sin^2 \theta \cos^2 \theta$$

$$\bar{Q}_{12} \simeq 0.5\, E_2 + E_1 \sin^2 \theta \cos^2 \theta$$

$$\bar{Q}_{16} \simeq E_1 \sin \theta \cos^3 \theta$$

$$\bar{Q}_{26} \simeq E_1 \sin^3 \theta \cos^3 \theta$$

[5.49]

Some of the more peculiar aspects of flexible composites, not observed in rigid materials, can be explored through the above equations. As with all off-axis lamina there is coupling between applied normal loads and shear strain and this can be shown to be quantified by:

$$\gamma_{xy} = -2\frac{\sin\theta\cos^3\theta}{E_2}(2 - \tan^2\theta)\sigma_x \qquad [5.50]$$

As can be seen from examination of equation 5.50 the stretching-shear coupling vanishes at $\tan^2\theta = 2$, i.e. 54.7°, and for other orientations $\gamma_{xy} < 0$ for $\theta < 54.7°$ and $\gamma_{xy} > 0$ for $\theta = 54.7°$.

Laminates can be treated in a similar way where the approximations for basic properties can be used in the calculation of the stiffness matrix, $[A]$. Applying this approach to simple angle-ply laminates, i.e. those of the $\pm\theta$ configuration, yields:[58]

$$E_x = E_f v_f \cos^4\theta + 4G_m/(1 - v_f) - \frac{[E_f v_f \sin^2\theta\cos^2\theta + 2G_m/(1 - v_f)]^2}{[E_f v_f \sin^4\theta + 4G_m/(1 - v_f)]}$$

$$E_y = E_x(\pi/2 - \theta) \qquad [5.51]$$

$$G_{xy} = E_f v_f \sin^2\theta\cos^2\theta + G_m/(1 - V_f)$$

$$v_{xy} = \frac{[E_f v_f \sin^2\theta\cos^2\theta + 2G_m/(1 - v_f)]}{[E_f v_f \sin^4\theta + 4G_m/(1 - v_f)]}$$

$$v_{yx} = v_{xy}(\pi/2 - \theta)$$

Applying the condition that the cords are inextensible, i.e. $E_f \to \infty$, provides some further simplification:

$$E_x = 4G_m(1 - v_f)(\cot^4\theta - \cot^2\theta + 1)$$

$$E_y = E_x(\pi/2 - \theta)$$

$$G_{xy} = E_f v_f \sin^2\theta\cos^2\theta + G_m(1 - v_f) \qquad [5.52]$$

$$v_{xy} = \cot^2\theta$$

$$v_{yx} = \tan^2\theta$$

Figure 5.57 shows the results of analytical predictions, in this case those for angle-ply tensile modulus, together with experimental results.[58] These are based on values of E_1 and E_2 of 1440 MPa and 6.9 MPa respectively.

The treatment of flexible composites in this manner is adequate as a first approximation. However, there are a number of factors which limit general applicability. Of significance amongst these is material nonlinearity, but also of importance are finite strains in the cord, interlaminar deformations, viscoelastic behaviour and non-unidirectional behaviour of the cord/rubber interfacial bond. A simple approach to treat nonlinear constitutive behaviour is based on an incremental analysis using superposition of infinitesimal linear elastic deformation. Although this is clearly approximate it does provide a

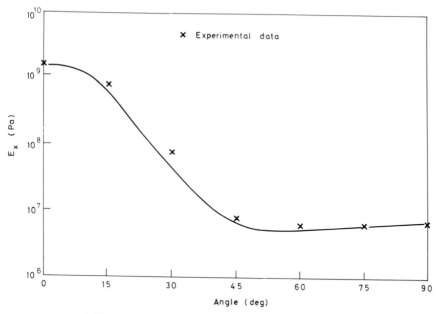

5.57 Young's modulus E_x plotted against cord angle
for a two-ply laminate.

convenient tool for predicting basic properties. A more accurate analysis
based on concepts of strain energy density can also be used to good effect.[59]
Considering an off-axis lamina it can be shown that the following constitutive
equations apply:

$$e_1 = S_{11}\sigma_{11} + S_{111}\sigma_1^2 + S_{111}\sigma_1^3 + S_{12}\sigma_2 + S_{166}\sigma_6^2$$
$$e_2 = S_{22}\sigma_2 + S_{222}\sigma_2^2 + S_{2222}\sigma_2^3 + S_{12}\sigma_1 + 2S_{2266}\sigma_2\sigma_6^2 \qquad [5.53]$$
$$e_6 = S_{66}\sigma_6 + S_{6666}\sigma_6^3 + 2S_{166}\sigma_1\sigma_6 + 2S_{2266}\sigma_2^2\sigma_6$$

where $e_1 = e_{11}, e_2 = e_{22}$ and $e_6 = 2e_{12}. e_1. e_2, e_{16}$ are Eulerian strains and can
be related to the engineering strain as follows:

$$e_{11} = \tfrac{1}{2}[1 - (1 + \varepsilon_1)^{-2}]$$

$$\qquad [5.54]$$

$$e_{22} = \tfrac{1}{2}[1 - (1 - \varepsilon_2)^{-2}]$$

Of the compliance terms in equation 5.53, S_{11}, S_{22}, S_{12} and S_{66} are required
for linear deformation, S_{111} and S_{222} represent bimodular behaviour in the
axial and transverse direction and S_{1111}, S_{2222} and S_{6666} are nonlinear terms.
The two terms S_{166} and S_{2266} are necessary to represent coupling between
normal and shear deformation. In order to determine the compliance
constants, curve fitting is required. For example, the stress/strain curve from a

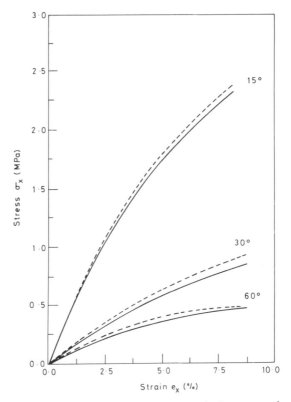

5.58 Comparisons between theoretical curves and experimental results for tyre cord/rubber composite laminates.

unidirectional test is given by:

$$e_1 = S_{11}\sigma_1 + S_{111}\sigma_1^2 + S_{1111}\sigma_1^3 \qquad [5.55]$$

S_{11} is determined from the initial slope of the cure and S_{111} and S_{1111} are obtained from curve fitting. The other constants can be determined from similar tests on off-axis specimens. Comparisons between theory and experiment are shown in Fig. 5.58 and 5.59 and good agreement can be observed.[59] The calculated stress/strain plots in the figures are based on the data in Table 5.6.

An interesting application of the novel properties of elastomeric composites is that of pressure tubes.[60] When tubes of angle-ply construction are axially loaded, the apparent reinforcing angle changes. This in turn causes a change in internal volume (Fig. 5.60). For angles $< 54.7°$ the volume decreases with stretching of the tube. Figure 5.61 shows typical results taken from measurements from a tube with $\pm 45°$ aramid reinforcement. The characteristics of the

Table 5.6. Elastic properties of elastomeric composite laminates

Stiffness parameter	Tyre cord/rubber	Kevlar-49/elastomer ($v_f = 9\%$)
S_{11} (MPa)$^{-1}$	0.165×10^{-3}	0.114×10^{-3}
S_{1111} (MPa)$^{-3}$	0	0
S_{12} (MPa)$^{-1}$	-65.9×10^{-6}	-69.9×10^{-6}
S_{22} (MPa)$^{-1}$	0.121	0.306
S_{2222} (MPa)$^{-3}$	51.4×10^{-3}	0.563
S_{66} (MPa)$^{-1}$	0.408	0.387
S_{6666} (MPa)$^{-3}$	0.183	77.5×10^{-3}
S_{166} (MPa)$^{-2}$	0.131×10^{-3}	3.43×10^{-6}
S_{2266} (MPa)$^{-3}$	0.469	56.3×10^{-3}

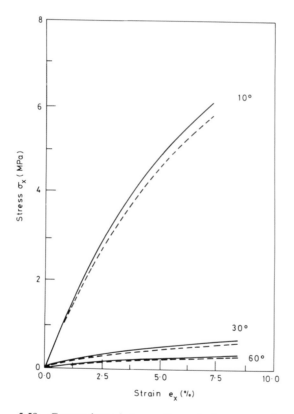

5.59   Comparisons between theoretical curves and
experimental results for aramid/silicone elastomer
laminates.

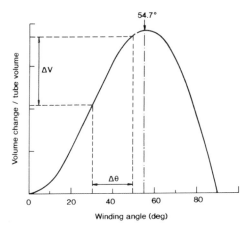

5.60   Predicted relationship between reinforcing angle
and volumetric strains in reinforced rubber tubes.

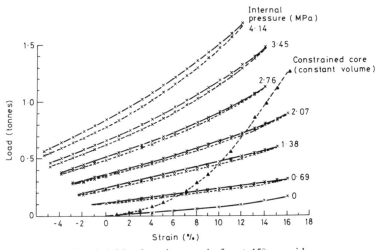

5.61   Axial load against strain for ±45° aramid
reinforcement.

tube deformed at both constant pressure and constant volume are also shown.
The stiffness of the tube, i.e. the slope of the curves shown in Fig. 5.61, is
controlled both by the elastic properties of the material and on the extent to
which volume change is opposed by the internal pressure. In the extreme, the
stiffness characteristics can be adjusted from extreme compliance, when there
is no constraint on volume change, to high stiffness where there is effectively
total constraint. As a result, a range of mechanical responses can be obtained
simply by controlling internal pressure. Such a device could be used as an
active mooring line where stiffness could be adjusted to suit conditions or,

conversely, the volume contractions which occur on stretching could be used to provide a pumping action.

Environmental effects

There is a wide-ranging array of applications where static loading at or near ambient conditions represents the governing design condition. It is often the case, however, that the operating environment is somewhat more arduous than this and may involve cyclic stresses, elevated temperatures or exposure to media other than air. It may be possible in an initial scoping study to carry out preliminary calculations using data obtained at, say, room temperature and then to exercise judgement based on past experience, but at some point in the analysis due account must be taken of the influence of other effects. This is not to say that composites perform badly in these respects when compared with alternative materials, as in a number of instances they are selected in preference to others for reasons concerned with their good behaviour. For example, GRP is noted for its corrosion resistance and is used in preference to costly alloy or lining systems and the fatigue response of CFRP is regarded as being excellent. At the detailed level the mechanisms that control behaviour, property change and, ultimately, design life are fundamentally different from those that occur with metals and an appreciation of the issues concerned is important in the evaluation of component performance.

Chemical media

The degradation of composite materials under the influence of an aggressive environment can result from a number of factors:

* Loss of strength of reinforcing fibres by stress corrosion.
* Loss of bond strength through degradation of the interfacial fibre/matrix bond.
* Chemical degradation of the matrix.
* Accelerated degradation caused by the combined action of temperature and chemical environment.

As a result of each of these factors, acting singly or in combination, mechanical properties can be adversely affected. The nature of corrosion phenomena is entirely different from that observed with metals where electrolytic effects may predominate. For composites the physical processes of diffusion and osmosis are often the more important mechanisms to be considered.

With regard to the effects of environment on the reinforcing fibres it has been demonstrated that under certain circumstances delayed failure under load or 'static fatigue' can have a pronounced effect. This features particularly

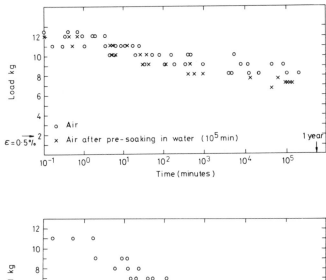

5.62   Stress rupture of polyester impregnated E-glass.

with glass fibres where it has been concluded that initially non-critical cracks grow under the influence of stress and the reactive environment until they are sufficient to cause failure. Figure 5.62 shows the stress rupture behaviour of E-glass strands in air, water and distilled water and as can be seen the fall off in strength is considerable, although it should be noted that at normal design levels (design strains < < 0.5%) it may not be a significant issue.[60] The behaviour of other fibres is shown in Fig. 5.63. Carbon fibres are essentially unaffected, whereas aramids fall somewhere between the two extremes.[60]

Because of their nature, polymer matrix materials can have a comparatively rapid uptake of environmental agents such as water due to diffusion processes. The most common technique for the modelling of diffusion is to consider Fick's law,[61] i.e.:

$$\frac{\partial c}{\partial t} = D_x \frac{d^2 c}{dx^2} \qquad [5.56]$$

where dc/dx is the concentration gradient and D_x is the diffusion gradient in direction x. It can be shown that the mass absorbed in time t, M_t, can be

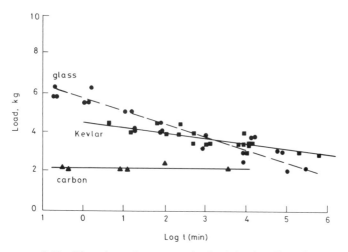

5.63 Time-dependent strength of reinforcing fibres in distilled water.

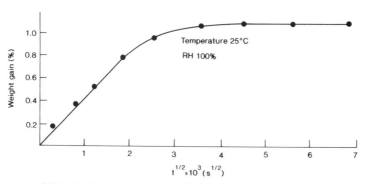

5.64 Moisture absorption of glass reinforced epoxy.

expressed as:[62]

$$\frac{M_t}{M_\infty} = \frac{4}{h}\left(\frac{Dt}{\pi}\right)^{1/2}$$ [5.57]

where M_∞ is the mass absorbed at saturation and h is the thickness of the plate of concern. Figure 5.64 shows the absorption cure for water uptake for a glass reinforced epoxy indicating classical behaviour as given by equation 5.57.

By assuming fibre packing arrangements and using the analogy with thermal conductivity, approximate relationships for the diffusion coefficient for a unidirectional composite can be obtained:

$$D_{11} = (1 - v_f)D_m$$ [5.58]

$$D_{22} = [1 - 2(v_f/\pi)^{1/2}]D_m$$

where D_{11} and D_{22} are the diffusion coefficients parallel and transverse to the fibre direction and D_m is the value for the matrix. The diffusivity at angle θ is then given by:

$$D_\theta = D_{11} \cos^2 \theta + D_{22} \sin^2 \theta \qquad [5.59]$$

Generally, Fickean behaviour is found to be representative of a number of systems at the onset of the absorption process. Derivations can be observed as saturation is reached or if degradation of the material occurs in response to the environment. In these circumstances more complex models of behaviour need to be used to accommodate interactions between the various reactants concerned.

One of the most marked effects of absorption of a diluent such as water into a polymer is the reduction of its glass transition temperature. This plasticizing effect can be described by:

$$T_g = \frac{\alpha_m T_{gm}(1 - v_f) + \alpha_f v_f T_{gf}}{\alpha_m(1 - v_f) + \alpha_f v_f} \qquad [5.60]$$

where T_{gm} and T_{gf} are the glass transitions of the matrix and fibre respectively, α_m and α_f are expansivity values and v_f is the volume fraction of the fibre. The fall of T_g can be considerable and can be of the order of several tens of degrees.

Interfacial bond strength can also be affected by ingress of moisture and other reagents. In fact, capillary action along the fibres can account for a significant proportion of uptake in the first instance. In general terms, the effect of moisture is to cause hydrolytic breakdown of the fibre matrix bond which will in turn affect the efficiency of load transfer between the phases. In some cases this effect is reversible, whereupon on drying out, the composite properties return to their original values. Figure 5.65 shows the load deflection curves for three-point notched bending tests carried out during/after boiling and dried after boiling.[63] The effects of drying are to return the material almost to its original state. This behaviour is not repeated over a large number of cycles owing to progressive irreversible damage of the laminate.

It is difficult to generalize with respect to the overall response effects of composites under the action of chemical media because of the vast number of material and reagent combinations possible. In Fig. 5.66 the results of stress corrosion tests of GRP filament wound pipe under acid conditions are plotted. Results are given for both unidirectional (hoop) and angle-ply laminates ($\pm 55°$) and are expressed in terms of the fibre stress σ_1. Plotted in this way, there is a common relationship between fibre stress and time to failure. The displacement of the plots is thought to be due to initial mechanical damage which allows more rapid ingress of the acid.[64] Figure 5.67 shows times to failure of a different laminate type, chopped strand mat (CSM), for a range of chemical media and, as can be seen, the picture is not straightforward.[65] Also

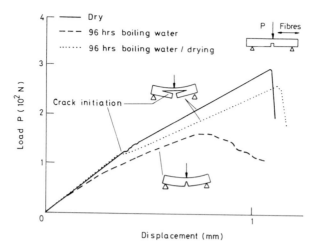

5.65   Load deflection curves for notched
unidirectional GRP.

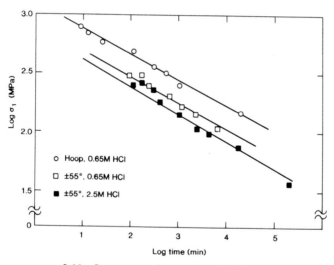

5.66   Stress corrosion tests on GRP pipe.

plotted is the maximum tensile strain experienced during the test and it is worth noting that typical design values lie between 0.2% and 0.4%. The chemical process plant industry, for example, uses 0.2% as a norm.

High temperature performance

For the vast majority of high performance structural applications, epoxy resins are used as the matrix material. There is a great variety of systems to

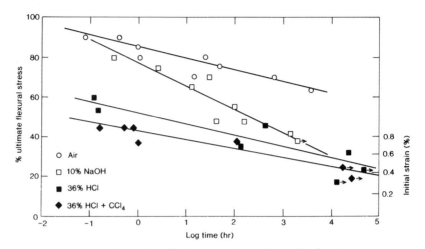

5.67   Effect of applied stress and strain on the time to
failure of CSM.

choose from and their detailed behaviour depends on the formulation used, curing schedule, reinforcement details, stress conditions and exposure time. Generally, however, standard epoxies have a maximum upper working temperature in the range 130–180 °C. Also the toughness of epoxies can be low and moisture uptake can cause a reduction in mechanical properties under hot wet conditions. As a result the development of systems which can sustain exposure at high temperature has been of particular interest.

The ability to use a polymer decreases as the temperature approaches its glass transition temperature (T_g). At this point stiffness and strength properties fall off rapidly and the tendency for creep increases significantly. Assessing the working temperature of a polymeric material is not a straightforward matter, as determination of, for example, T_g is strongly influenced by the test method used. Other methods have been developed in an attempt to produce pragmatic and reproducible measurements. The most widely used is the measurement of heat distortion temperature (HDT). Here the deflection of a specimen in three-point bending is measured as a function of temperature; typically the test is carried out in an oil bath and the temperature is increased at 2 °C/min. The temperature at which the beam deflection reaches a predetermined level is defined as the upper working point.[66] An even simpler approach is to determine the length of exposure which a material can withstand before its properties are degraded by a set amount – say 50%. All of these test methods are dependent on the type, quantity and orientation of reinforcement so individual measurements are very specific to the sample of concern. In general terms only broad guidelines or a ranking of materials can be given.

Table 5.7. Estimated continuous service temperatures

Material	Temperature (°C)
Epoxy	180
Phenolic	260
Poly styryl pyridene (PSP)	400
Poly bismaleimide (BMI)	250
Polyimide (PI)	300
Polyether sulphone (PES)	180
Polyether ether ketone (PEEK)	250
Polyether imide (PEI)	170
Polyamide imide (PAI)	230-260
Polyether sulphone (PPS)	200-240

Figures depend on grade, post cure, presence of reinforcement, applied load and environment.

As apparent temperature capability is increased, care must be taken to ensure that other properties are not seriously affected. Two characteristics that tend to reduce with thermal stability are processability and toughness. High temperature systems generally have large viscosity values and therefore can only be processed wet with difficulty or in prepreg form. Additionally cure cycles can be onerous, involving complex temperature profiles over extended periods. High temperature capability is often associated with an increasing density of cross-links and this in turn tends to increase brittleness. The incorporation of reactive rubber particles or a fine dispersion of a thermoplastic such as polysulphone or polyether sulphone, or the use of thermoplastics themselves, can help to increase matrix toughness, but these benefits are not always translated into improved composite properties.

Examples of high temperature systems are given in Chapter 2 and specific examples of thermosets include phenolics, epoxies, bismalemides, polystrylpyridene and polyimides, whilst the thermoplastics available include polyphenylene sulphide, polyaryl sulphide, polyether imide and PEEK. Estimated service temperatures for composites made from these systems are given in Table 5.7. Instantaneous mechanical properties as a percentage of the corresponding room temperature values are listed in Table 5.8 and the effect of prolonged exposure at temperature is shown in Table 5.9.[67,68]

Fatigue

The behaviour of composites under cyclic loading conditions is fundamentally different from that of traditional materials. For metals, fatigue damage tends to be localized, with failure being initiated by the progressive growth of a single flaw. Linear elastic fracture mechanics is often used to predict behaviour. In

Table 5.8. Properties of composites at elevated temperatures

Resin	Fibre	Modulus (%)	Strength (%)	$T(°C)$
Epoxy	Glass	90	60	200
Phenolic	Glass	–	30	300
BMI	Carbon	79	71	250
BMI	Glass	97	80	175
PSP	Carbon	83–100	50–63	350
PSP	Glass	104	90	400
PI	Carbon	81	57	343
PPS	Glass	21	26	250
PEEK	Glass	–	58	177
PEEK	Glass	27	19	300

Modulus, strength values expressed as percentage of room temperature value.

Table 5.9. Properties of composites after exposure to temperature

Resin	Fibre	Time (h)	Exposure temperature (°C)	Modulus (%)	Strength (%)	Test temperature (°C)
Epoxy	Carbon	030 000	150	100	107	150
Epoxy	Glass	10 800	200	–	89	200
Phenolic	Glass	100	250	–	59	250
Phenolic	Glass	1	500	–	31	300
BMI	Glass	10 000	220	97	21	20
BMI	Glass	2000	250	23	77	200
PSP	Glass	1000	250	98	104	250
PI	Carbon	8600	260	71	63	260
PI	Carbon	500	343	59	59	343
PPS	Glass	10 000	246	–	50	20
PES	Glass	7000	230	50	50	20
PES	Glass	19 000	210	50	50	20

Modulus, strength values are expressed as percentage of pre-exposure reading at same temperature.

composites, however, damage is much more widespread and involves a number of microstructural mechanisms including fibre and matrix cracking, delamination and debonding. At stress concentrations complex damage zones are generated which have the effect of redistributing peak stress values. The net result is that fatigue behaviour is generally better with ratios of fatigue strength at high cycles to static strength very much greater than with metals. Damage mechanisms are generally related to matrix-dominated properties and therefore matrix strain is a key factor. Figure 5.68 shows typical plots of peak tensile stress against cycles to failure for unidirectional laminates.[69] As would be expected, the ratio of fatigue strength to static strength increases with the fibre modulus.

At low cyclic stress levels composites tend to sustain the full range of damage

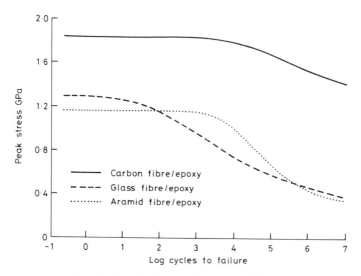

5.68   *S–N* tensile fatigue data for unidirectional
composite materials.

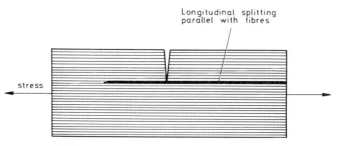

5.69   Longitudinal splitting in unidirectional
composites.

mechanisms throughout the loaded region. During this time stiffness properties can be degraded, but the effects on strength are more complex. Strength
reductions can, in part, be offset by strength increases due to changes in fibre
orientation or time-dependent deformations. Depending on the type of
composite material, crack propagation may or may not develop from an
existing flaw. For stiff, high volume fraction systems, cracks normal to the
fibres can be arrested by the reinforcement and be changed into a longitudinal
splitting type of damage (Fig. 5.69).

In multi-ply laminates the position becomes dominated by the interactions
between individual layers. With cross-plies, for example, cracking in the
transverse plies occurs after a comparatively short number of cycles and this is
accompanied by a significant stiffness reduction. Once this has occurred fibre

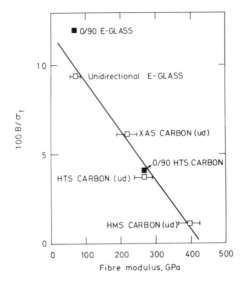

5.70   Linear relationship between fibre modulus and
fatigue life.

characteristics dominate the response and there may not be further significant falls in stiffness until immediately prior to failure.[70]

In principle, unidirectional materials where the reinforcement fibre is of a brittle nature should possess a large degree of fatigue resistance, and, indeed, this is the case with carbon reinforced systems which are of high modulus (Fig. 5.68). With an increasing number of cycles, however, there will be a tendency for small deformations in the matrix which will result in some local redistribution of stress. A degree of fibre damage could then ensue. As these mechanisms proceed, the damage level will accumulate to some critical state until the residual composite strength becomes less than the peak cycle stress and failure occurs. For a given resin system the proportion of load carried by the matrix is a function of fibre stiffness; the higher the modulus and the lower the failure strain of the fibre, the less the matrix fatigue damage. Hence the differences in response between glass, aramid and carbon reinforced systems.

Given this rationale for establishing a common mechanism for fatigue effects across a range of materials it has been possible to express maximum cyclic stress as a function of fibre stiffness by a single relationship:[71]

$$\frac{\sigma}{\sigma_f} = 1 - \left(\frac{B}{\sigma_f}\right)\log n \qquad [5.61]$$

where σ is the maximum cyclic stress, n is the cycle life at that stress level and σ_f is the failure stress.

Figure 5.70 shows that the linear relationship indicated by equation 5.61 is

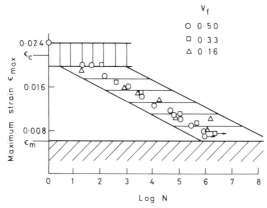

5.71   Fatigue life diagram for unidirectional GRP.

in fact reflected by experimental data. The slope of the curve is approximately 10. Even for cross-ply laminates,[72] where to a large extent response is dominated by the 0° plies and hence the fibres, the results fall close to the unidirectional values. Correlation is less close for the GRP cross-plies due to the lower modulus of the glass and interaction between adjacent layers. The principle that can be derived from presenting the results in this way is that fatigue is primarily governed by the gradual deterioration of load-bearing fibres and for any given laminate, lifetime is determined by other mechanisms, usually matrix dominated, which affect the rate of damage accumulation. These guidelines appear to be adequate for the brittle reinforcement types such as glass and carbon, but the approach tends to break down where the fibres possess a certain degree of resilience. Aramid fibres, for example, behave well at high stress levels but performance tends to fall away at large numbers of cycles.[72] The transition corresponds to changes of behaviour in the reinforcement itself; initially the response corresponds to a tough material and this then becomes overtaken by the effects of internal structural flaws within the fibres.

One means of providing an understanding of cyclic behaviour is through the use of fatigue life diagrams.[73] Figure 5.71 shows an example for unidirectional GRP. Note that strain is plotted, not stress. The lower limit to the diagram is given by the fatigue limit of the matrix and the upper limit to the diagram is given by the strain to failure of the reinforcement. For a given matrix it would be expected that the fatigue limit of any two composites would be the same, but that the upper limit would be dependent on the fibre. In Figure 5.71 these values are 0.6% and 2.2% respectively. In the intermediate region, whilst a level of fibre breakage does occur, it is not sufficient for ultimate failure and other mechanisms such as matrix cracking occur. For carbon fibres (Fig. 5.72), the picture is somewhat different. Here the static fibre failure strain is of the same order as that of the resin fatigue limit and fatigue damage is

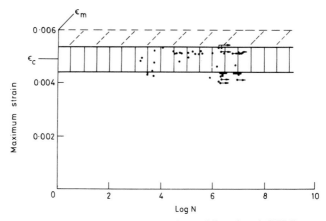

5.72   Fatigue life diagram for unidirectional CFRP.

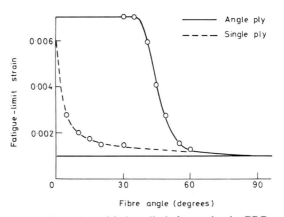

5.73   Variation of fatigue limit for angle-ply GRP.

essentially suppressed. As can be seen, this is borne out by the experimental results.

When load is applied at an angle to the fibre direction the premise regarding lower limit strains is no longer valid. Here fatigue damage is dominated by matrix cracking even for orientations where the load direction deviates only slightly from the fibre orientation. The fatigue limit reduces rapidly to about 0.10% for the transverse loading case (Fig. 5.73). Also shown in Fig. 5.73 are fatigue limit strains for angle-ply laminates. The performance of these layers is somewhat superior to single plies due to the constraint of damage progression provided by adjacent layers.

If the fatigue regime takes the material into a compressive or shear state, additional problems can arise. The relatively poor behaviour of composites in compression means that fully reversed fatigue can be a severe loading

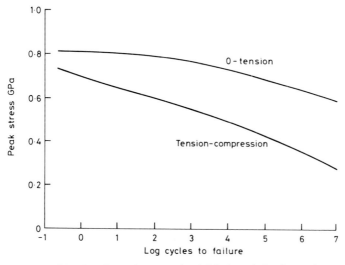

5.74 Tensile and reversed axial fatigue behaviour of $(\pm 45°, 0_2°)_s$ CFRP material.

condition. Any delamination, which may be relatively unimportant in tension, can be deleterious in compression owing to instability of individual plies. Figure 5.74 shows the effect in this case for quasi-isotropic CFRP laminates.

In terms of design the degree to which fatigue presents a problem depends on the nature of the application. Where a structure is stiffness-limited the low stresses may not be significant with respect to those where fatigue-related damage will be apparent. For stress-limited designs full consideration of fatigue effects must be made, although in some circumstances it may be found that other issues reduce design allowables below those at which fatigue will be a problem.

Residual stress

The cure of a composite material is a complex process often involving the application of both temperature and pressure and this results in an irreversible combination of physical and chemical changes within the material. The chemical and thermal phenomena which occur can lead to a number of dimensional changes in the constituents at both the microscopic and macroscopic levels. On a larger scale, when laminae are combined within a laminate, the resulting mismatch in thermal expansivity can give rise to residual stresses which can be high enough to cause cracking and therefore affect structural integrity. Dimensional stability can also be adversely affected, in that high expansion rates and the effects of laminate symmetry can result in distortions which can prejudice required tolerance levels.

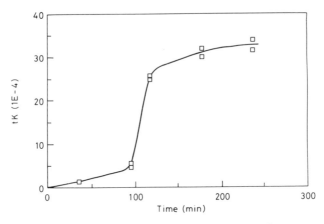

5.75   Change of curvature (κt) with cure time for an
asymmetric laminate.

The simplest manifestations of the effects of residual stress are seen in the
bending or warping of asymmetric laminates, analogous to the bimetallic strip
effect. For an asymmetrical cross-ply laminated beam, cooling from any
temperature will result in curvature with the transverse plies at the inner radius
because of their higher expansivity. The development of residual stresses can
be evaluated by stopping the process at intervals and measuring the curvature
on cooling to room temperature. Intuitively it follows that residual stresses
become progressively 'locked into' the material as chemical cross-linking
proceeds and therefore it would be expected that residual stress and viscosity
are related. For a simple cross-ply strip the curvature can be calculated
approximately from:

$$\kappa = \frac{24\,(\alpha_1 - \alpha_2)\Delta T}{[14 + (E_2/E_1) + (E_1/E_2)]\cdot t} \qquad [5.62]$$

where α_1, α_2 and E_1, E_2 are the expansivities and moduli of a unidirectional
lamina, ΔT is the difference between the stress-free temperature and ambient
and t is the thickness of the laminate.

Figure 5.75 shows typical results for CFRP and indicates a rapid increase in
curvature at a time which corresponds to the resin gel point.[74] The
assessments of the degree of cure and changes in resin viscosity during the
process (Fig. 5.76), follow similar patterns.

Clearly the accuracy of the calculation is dependent on the selection of an
appropriate stress-free temperature and this is usually assessed using relatively
simple experiments, e.g. evaluating the temperature at which an asymmetric
curved beam becomes flat. Alternatively the cure temperature can be used as a
first approximation. The position with thermoplastics composites is somewhat similar. For amorphous materials the stress-free temperature is usually

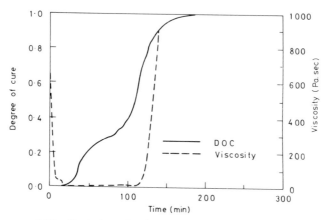

5.76 Variation of resin viscosity and degree of cure
(DOC) with time.

taken as the glass transition T_g. In semi-crystalline polymers matters are more
complex as the stress-free temperature is dependent on the details of polymer
morphology and the point at which the material assumes a solid-like
character. This in turn can be a function of processing conditions.

On the laminate level, the presence of residual stress can be evaluated using
the classical plate theory described in Chapter 3 by incorporating the thermal
strains into the analysis. For plane stress of an orthotropic lamina in the
principal material directions:

$$
\begin{bmatrix} \sigma_1 \\ \sigma_2 \\ \tau_2 \end{bmatrix} = \begin{bmatrix} Q_{11} & Q_{12} & 0 \\ Q_{12} & Q_{22} & 0 \\ 0 & 0 & Q_{66} \end{bmatrix} \begin{bmatrix} \varepsilon_1 & - \alpha_1 \Delta T \\ \varepsilon_2 & - \alpha_2 \Delta T \\ \varepsilon_3 & \end{bmatrix}
\tag{5.63}
$$

Transformation of equation 5.63 yields:

$$
\begin{bmatrix} \sigma_x \\ \sigma_y \\ \tau_{xy} \end{bmatrix} = [\bar{Q}] \begin{bmatrix} \varepsilon_x & - \alpha_x \Delta T \\ \varepsilon_y & - \alpha_y \Delta T \\ \varepsilon_{xy} & - \alpha_{xy} \Delta T \end{bmatrix}
\tag{5.64}
$$

The presence of the term α_{xy} should be noted. This effectively represents a
coefficient of thermal shear and arises due to coupling between normal and
shear strains. Integration across the thickness of a laminate results in a series of
equivalent thermal forces and moments:

$$
[N^T] = \int [\bar{Q}][\alpha] \Delta T \, dz
$$
$$
[M^T] = \int [\bar{Q}][\alpha] \Delta T z \, dz
\tag{5.65}
$$

These thermal loads can be treated in the same way as mechanical loads,

Table 5.10. Comparison of pressure and residual stresses in a thick wall GRP cylinder

		σ_{11}(MPa)	σ_{22} (MPa)	σ_{33} (MPa)	σ_{12} (MPa)
Pressure stresses for 1 MPa					
Inner layer	inner surface	2.46	2.35	−1.00	± 1.08
	interface	1.83	1.89	−0.35	± 0.84
Outer layer	interface	4.97	1.18	−0.35	± 0.96
	outer surface	4.52	1.14	0.00	0.87
Thermal stresses for $\Delta T = 10\,°C$					
Inner layer	inner surface	3.71	− 0.26	0.09	± 0.04
	interface	3.79	− 0.19	0.11	± 0.05
Outer layer	interface	4.02	− 0.23	0.14	± 0.06
	outer surface	4.07	− 0.22	0.11	± 0.07

although care must be taken to ensure that the nature of the laminate boundary conditions is taken into account. For example, free thermal expansions which do not generate forces and moments should be deducted. The constitutive relationships incorporating both thermal and mechanical loading can then be written as:

$$\begin{bmatrix} \bar{N} \\ \bar{M} \end{bmatrix} = \begin{bmatrix} A & B \\ B & D \end{bmatrix} \begin{bmatrix} \varepsilon \\ \kappa \end{bmatrix} \qquad [5.66]$$

where

$$[\bar{N}] = [N] + [N^T]$$

$$[\bar{M}] = [M] + [M^T]$$

Individual stress and strain distributions within each lamina can now be calculated and taken into account in a design assessment.

The significance of residual stress with respect to the other applied loadings depends on the magnitude of the temperature change, the laminate details, the geometry of the component and processing conditions.[75-77] In Table 5.10 stress distributions for a thick-wall cylinder under internal pressure and a temperature change of 10 °C are shown.[78] The cylinders are 500 mm inside diameter and wall thickness is 150 mm. The latter was constructed from two GRP layers; 100 mm of ± 30° and 50 mm of ± 50°. The stresses given are in the local material direction. As can be seen, the thermal stresses, even for the modest temperature rise of 10 °C, are of the same order as the externally applied stress.

The analysis presented in equations 5.63–5.66 is adequate for examining the influence of in-plane properties on residual stress distributions. Where the thickness of the component is comparatively large, attention must also be given to out-of-plane properties. Figure 5.77 shows the effects of laminate layup for thick-walled GRP cylinders.[79] The magnitude of residual stress

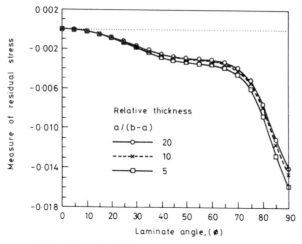

5.77 Effect of reinforcement angle on the residual
stress in angle-ply laminates.

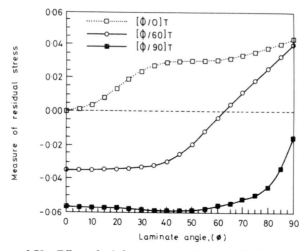

5.78 Effect of reinforcement angle on residual stress
in dual laminates.

levels increases with winding angle. This also corresponds to an increasing
ratio of hoop to radial stiffness. By using a combination of laminate angles, a
degree of control can be exercised on the level of residual stress in the
cross-section. Figure 5.78 shows the effect of a dual laminate construction
where the outer layer is kept constant and the inner is varied. Even a net stress
level of zero is possible for certain laminate combinations. Generally, by
tailoring the laminate it can be shown a gradual reduction in stiffness from the
inner to outer diameters has the effect of reducing stress levels.

Processing conditions can also have a significant influence on residual stress distribution. Tension applied to tows during winding is a common method of achieving consolidation of a laminate. However, depending on how the tension is controlled and the properties of the mandrel, induced stress can be high. The effect of winding is to prestress each layer with a radial compressive stress which is cumulative as winding proceeds.[80,81] The equivalent radial pressure, P_t, generated by a winding tension is given by:

$$P_t = \frac{Ft}{a} \qquad [5.67]$$

where t is the thickness of the ring and a is its inner radius. For a cylindrical ring with orthotropic properties subjected to uniform internal and external pressures, p and q respectively and a uniform temperature change ΔT, the radial and circumferential, σ_r and σ_h, stresses are given by:

$$\sigma_r = \frac{pc^{k+1} - q}{c^{2k} - 1}\left(\frac{r}{b}\right)^{k-1} - \frac{p - qc^{k-1}}{c^{2k} - 1}c^{k+1}\left(\frac{b}{r}\right)^{k+1}$$

$$+ \frac{E_h(\alpha_r - \alpha_h)\Delta T}{1 - k^2}\left[\frac{1 - c^{k+1}}{c^{2k} - 1}\left(\frac{r}{b}\right)^{k-1} + \frac{1 - c^{k-1}}{c^{2k} - 1}c^{k+1}\left(\frac{b}{r}\right)^{k+1} + 1\right]$$

$$[5.68]$$

$$\sigma_h = \frac{pc^{k+1} - q}{c^{2k} - 1}k\left(\frac{r}{b}\right)^{k-1} + \frac{p - qc^{k-1}}{c^{2k} - 1}kc^{k+1}\left(\frac{b}{r}\right)^{k+1}$$

$$+ \frac{E_h(\alpha_r - \alpha_h)\Delta T}{1 - k^2}\left[\frac{1 - c^{k+1}}{c^{2k} - 1}k\left(\frac{r}{b}\right)^{k-1} - \frac{1 - c^{k-1}}{c^{2k} - 1}kc^{k+1}\left(\frac{b}{r}\right)^{k+1} + 1\right]$$

where $k = (E_h/E_r)^{1/2}$ and $c = a/b$. E_h and E_r are the circumferential and radial modulus values respectively and α_h and α_r are the corresponding expansivities.

The corresponding strain components are given by:

$$\varepsilon_r = \frac{\sigma_r}{E_r} + \frac{v_{hr}}{E_h}\cdot\sigma_h + \alpha_r\Delta T \qquad [5.69]$$

$$\varepsilon - h = \frac{\sigma_h}{E_h} - \frac{v_{rh}}{E_r}\sigma_r + \alpha_h\Delta T$$

The latter terms in equations 5.69 represent the thermal expansion of the composite due to ΔT.

As the winding proceeds, new layers will alter the stress state in the underlying lamina and mandrel due to winding tension and thermal expansion. The contribution of each layer is summed as the winding process

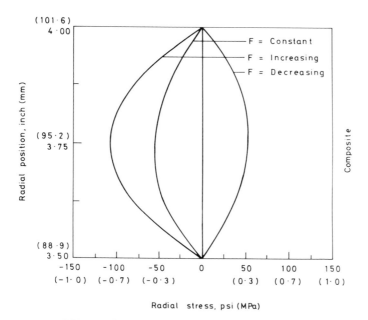

5.79   Radial stress distributions for various winding
tension histories.

continues through to completion. Depending on the sophistication of control, winding tension can be varied throughout the fabrication schedule. On removal of the mandrel, induced radial stresses at the outer surfaces must become zero and this gives rise to a complex through-thickness distribution of stresses. Figures 5.79 and 5.80 show final stress distributions as a result of a number of winding tension scenarios. Table 5.11 gives details of the material properties used for this calculation. Radial stresses remain compressive for a constant or increasing tension regime whilst the opposite is true for a decreasing fibre tension. The slope of the circumferential stress distribution is strongly influenced by the nature of the applied tension and it is clear that the winding tension can be effective in controlling the magnitude and sign of the residual stress distribution.

A further important consideration is the means by which temperature is applied during processing.[80,81] The effect of temperature is shown in Fig. 5.81 and 5.82. In this case the temperature change was taken as 315 °C and two calculations are presented. The first is where the component is wound and temperature applied at the end of the process and the second where the temperature is applied after the application of each layer. This latter case is analogous to staged cure in thermosets or an on-line heating process for thermoplastics. The potential benefits of the second method of processing is clear.

Table 5.11. Parameters for residual stress calculation

Composite material (CFRP)	
Circumferential modulus	135 GPa
Radial modulus	9 GPa
Poisson's ratio	0.3
Circumferential coefficient of thermal expansion	$-5 \times 10^{-7}/°C$
Radial coefficient of thermal expansion	$3 \times 10^{-5}/°C$
Mandrel	
Modulus	200 GPa
Poisson's ratio	0.25
Coefficient of thermal expansion	$2 \times 10^{-6}/°C$
Component dimensions	
Mandrel thickness	13 mm
Inner radius of composite ring	89 mm
Outer radius of composite ring	102 mm
Composite layer thickness	0.13 mm

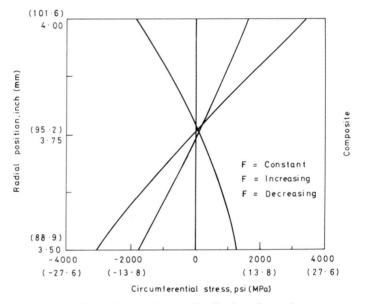

5.80   Circumferential stress distributions for various winding tension histories.

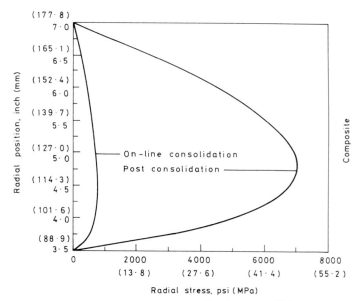

5.81   Effect of temperature application on radial stress
distribution.

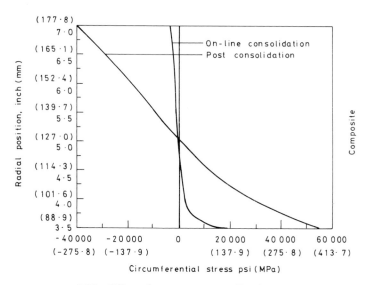

5.82   Effect of temperature application on
circumferential stress distribution.

Table 5.12. Weibull modulus for unidirectional composites under tension

Material	No. of test specimens	Weibull modulus
Glass fibre/epoxy	20–25	30
High strength carbon fibre/epoxy	30–35	12
Ultra-high modulus carbon fibre/epoxy	40–45	9

Statistical approach to design

In Chapter 1 the use of a statistical approach to the design of brittle materials was outlined. The probability of failure, P_f, for a volume of material under stress was given by (equation 1.2):

$$P_f = 1 - \exp\left\{ - \int_v \left(\frac{1}{m}!\right)^m (\sigma/\bar{\sigma}_f)^m \, dv \right\} \qquad [5.70]$$

where $\bar{\sigma}_f$ is the mean value of failure stress, σ is the applied failure stress and m is the Weibull modulus. For conditions of uniaxial tension where the stress system is uniform, equation 5.70 becomes:

$$P_f = 1 - \exp\left\{ - \left(\frac{1}{m}!\right)^m (\sigma/\bar{\sigma}_f)^m \right\} \qquad [5.72]$$

In order to estimate the parameters for the Weibull analysis a linear regression can be used with testing data. The values of P_f can be derived by ranking the data in order of magnitude and ascribing each with a probability value. There are a number of accepted methods,[82] typical of which is the mean rank probability, where:

$$P_{f,i} = i/(n + 1) \qquad [5.72]$$

where $P_{f,i}$ is the mean rank probability of the ith measurement and n is the total number of measurements. Examining equation 5.71 it can be seen that a plot of $\ln [1/(1-P_f)]$ against $\ln \sigma$ will be linear with a slope of m, the Weibull modulus, and an ordinate σ_f. The lower the value of the Weibull modulus, the greater the scatter in the data and the more relevant is the ability to treat the design with a statistical perspective. Table 5.12 shows a typical value of m for a range of unidirectional materials under tension.[83] The differences between GRP and CFRP are as expected given comparative properties of toughness, etc.

A feature of the behaviour of brittle materials is the size dependence of strength, the larger the component or specimen, the greater the probability there is of being flaws of critical size. Assuming that the flaw distribution is uniform then:

$$\frac{\bar{\sigma}_f}{\bar{\sigma}_{fv}} = \left(\frac{v}{V}\right)^{\frac{1}{m}}$$ [5.73]

$\bar{\sigma}_{fv}$ is the mean failure strength associated with volume v and σ_f that associated with volume V. Using equation 5.73 and with a knowledge of values for the Weibull modulus, the effect of size can be readily quantified. It is often convenient to equate v to unity thereby defining σ_{fv} as the mean failure strength of unit volume. Substituting equation 5.73 into equation 5.70 gives:

$$P_f = 1 - \exp\left\{1\left(\frac{1}{m}!\right)^m\left(\frac{1}{\bar{\sigma}_{fv}}\right)^m\frac{1}{v}\int_v \sigma^m\,dv\right\}$$ [5.74]

In many cases a stress system may contain both tensile and compressive components and allowance must be made for the fact that the two strengths are different. Equation 5.74 can be rewritten as:

$$P_f = 1 - \exp\left\{-\left(\frac{1}{m}!\right)^m\left(\frac{1}{\bar{\sigma}_{fv}}\right)^m\frac{1}{v}\int_v\left(\frac{\sigma}{H(\sigma)}\right)^m\,dv\right\}$$ [5.75]

$H(\sigma)$ is a step function defined by: $H(\sigma) = 1$ for $\sigma \geq 0$ (tension) and $H(\sigma) = \alpha$ for $\sigma < 0$ (compression), where α is the ratio of compressive and tensile strengths.

For multi-axial stress systems, equation 5.75 becomes:

$$P_f = 1 - \exp\left\{-\left(\frac{1}{m}!\right)^m\left(\frac{1}{\bar{\sigma}_{fv}}\right)^m\frac{1}{v}\int_v\left[\left(\frac{\sigma_1}{H(\sigma_1)}\right)^m + \left(\frac{\sigma_2}{H(\sigma_1)}\right)^m\right.\right.$$
$$\left.\left. + \left(\frac{\sigma_3}{H(\sigma_3)}\right)^m\right]dv\right\}$$ [5.76]

Further expansion can be carried out whereby σ_{fv} is brought into the integral and given different values to take into account strengths in different directions and the fact that each term is likely to have a different value of m. For a complex component where the stress distribution varies with position, the structure must be divided into elements and the survival probability of each element evaluated. The response of the total structure would then be determined by the product of those for the individual elements.

As an example consider a thick-wall cylinder. For an isotropic component the stress distributions would be as per those calculated by the well-known Lamé equations:

$$\sigma_h = \frac{pr_i^2}{r_o^2 - r_i^2}\left[1 + \left(\frac{r_o}{r}\right)^2\right]$$

$$\sigma_a = \frac{pr^2}{r_o^2 - r_i^2}$$ [5.77]

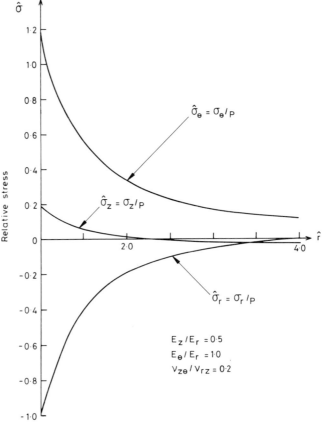

5.83 Stress distribution in a thick-wall orthotropic cylinder.

$$\sigma_r = \frac{pr_i^2}{(r_o^2 - r_i^2)}\left[1 - \left(\frac{r_o}{r}\right)^2\right]$$

where σ_h, σ_a and σ_r are circumferential, axial and radial components of stress at radius r and under pressure p; r_o and r_i are the outer and inner radii of the cylinder. Solutions are available for an orthotropic material and these can be used in a probability assessment.[84] Figure 5.83 shows the stress distribution in an orthotropic cylinder under internal pressure.[85] Figure 5.84 shows the calculated probability of failure as a function of internal pressure. As would be expected, the breadth of the probability curve is greater with reducing Weibull modulus.

As a further example, consider the case of a turbine blade (Fig. 5.85).[85] Not only can a statistical analysis provide a means of determining the probability

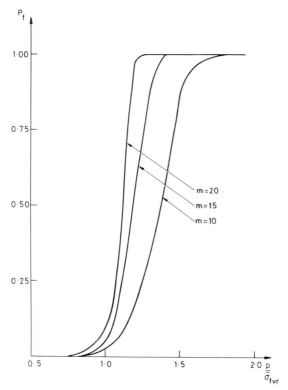

5.84   Dependence of failure probability on pressure.

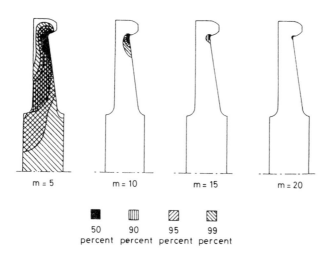

5.85   Probability of failure of a turbine wheel.

of failure for a given material and rotational speed, but it can be taken further to provide useful information on how to modify the structure to improve its performance. As is often the case, particularly when stress distributions are complex, the stresses in a relatively small region contribute predominantly to the stress volume integral. This is of importance as the percentage contribution to the stress volume integral is approximately equal to the probability of the total failure initiating from that volume. Figure 5.85 demonstrates this effect by indicating the volumes contributing for different Weibull moduli. A number of important design conclusions may be drawn:

- For high values of m an increased fillet radius at the outer tip would have a significant influence on component strength, whereas for low values of m a more radical change in design would be necessary to have the same effect.
- For high values of m inspection procedures may be concentrated in the fillet area, whereas for lower values of m the whole component would need to receive rigorous examination.

From this type of assessment the next stage of the design may be undertaken. This may include not only the changes in geometry cited above, but also modification of the detailed design.

References

1 Kinloch A J, *Adhesion and Adhesives, Science and Technology*, Chapman and Hall, London, 1987.
2 Adams R D and Wake W C, *Structural Adhesive Joints in Engineering*, Elsevier Applied Science, London, 1984.
3 Hart-Smith L J, *Adhesive-bonded Double-lap Joints*, Report CR-112235, NASA, Washington, DC, 1973.
4 Hart-Smith L J, *Adhesive-bonded Scarf and Stepped-lap Joints*, Report CR-112237, NASA, Washington, DC, 1973.
5 Hart-Smith L J, *Adhesive-bonded Single-lap Joints*, Report CR-112236, NASA, Washington, DC, 1973.
6 *Adhesives Bonding Handbook*, BNF–Fulmer Materials Centre, Wantage, Oxon.
7 'ADHESYS expert system', AEA Technology, Harwell Laboratory, Oxon.
8 Lee R J and McCarthy J C, 'An overview of the composite to metal jointing project', in Proceedings of the Conference on Advances in Joining Plastics and Composites, Bradford, 1991. Abington Publishing, Cambridge, England.
9 Eckold G C, Lee R J and McCarthy J C, 'Design of composite to metal joints', 10th International Conference, OMAE, ASME, Stavanger, 1991.
10 Davidson R, 'The study of strain fields in adhesive joints by laser moiré interferometry', in Proceedings of the 23rd Annual Conference on *Adhesion and Adhesives*, City University, Elsevier Applied Science, London, 1988.
11 Lee R J, Davidson R and McCarthy J C, 'Composite to metal jointing for transport applications', in Proceedings of the 23rd Annual Conference on, *Adhesion and Adhesives*, City University, Elsevier Applied Science, London, 1988.

12 Matthews F L, (Ed) *Joining Fibre-Reinforced Plastics*, Elsevier Applied Science, London, 1986.

13 Hart-Smith L J, 'Design and empirical analysis of bolted or riveted joints', in *Joining Fibre-Reinforced Plastics*, Elsevier Applied Science, London, 1986.

14 Matthews F L, *Proceedings of Conference on New Materials and their Applications*, University of Warwick, The Institute of Physics, London, 1987.

15 Collings T A, 'Experimentally determined strength of mechanically fastened joints', in *Joining Fibre-reinforced Plastics*, Ed F L Matthews, Elsevier Applied Science, London, 1986.

16 Taylor N S and Jones S B, 'The challenge of fabricating thermoplastic composite materials', Proceedings on Use of Plastic Composites for Advanced Engineering and High Technology Commercial Application, Rotterdam, TWI, Cambridge, 1989.

17 Greszczuk LB, 'Response of isotropic and composite materials to particle impact', *Foreign Object Impact Damage to Composites*, ASTM STP 568, in American Society for Testing and Materials, Philadelphia, PA, 1975.

18 Zukes J A, Nicholas T, Swift H F, Greszczuk L B and Curran D R, *Impact Dynamics*, Wiley, Chichester, 1982.

19 Timoshenko S, *Theory of Elasticity*, McGraw-Hill, New York, 1934.

20 Choi H Y, Wu H Y T and Chang F K, 'A new approach toward understanding damage mechanisms and mechanics of laminated composites due to low velocity impact, Part II – Analysis', *J Comp Mat*, **25**, 1012–1038, 1991.

21 Grady J E and Sun C T, 'Dynamic delamination crack propagation in a graphite/epoxy laminate', in *Composite Materials: Fatigue and Fracture*, ASTM STP 907, Ed HT Hahn, American Society for Testing and Materials, Philadelphia, PA, 1986.

22 Gosse J H and Mori P B Y, 'Impact damage characterization of graphite/epoxy laminates', in Proceedings of the American Society for Composites, 3rd Technical Conference on Composite Materials, American Society for Composites, 1988.

23 Finn S R and Springer G S, 'Delamination in composite plates under transverse impact loads – a model', *Comp Struct*, **23**, 177–190, 1993.

24 McCarthy J C, AEA Technology, private communication.

25 Clark G, 'Modelling of impact damage in composite laminates', *Composites*, **20**, 1989.

26 Adams D F and Miller A K, 'An analysis of the impact behaviour of hybrid composite materials', *Mat Sci Eng*, **19**, 245–260, 1975.

27 Cantwell W J and Morton J, 'The impact resistance of composite materials – a review', *Composites*, **22**, 342–361, 1991.

28 Chamis C C and Sinclair J H, 'Impact resistance of fibre composites: energy absorbing mechanisms and environmental effects', in *Recent Advances in Composites in the United States and Japan*, ASTM STP 864, Ed JR Vinson and M Taya, American Society for Testing and Materials, Philadelphia, PA, pp 326–345, 1985.

29 Hirschbuehler K R, 'A comparison of several mechanical tests used to evaluate the toughness of composites', Toughened Composites, ASTM STP 937, Ed NJ Johnston, ASTM, Philadelphia, pp 37–60, 1987.

30 Dorey G, 'Damage tolerance and damage assessment in advanced composites', in *Advanced Composites*, Ed IK Partridge, Elsevier Applied Science, London, pp 369–398, 1989.

31 Cantwell W J, Curtis P T and Morton J, 'Low velocity impact damage tolerance in CFRP laminates containing woven and non-woven layers', *Composites*, **14**,

301–305,1983.

32 Johnson W and Reid S R, 'Metallic energy dissipation systems', *App Mech Rev*, **31**, 277–288, 1978.

33 Hull D, 'Energy absorbtion of composite materials under crush conditions', ICCM-IV, Tokyo, pp 861–870, 1982.

34 Thornton P H, 'Energy absorbtion in composite structures', *J Comp Mat*, **13**, 247–262, 1979.

35 Pipes R B and Daniels I M, 'Moiré analysis of interlaminar shear edge effects in laminated composites', *J Comp Mat*, **5**, 255–259, 1971.

36 Pipes R B and Pagano N J, 'Interlaminar stresses in composite laminates under axial extension', *J Comp Mat*, **4**, 538–548, 1970.

37 Conti P and De Paules A, 'A simple model to simulate the interlaminar stresses generated near the free edge of a composite laminate', ASTM STP 876, Ed W S Johnson, ASTM STP 876, American Society for Testing and Materials, Philadelphia, PA, 1983.

38 Whitney J M, Interlaminar stresses and free edge effects in the characterization of composite materials, in 'Analysis of the Test Methods for High Modulus Fibers and Composites', ASTM STP 521, American Society for Testing and Materials, Philadelphia, PA, p 167, 1973.

39 Wang A S and Crossman F W, 'Some new results on edge effects in symmetric composite laminates', *J Comp Mat*, **11**, 92–106, 1977.

40 Pagano N J, 'Stress fields in composite laminates', *Int J Solids Struct*, **14**, 385–400, 1978.

41 Pagano N J and Pipes R B, 'The influence of stacking sequence of laminate strength', *J Comp Mat*, **5**, 50–57, 1971.

42 Whitney J M, 'Free edge effects in the characterization of composite materials', in *Analysis of Test Methods for High Modulus Fibres and Composites*, ASTM STP 521, American Society for Testing and Materials, Philadelphia, PA, 1973.

43 Wells H and Hancox N L, 'Izod impact testing of carbon-fibre-reinforced plastics', *Composites*, **2**, 41–45, 1971.

44 Hancox N L and Wells H, 'Izod impact properties of carbon-fibre sandwich structures', *Composites*, **4**, 26–30, 1973.

45 Hancox N L, 'Introduction to fibre composite hybrids', in *Fibre Composite Hybrids*, Ed N L Hancox, Applied Science, London, 1981.

46 Timoshenko S P and Gere J M, *Mechanics of Materials*, Van Nostrand, New York, 1972.

47 Harris B and Bunsell A R, 'Dynamic elastic moduli and toughness of dough moulding compounds', *Composites*, **6**, 25–29, 1975.

48 Phillips M G, 'Composites parameters for hybrid composite materials', *Composites*, **12**, 113–116, 1981.

49 Harris B, 'Engineering Composite Materials', Institute of Metals, London 1986.

50 Parratt N J and Potter K D, PERME Report TR81, 1978.

51 Phillips M G, 'Fracture and fatigue of hybrid composites', in *Fibre Composite Hybrid Materials*, N L Hancox, Applied Science, London, 1981.

52 Lubin G, *Handbook of composites*, Van Nostrand, New York, 1982.

53 Roark R J and Young W C, *Mechanics of Materials*, Von Nostrand Reinhold, New York, 1972.

54 Chou T W, 'Review of flexible composites', *J Mat Sci*, **24**, 761–782, 1989.

55 Clark S K, 'The role of textiles in pneumatic tyres', in *Mechanics of Flexible Fibre Assemblies*, Eds J W S Hearle, J J Thwaites and J Amirbayat, The Netherlands, 1980.

56 Akasaka T, 'Flexible composites', in *Textile Structural Composites*, Ed T W Chou and F Ko, Elsevier Science, The Netherlands, in press.

57 Akasaka T and Hirano M, 'Approximate elastic constants of fiber reinforced rubber sheet and its composite laminate', *Comp Mat Struct* 1, 1972.

58 Clark SK, 'Plane elastic characteristics of cord-rubber laminates', *Textile Res, J* 33, (4), 295–313, 1963.

59 Davidson R and Phillips D C, 'Spiral moorings and tube pumps', in *Rubber in Offshore Engineering*, Ed A Stevenson, Adam Hilger, Bristol, 1983.

60 Aveston J, Keely A and Silkwood J M, in *ICCM3, Advances in Composite Materials*, 1, Bunsel, Pergamon, Oxford, 1980.

61 Crank J, *The Mechanics of Diffusion*, Clarendon, Oxford, 1956.

62 Springer G S, 'Moisture absorbtion and disorption of composite materials', in *Developments in Reinforced Plastics*, Vol. 2, Ed. G Pritchard Applied Science, London, 1982.

63 Bunsell A R, 'Long term degradation of polymer matrix composites', in *Encyclopedia of Composite Materials*, Ed A Kelly, Pergamon, Oxford, 1989.

64 Hogg R J, Hull D and Legg M J, 'Failure of GRP in corrosive environments', in *Composite Structures*, Ed I M Marshall, Applied Science, London, pp 106–122, 1981.

65 Roberts R C, 'Design, strain and failure mechanism of GRP in a chemical environment', Reinforced Plastics Congress BPF, 1978.

66 BS2782, 'Methods of testing plastics, Method 121A Determination of temperature of deflection under a bending stress of 1.8 MPa of plastics and ebonite', BSI, London, 1991.

67 Hancox N L, 'The production and properties of bismaleimide composites', *Materials and Design*, 12, 317, 1991.

68 McLaren J R and Hancox N L, 'Carbon fibre reinforced bismaleimide and polystyryl pyridine laminates', Institute of Physics Conference, Ser. No. 89: Session 1, Warwick, 1987.

69 Curtis P T, 'Fatigue of organic matrix composites', in *Advanced Composites*, Ed I K Partridge, Elsevier, London, pp 331–368, 1989.

70 Reifsnider K L and Jamison R D, 'Fracture of fatigue-loaded composite laminates', *Int J Fatigue*, 4, 187–198, 1982.

71 Harris B, *Engineering Composite Materials*, Institute of Metals, London, 1986.

72 Jones C J, Dickson R F, Adam T, Reiter H and Harris B, *Proc Roy Soc*, A396, 313–335, 1984.

73 Taireja R, 'A continuum mechanics characterisation of damage in composite materials', *Proc R Soc*, A378, 195–216, 1981.

74 Kim K S and Hahn H T, 'Residual stress developing during processing of graphite/epoxy composites', *Comp Sci Technol* 36, 121–132, 1989.

75 Nairn J A and Zoller P, 'Matrix solidification and the resulting residual thermal stresses in composites', *J Mat Sci* 20, 355–367, 1985.

76 Kim K S, Hahn H T and Croman R B, 'The effect of cooling rate on residual stress in a thermoset composite', *ASTM J Comp Technol Res* 11, 121–132, 1987.

77 Fourney W L, 'Residual strain in filament wound rings', *J Comp Mat*, 2, 47–52, 1968.

78 Wuthrich C, 'Thick walled composite tubes under mechanical and hydrothermal loading', *Composites*, **23**, 407–413, 1992.

79 Roy A K, 'Response of thick laminated composite rings to thermal stresses', *Composite Structures*, **18**, 125–138, 1991.

80 Cirino M and Byron-Pipes R, 'In-situ consolidation for the thermoplastic composite ring residual stress state', *Composite Manufacturing*, **2**, 105–113, 1991.

81 Filiou C, Gauotis C and Batchelder D N, Residual stress distribution in carbon fibre/thermoplastic matrix pre-impregnated composite tapes, *Composites*, **23**, 28–38, 1992.

82 Robinson E Y, 'Estimating Weibull parameters for materials', Technical Memorandum 33-580, Jet Propulsion Laboratory, California Institute of Technology, Pasadena, California, 1972.

83 King R L, 'Statistical derivation of design data for composite materials', in *Computer Aided Design in Composite Material Technology*, Ed C A Brebbis, Computational Mechanics Publications, Springer Verlag, 1988.

84 Stanley P, Sivill A D and Fessler H, 'Design exercise for a ceramic turbine disc subjected to centrifugal stresses', *J Strain Analysis*, **13**, 103–113, 1978.

85 Stanley P and Margetson J, 'Failure probability analysis of an elastic orthotropic brittle cylinder subjected to axisymmetric thermal and pressure loading', *Int J Fracture*, **13**, 787–806, 1977.

6

MANUFACTURE

For metallic materials that are homogeneous and isotropic the choice of manufacturing method is often taken after considerations of material selection and part design. With composites, however, the constraints and characteristics of the process have a direct influence on fibre volume fractions and their distributions, fibre orientations and details of laminate layup. The main issues which affect component performance are therefore intimately linked with manufacture method, and failure to acknowledge this early in the component development cycle could lead to missed opportunities for parts integration and optimization or even result in the objectives of the programme not being met in either technical or commercial terms.

The driving factors behind manufacturing considerations for monolithic materials are primarily cost effectiveness, the minimization of scrap, the control of assembly operations and the sourcing of standard parts. Whilst these are equally applicable to composite materials options, they do not include issues such as component integration or hybrid combinations of materials. Furthermore, products of nominally the same form, but manufactured through different routes, could have markedly different properties. This not only affects mechanical considerations such as stiffness and strength, but also other attributes such as surface finish, chemical resistivity and internal damping, as well as electrical and thermal properties, each of which may be key to a particular application.

The analysis of unit cost of a given component or assembly is also linked to individual processes. The breakdown of cost elements between equipment, tooling, labour and materials varies greatly – see Fig. 6.1.[1] The balance between achieving the maximum level of performance, e.g. highest possible reinforcement volume fraction, lowest component cost and fastest production rate, is finely drawn and will need to satisfy the requirements of different industry sectors. The automotive industry, for example, demands unit costs and production rates commensurate with large volumes. Single stage manufacture and low cost materials are therefore a prerequisite. Higher cost capital investments in plant can be tolerated through amortization over large production runs. Space and process plant applications are more likely to

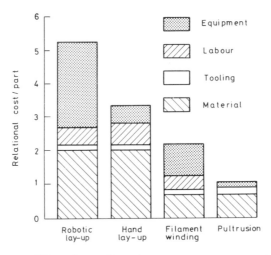

6.1   Effect of manufacturing process on component
cost.

require individually designed structures and as a consequence are less likely to
employ costly production processes geared to replication. They will, however,
occupy different extremes of the material cost spectrum so techniques and
methods will need to be tailored to suit.

The over-riding message when considering a composite material option for
a component is that consideration of manufacture method cannot be divorced
from material selection, component specification and component design.

Continuous reinforcement processes

For structural applications it is often the case that recourse is first made to the
use of continuously reinforced materials when at the initial stages of design.
The concept of an optimum level of material design is clearly an attractive
option. Manufacture processes must therefore reflect this desire and be
capable of dispensing material accurately and uniformly as per the design
specification. The effects of anisotropy, a consequence of which are the high
but directional mechanical properties that feature in the design, can be both a
benefit and a drawback. Such materials can be unforgiving and the sensitivity
of the design must be robust with respect to tolerances and potential variations
due to manufacture. Also, the use of continuous unidirectional systems can be
restrictive in terms of product form. By definition, material precursers, i.e.
fibre or prepreg, are of high aspect ratio and this lends itself to certain types of
processes. However, while there are limitations, there are also advantages.
Many of the available techniques have their origin in the textile industry and

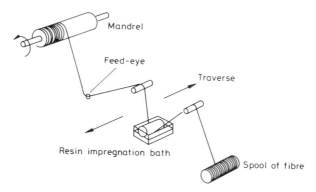

Mandrel

Feed-eye

Traverse

Resin impregnation bath

Spool of fibre

6.2   Schematic of a two-axis winding machine.

production rates in terms of material conversion are perhaps higher than for any other range of fabrication methods.

Filament winding

In this process, strong, stiff shell structures can be made by winding bands of continuous fibres impregnated with resin onto a former or mandrel which can be withdrawn after the resin has been cured.[2,3] The method is analogous to the technique of wire wrapping for the reinforcement of steel pressure vessels and gun barrels. Filament winding machines, in their simplest configuration, are similar in principle to a lathe. A mandrel is rotated between supports while fibre is drawn from a carriage which travels along a path parallel to the axis of rotation. The helix angle can vary from near hoop to axial and is determined by the relative speed of mandrel rotation and carriage translation. A linear winding machine, where the relative speed of the two axes is constant, is only capable of winding at a single helix angle on a surface of fixed diameter. Variations in helix angle and limited variations in diameter can be accommodated by machines with a nonlinear capability, where the relative speed may be varied during a circuit of the mandrel. This flexibility can be accomplished by mechanical means, e.g. through the use of cams, or numerical means, e.g. software control of independent drives. Two axis winding machines of either linear or nonlinear type are widely available and in terms of the many types of applications are often all that are required. Figures 6.2 and 6.3 illustrate details of the configuration of typical winding equipment. In addition to the helical winding machines there is also equipment that is capable of deploying material in a polar fashion. Examples are shown in Fig. 6.4.[4] On a larger scale, machines have also been developed to allow the on-site winding of large components where problems of transportation would otherwise have precluded the technique as a method of manufacture.[5]

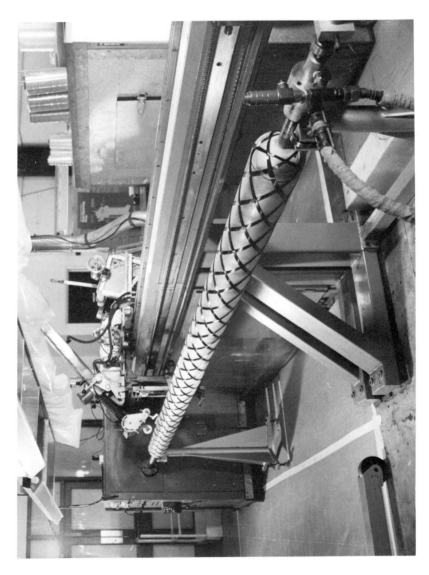

6.3   Typical winding machine.

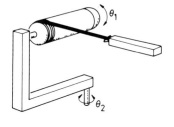 

6.4   Polar winding machines.

Despite apparent limitations to cylindrical or near-cylindrical form the process has found wide application in a number of industrial sectors:

- **Process**: pipework, vessels, storage tanks.[6-8]
- **Aerospace**: fuel tanks, helicopter blades, nose cones, driveshafts.[9-12]
- **Rocket motor casings**: strategic delivery systems, shuttle boosters, pressure bottles.[13,14]
- **Defence**: launchers, portable bridge struts.[13,15]
- **Offshore**: pipework, risers, casings.[14]
- **Transportation**: draftshafts, stiffened monocoques, suspension units.[13-15]

During winding, fibre impregnation is achieved in one of two ways: preimpregnation and wet winding. In the first, conventional unidirectional prepreg in tape form is wound on to the mandrel under tension. The radial component of the applied tension is normally sufficient to effect adequate consolidation, but in some cases an intermediate compaction stage may be necessary. Wet impregnation of reinforcement is achieved by pulling the fibre through a resin bath or over a roller that contains a metered volume of resin. Figure 6.5 shows an early design of resin bath.[16] The resin/fibre ratio in such a system is governed by resin viscosity, temperature, fibre tension, fibre speed and the mechanical arrangement of rollers and supports which guide the fibre on to the mandrel. Thermoplastic materials can also be processed by winding, but in this case the pre-impregnated tow must be heated to melt the polymer prior to application on the mandrel. Cooling and consolidation take place after initial placement. An example of a scheme for thermoplastics winding is shown in Fig. 6.6 where heating is provided by a hot gas gun. The fabrication of re-entrant shapes is possible by forcing the material on to the mandrel surface where it will remain on freezing of the matrix. Speeds for winding thermoplastic materials are currently much less than for thermosets because of the time necessary to impart sufficient heat into the tow.

Mandrel materials vary according to the application, required dimensional tolerances and the need for robustness. Metal tooling, usually steel, with a ground surface is appropriate if high levels of accuracy are necessary and

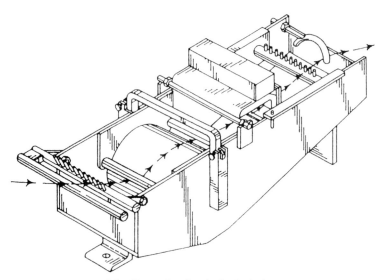

6.5   Example of resin bath design.

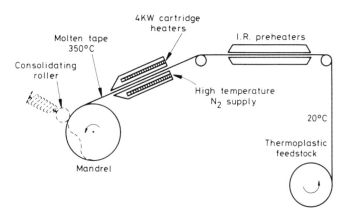

6.6   Scheme for thermoplastic winding.

components are to be large in number. For limited volumes, and when dimensions are less critical, wood, GRP or thermoplastics are all suitable materials. Extraction of mandrels is usually carried out through pulling the composite component axially against a collar. For carbon laminates where expansion coefficients are small the forces required are low. For glass, on the other hand, shrinkage on cooling from cure means that a significant end load may need to be applied. Where simple axial extraction is not possible, e.g. where windings have been continued over an end closure, a different approach is necessary. There are a number of options, including demountable metal fabrications, water-soluble sand and plaster systems as well as friable foams.

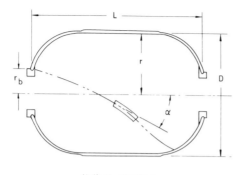

Helical winding

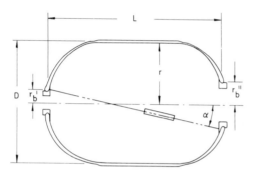

Polar winding

6.7   Types of winding pattern.

With winding there are three patterns which may be adopted. In the simple helical winding already described, the pattern is characterized by a series of fibre crossovers along the mandrel length. After the first circuit of the mandrel the fibres do not lay adjacent to one another and a series of circuits, the number depending on band width and component length, are required before the pattern is such that material is applied adjacent to previous winding. To ensure that the fibre does not slip whilst being wound under tension the geodesic is normally followed. This is essentially the shortest distance over a surface, given a starting position and helix angle. It can be shown that for an axisymmetric shape the geodesic must satisfy the relationship:

$$k = r \sin \theta \qquad [6.1]$$

where r is the mandrel radius at the point of concern, θ is the winding angle at radius r and k is a constant.

Deviations from the geodesic are possible, but are limited by the level of friction which is available to stop the fibre slipping. Figure 6.7 shows the concept of geodesic winding.[2] For the component shown, the need to wind around the boss in the end closure essentially defines the winding path. At the boss, reinforcement will be tangential, i.e. $\sin \theta = 1$. Therefore, the fibre angle in the cylinder, α, is given by:

$$\alpha = \sin^{-1}\left(\frac{r_b}{r}\right) \qquad [6.2]$$

where r_b is the radius of the boss. Note that the openings in the ends must be of equal size in order to complete the winding.

A second form of pattern is polar winding. Here the fibre passes tangentially between polar openings. The windings are applied by the polar arm which describes a great circle about the mandrel. Figure 6.7 also shows the polar winding technique. End closure openings of different size can be accommodated. In this case the cylinder winding angle is given by:

$$\alpha = \tan^{-1}\frac{r_b' - r_b''}{L} \qquad [6.3]$$

The polar winding technique is a pragmatic approach and simple in principle. Fibre slippage can be a problem, however, and generally length to diameter (L/D) ratios of wound components should be limited to approximately 2.

The third pattern in common use is hoop winding, where the fibre is placed circumferentially along the mandrel. Owing to the problems of slippage, such winding methods can only be applied to prismatic sections, i.e. those that are constant in the axial direction. Usually they are applied in conjunction with other windings in order to achieve a good balance of mechanical properties.

Most winding equipment is of the simple two-axis type. However, the addition of further degrees of freedom greatly enhances the flexibility of the system:

• The incorporation of a third axis allows the machine to handle mandrels of greatly varying diameter. With three axes the machine is capable of positioning the fibre deployment mechanism in any position in space with respect to the mandrel surface, and hence in principle a component of any geometry can be wound with such a device.
• When orientation with respect to the mandrel surface is important, e.g. when winding tapes or a wide band of fibre tows, a fourth axis, namely rotation of the feed eye, is required.
• For very complex structures where there may be problems associated with clash or shadowing, five or six degrees of freedom may prove to be necessary.

Figure 6.8 illustrates the basic motions of a five-axis winding machine.

The winding of complex shapes that are not rotationally symmetric is not, in

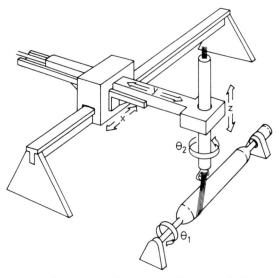

6.8   Basic motions of a five-axis filament winding
machine.

general, amenable to description by simple expressions such as equations 6.1
or 6.2. Control data can be determined experimentally by stretching a fibre
over a surface and following its path with a joystick control. The path followed
by the machine is then stored by the machine control system. Increasing use is
being made of simulation systems[17-20] which employ graphical or mathematical representations of the surface to be wound, and the equations that describe
the fibre path, either based on the geodesic or using friction, are solved
numerically.

Use of friction effects allows much more flexibility in that the steering of the
reinforcement over the surface becomes feasible. The degree of departure from
the geodesic which can be wound without slippage depends not only on the
friction coefficient, but also on the geometry of the surface. The greater the
local curvature, the greater the normal reaction force generated due to the
fibre tension. This in turn causes the available frictional force, which is
instrumental in preventing slippage, to increase.

Coverage of surfaces of the more complex type, particularly those with
asymmetric geometry, can pose a number of difficulties which are not
apparent when a component has an axis of symmetry. For these simpler
geometries all that is necessary are increments of mandrel rotation between
consecutive circuits to effect coverage in a controllable manner. In asymmetric
structures the concept of a circuit cannot be defined in the same way, since,
even if two paths had only a slight relative displacement, they would ultimately
be divergent when they encounter areas of the surface that are not identical.

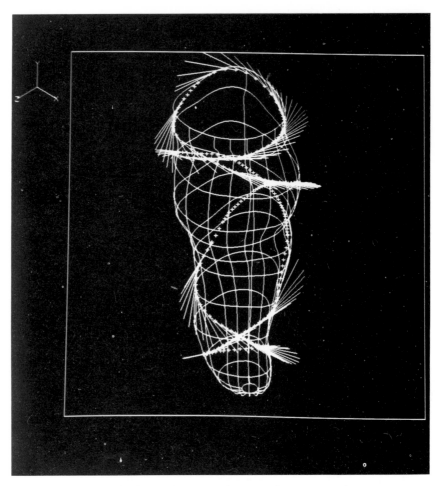

6.9   Winding path generated on a surface representing
a prosthesis.

Filament winding such a shape requires the generation of a large number of
related winding paths which, when combined, cover the mandrel surface in the
designated manner. All of these will be unique, unlike an axisymmetric
product where they are merely displaced round the axis from the initial circuit.
To facilitate asymmetric winding it is necessary to generate the winding
machine instruction set from one or more reference paths that can be
transformed from their initial positions into the other winding circuits
required for manufacture.

Figure 6.9 shows an example of the scope of filament winding technology.
The surface of concern is a leg prosthesis. These are already manufactured
using reinforced plastics in the form of an impregnated fabric draped around a

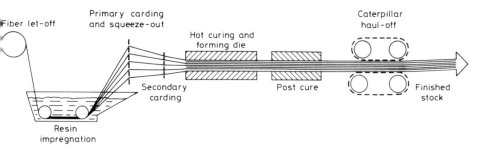

6.10 Schematic of pultrusion process.

mould. This offers a light structure with the necessary stiffness, but there is little opportunity to provide additional strength in specific areas. The fibre path shown indicates a winding which could preferentially reinforce the structure.

Pultrusion

Pultrusion is a process which is analogous to the extrusion of plastics and nonferrous metals in that the profile is shaped by continuous passage of the feedstock through a forming die, but in the case of pultrusion the reinforcing fibres are used to pull the material through the die.[21] Figure 6.10 shows a typical arrangement. Reinforcements of various types, but usually in a continuous form, are drawn from spools into the pultrusion die. A series of guides or bushings is generally used in front of the die to preform the reinforcement to the desired shape. Within the die, impregnation and consolidation occur as liquid resin is injected under pressure. The materials are shaped and curing is initiated as the section progresses through the heated part of the tool. The rate of throughput is controlled so that curing is essentially complete by the time the product emerges from the die. This is necessary because the material must have sufficient strength to resist the forces exerted in the traction stage. After passing through the pullers, sections are cut to the desired length.

Glass fibres and polyester resins are the most commonly employed materials because of their relatively low cost and good processability. Pre-heating by radio frequency methods can be used to increase the throughput, and production rates of several metres per minute can be achieved depending on section thickness. The process is ideally suited to the production of structural elements of comparatively high strength and stiffness, owing to the presence of axial fibres necessary to sustain the large tractive forces required to pull the fibres through the die. It is also possible to incorporate varying geometry, although the cross-sectional area must remain constant. Most pultrusions employ unidirectional reinforcement, but it is also possible to process fibres of different form provided that they have sufficient strength to sustain the applied tractive loads.

Commercial pultrusion operations use heated steel dies to form and shape the products. The working surfaces are usually chromium plated and polished to reduce wear and to improve the surface finish of the product. Precise machining tolerances are required over the critical curing section of the die in order to maintain parallelism and hence reduce internal friction. Resin injection cavities and resin distribution channels need to be designed to avoid stagnant areas where resin can gel or fibre debris accumulate.

Tubular sections are manufactured using a mandrel to form the inner surface. The mandrel generally runs the full length of the die and is rigidly supported at the input end by a bracket designed to allow the fibre reinforcement to move smoothly into the die. For large profiles a number of separately controlled heating zones are employed in order to heat the materials progressively and to reduce the peak exotherm temperatures which occur as the resin gels. In normal operation the cold materials entering the die cool the input end and prevent premature gelling of any resin which runs off the die entrance. Water cooling can be incorporated as an added precaution. When a resin injection cavity is machined in the die, the temperature in that region must be controlled to prevent the resin from gelling in the cavity.

The range of constant section profiles commonly produced is wide and essentially consists of open and closed structural sections.[22] Typical examples are illustrated in Figure 6.11. The simple solid rods and bars are mainly longitudinally reinforced, but the larger tubes, channels and structural shapes can be fabric reinforced. Tolerances that can be maintained in pultruded shapes depend on part complexity and on the materials used. Table 6.1 provides typical tolerance data.

In general, the detailed construction and form of a pultruded laminate is dependent on a number of process-controlled restraints. Thickness is limited by the time necessary to transmit sufficient heat to the centre of the section and also by possible problems with exotherm. Very thin sections are limited by the large tractive forces which must be applied during processing. Typical practical thicknesses are in the range 2–20 mm, although up to 50 mm is possible. The pulling capacity of pultrusion installations varies enormously (500 kg to 15 t is the normal range) and this determines the overall section size. Current maximums are around 1000×165 mm.[21]

Attempts to model pultrusion in order to optimize process parameters have centred around solutions of the heat balance equation in the die region.[23] Such expressions can be used to describe the extent of cure at any point in the modelling process. An example of such a relationship of empirical basis is as follows:

$$\frac{d\alpha}{dt} = (k_1 + k_2\alpha)^m (1 - \alpha)^n \qquad [6.4]$$

Table 6.1. Typical tolerance levels for pultruded profiles

Dimension	Typical tolerance
Thickness	$\pm 1\%$ minimum 0.15 mm (0.006 in.)
Width	$\pm 0.1\%$ minimum 0.15 mm (0.006 in.)
Diameter	$\pm 1\%$ minimum 0.15 mm (0.006 in.)
Wall thickness (tubes)	$\pm 10\%$ minimum 0.2 mm (0.008 in.)
Ovality	$\pm 1\%$ minimum 0.1 mm
Straightness	$\pm 0.2\%$
Length	± 0–2 mm

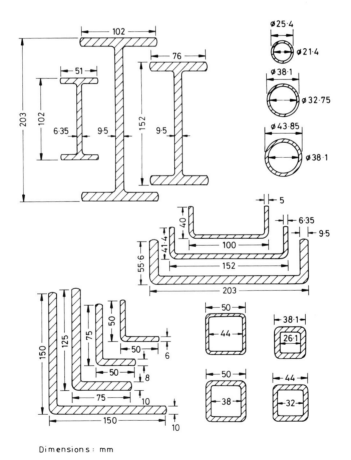

Dimensions : mm

6.11 Typical pultruded cross-sections.

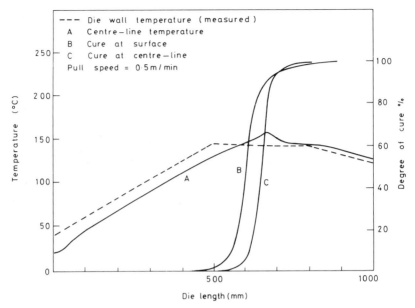

6.12   Calculated temperature and cure profile in
pultrusion die (5 mm diameter) for glass/polyester.

where α is the degree of cure, m and n are constants and k_1 and k_2 are thermally dependent rate constants determined by differential scanning calorimetry techniques. Figure 6.12 shows an example of a predicted temperature and cure profile through the die for a glass/polyester pultrusion. Agreement between the measured and calculated temperature profiles is reasonable. Thermal modelling such as this can be used in die design and to ensure adequate cure of the product as it leaves the die.

An important area that influences process considerations such as pull force and product quality and hence production speed is the interaction between the moving profile and the surface of the die.[23] The contact between the two results in an interfacial shear stress, the magnitude of which depends on the extent of cure, adhesion effects and thermal expansion. Figure 6.13 shows a comparison of the variation in shear stress with die position for unidirectional glass using three different resin systems. The stress distributions are characterized by three stress peaks. The first relates to viscous drag in the resin film between fibre and die. The stress reduces as the material is heated and viscosity falls and the relative magnitude of the stress peak increases with filler content in the resin. The second peak corresponds to the onset of cure where viscosity increases markedly with cross-linking and gelation. As the stress levels increase, debonding occurs between the product and the die surface. A key factor for process control is the second peak which provides the maximum

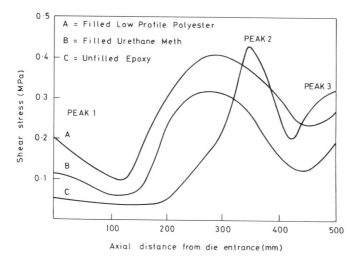

6.13   Results of die interfacial shear stress
measurements.

shear stress and hence traction load requirement. Reduction in friction effects
in this area could result in desirable enhancement in surface finish and
productivity. Finally, a third peak, of lower magnitude than the second, is
generated through Coulomb friction between the gelled (cured) section and
die wall.

Although originally developed for use with thermosetting resins, pultrusion
is now being extended to encompass thermoplastics materials.[24] Thermoplastic pultrusion offers potential productivity advantages over conventional
thermoset systems where production rates are limited by the kinetics of the
cure reaction. Thermoplastics processing, on the other hand, is determined by
heating and cooling and, in principle, speeds comparable to those encountered
in extrusion may be possible. Progress continues to be made in the
optimization of process parameters to establish a balance between composite
properties and speed of production. Heating and cooling rates, residence times
and consolidation forces have all been shown to have significant influence[25] –
Fig. 6.14 and 6.15 show the effect of processing temperature and consolidation
roll pressure on flexural strength of nylon/glass composites, and there have
been some developments in modelling the heat transfer conditions within the
system. There remains, however, some way to go before the apparent potential
in production rates is realized.

Other processes for continuous reinforcement

In addition to filament winding and pultrusion there are other methods which
are capable of handling continuous reinforcement.

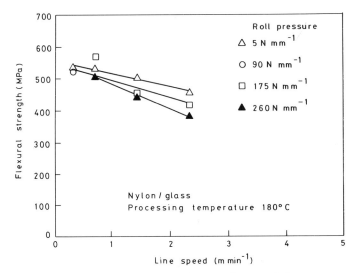

6.14   Flexural strength versus line speed for
nylon/glass pultruded at constant processing
temperature.

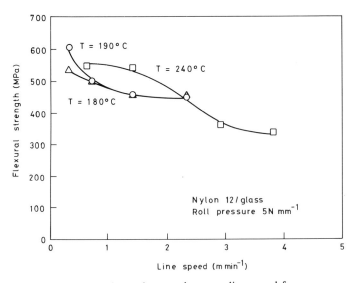

6.15   Flexural strength versus line speed for
nylon/glass pultruded at constant roll pressure.

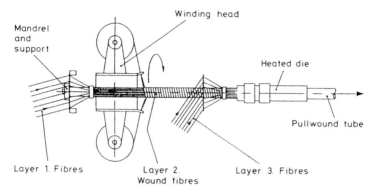

Mandrel
and
support

Winding head

Heated die

Pullwound tube

Layer 1. Fibres

Layer 2.
Wound fibres

Layer 3. Fibres

6.16   Schematic of pull-winding process.

One important technique is braiding which is a mechanized textile process dating from the last century. In the braiding operation, a mandrel is fed through the centre of the machine at a uniform rate, and fibres from moving carriers on the machine braid about the mandrel at a controlled angle. Fibre angles can include the whole range from 0° to hoop. The machine operates like a maypole with the carriers working in pairs to accomplish the over/under braiding sequence. This process also is applicable to channel sections and webs and, by exercising control over feed rates, fibre preforms of more complex shape can be produced.

Variants on the more established techniques have also been developed. The conventional pultrusion process can be considered as four in-line systems; in-feed of materials, curing in a heated die, continuous pulling system and a flying cut-off saw. The in-feed part of the process brings together all the reinforcing materials, impregnates them with resin and precisely guides them into the front end of the heated die. A development, the pull-winding process, employs a modification of this in-feed section and combines the techniques of pultrusion and filament winding.[26] Essentially the process consists of pultrusion equipment within which a rotating winding head is located (Fig. 6.16). There are a number of options available, but the general principle is to combine unidirectional fibres with fibres wound around them to give, essentially, combinations of 0° (unidirectional) and 90° (hoop) fibres in layers. Wind angles can be varied within the range 90° to 45°.

Another variation on the filament winding theme employs a pressure bag technique.[27] Continuous reinforced materials, in the form of prepreg, are wound on to a tubular elastic bag. This in turn is enclosed in a mould of the required section. This may be cylindrical but could also be of general prismatic form. On heating the resin, viscosity reduces and at the requisite time the bag is inflated, forcing the composite to unwind and be forced against the surface of

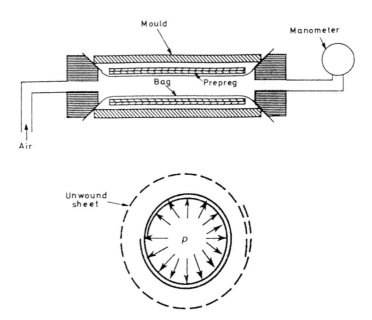

6.17   Schematic of pressure bag winding technique.

the mould. Figure 6.17 shows the arrangement.[28] The final thickness of the component depends on the degree of initial overlap in the prepreg winding.

Hand layup processes

Perhaps the oldest method of fabrication of structural composites is the hand layup or contact moulding technique. In this process components are normally produced on a male mould coated with a suitable release agent which is usually liquid poly(vinyl acetate) (PVA) or liquid wax, although poly(tetrafluoroethane) (PTFE)/silicone sprays or polyester films are sometimes used. The first step in the laminating process is the provision of a gel coat. This resin layer can be filled or pigmented, or reinforced with a glass tissue or synthetic veil. Subsequent layers are completed by applying resin to the surface followed by reinforcement. This sequence whereby resin is drawn through the reinforcement assists in wet out and the removal of air. Consolidation is carried out using brushes, rollers and 'squeegees' (Fig. 6.18).[29] Normally resin systems are unfilled, although thixotropic agents may be added to assist viscosity control. This can be important in the lamination of vertical surfaces where resin drainage after application can occur. Airless sprays or pressurized rollers may also be used to apply the resin. In most cases the curing reaction is an exothermic process and it is essential that laminating

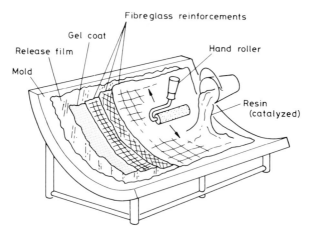

6.18   Hand layup process.

sequences are designed so as to limit the amount of heat generated during cure, otherwise damage may occur within the material. On completion, the component is removed from the mould and other operations, e.g. post-curing and, fitting of attachments, may be carried out as required.

Each laminate can contain layers of several different types of reinforcement, ranging from random mat to highly directional woven cloths. One of the features of the process is that the type and arrangement of the layers can be varied freely to suit particular requirements. The principal advantages of hand-lay processes are versatility – there is little limitation on the size and complexity of the moulding – and low capital cost, the only significant investment being that of tooling. For low volumes the tool material can be wood, GRP mouldings or even thermoplastic fabrications. Where the structure is large or where significant numbers are to be manufactured, tools of steel or aluminium are often used. The detailed design of the tool is a key element to the success of the moulding. Good external finish is necessary as the laminate will provide a reflection of this surface, adequate rigidity is required to maintain dimensional tolerance during fabrication and cure, and generous radii of curvature at corners must be provided so that laminates can be applied such that they do not 'lift off' after placement.

The major disadvantages of the technique are low fibre volume fractions and hence low mechanical properties. Typical values are 20–35% depending on the type of reinforcement. The random mats tend to be at the lower end of this range. Also, being a labour-intensive technique, provision must be made in the design for the greater degree of variability which occurs compared with an automated process. For components where structural integrity is a key issue, there are Codes of Practice which outline in considerable detail the layup procedure. Process equipment is an example where such documentation

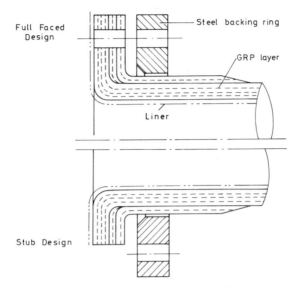

6.19   Laminated flange construction.

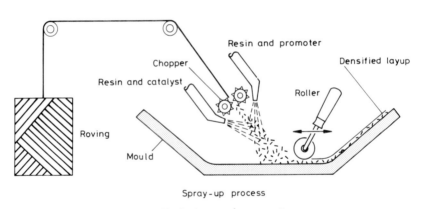

6.20   Typical spray layup system.

is available. Figure 6.19 shows an example of a layup sequence; in this case, the laminating details for a flange connection.[30] Both the thickness, and therefore number of layers, and the direction of reinforcement lay are specified.

Spray-up methods are usually classified in the same category as contact moulding. Here, chopped fibres and resin are deposited on to an open mould (Fig. 6.20).[29] The rovings are fed through a chopper and blown into a pre-cataiysed resin stream. A feature of the process is the very rapid rate at which material can be deposited, although this is in part offset by the low reinforcement volumes which are achieved. Usually after initial deposition the laminate is consolidated using hand rolling methods as per hand layup. Final product quality and thickness are very sensitive to operator technique and measures such as spray gun calibration and on-line measurement of glass/fibre utilization can be employed to effect a degree of control.

Moulding processes

In the first section (continuous reinforcement processes) some mechanical processes were described which provide a means of high volume production, but are generally limited to continuous reinforcement and prismatic sections. Whilst progress has been made in extending the range of these techniques to components of greater complexity, these developments have yet to have a commercial impact. The contact moulding process, on the other hand, offers a means of manufacturing large, complex, fabrications at relatively low capital cost, but using methods that are labour-intensive and therefore difficult to control in terms of quality and reproducibility.

Press moulding operations of one form or other where components are formed under pressure in a heated environment are the most widely used processes for the production of reinforced plastic components. There are many variants on the technique but they all entail the same steps:

- The basic materials, usually in the form of prepregs, are cut to shape and laid in the required orientations within a mould cavity.
- The mould is 'closed' and combinations of pressure and temperature are applied in accordance with curing schedules defined for the resin system of concern.
- The part/mould system is cooled to allow ejection of part and subsequent processing operations.

The basic techniques can be described as follows.

Matched-die moulding

Matched-die moulding involves the use of matching steel male and female dies that close to form a cavity of the shape of the component; Fig. 6.21 shows a schematic of the arrangement. The dies can be internally heated, if required, by electric elements or steam pipes. The fibre layers are placed over the lower mould section and the two halves of the mould are brought together in a press. The thickness of the part is usually controlled by lands built into the mould. Advantages of matched-die moulding are very high dimensional control, good surface finish produced on both surfaces, and reasonable production rates. However, the cost of the matching dies, which for large numbers may require hardened faces, is very high. Also, the size of the available hydraulic presses used to apply the closing pressure provides an overall limit on the size of parts

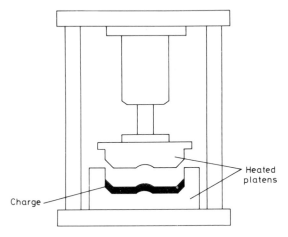

6.21   Simple press moulding arrangement.

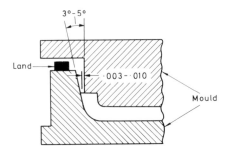

6.22   Example of detail of mould design.

that can be produced. One problem with the process is that fibres tend to move from their preset position as the mould is closed and fibre disposition in large component sections can be difficult to control. Adjusting the pressing speed on closure can be helpful in this respect. Gaps or vent holes need to be left to permit the escape of excess resin. Figure 6.22 shows example detail of mould design.[29]

Wet laminating procedures may be used, in which case the dry fibre is laid in the mould and the resin added subsequently. If random fibre mat is used, quite complex shapes can be produced and these can be preformed by spraying the fibres onto a male mandrel prior to placing the preform in the mould. The complexity of component form is limited, however, particularly with long or continuous fibre reinforcements. Bosses, ribs, shape changes in thickness and inserts are difficult. The process is ideally suited to planar/shell structures which require uniformity of section and properties. Cycle times are controlled primarily by the charging of the mould. Preheating the mould as an option is

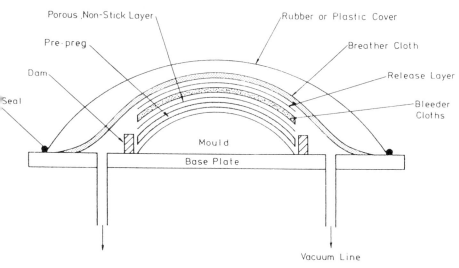

Porous ,Non-Stick Layer

Rubber or Plastic Cover

Pre-preg

Breather Cloth

Dam

Release Layer

Seal

Bleeder Cloths

Mould

Base Plate

Vacuum Line

6.23   Vacuum bag moulding assembly.

limited as pre-cure of the composite during filling must be prevented. Part removal is accomplished by compressed air or ejector mechanisms fitted in the tool.

Vacuum bagging

Bagging techniques have been developed for fabricating a variety of components and structures. Complex shapes, including double contours, and relatively large parts can be readily handled. The process is principally suited for those cases where higher pressure moulding cannot be used. The procedure involves the use of a flexible plastic membrane that is moulded over the surface of the layup to form a vacuum-tight bag. In vacuum bagging, the bag is simply evacuated and atmospheric pressure used to consolidate the layup against the surface of the mould. Tooling is less expensive than for matched-die moulding. In this process the tools must be able to withstand curing conditions without distortion and handling, but are not subjected to large pressure differentials. Release agents for bag-moulded composites include wax, aerosols containing wax or fluorocarbon and silicone sprays. In addition to the composite laminates and flexible bag, the other components in the system (Fig. 6.23), are:

- Bleeder cloths: a porous fabric to absorb excess resin as the consolidation occurs.
- Release layers: a porous release fabric will be placed between the laminated and bleeder cloths and a non-porous fabric.
- Breather cloth: a perforated nylon or PTFE-coated scrim to allow escape of volatiles.

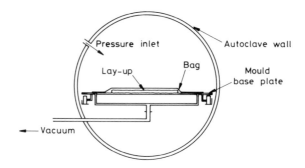

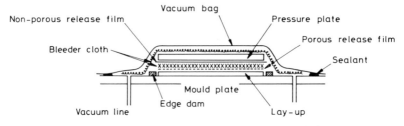

6.24   Autoclave layout and typical bag assembly for a
flat panel moulding.

Vacuum bagging is an inexpensive and versatile procedure. However, it can provide only limited consolidation pressure and may produce voided laminates due to the enlargement of the bubbles trapped in the resin in regions of low pressure.

Autoclave moulding

Autoclave moulding is similar to the vacuum bag process, except that the layup is subjected to greater pressures and denser parts are produced. Owing to the use of an autoclave, component dimensions can be somewhat limited compared with the vacuum bag technique. The bagged layup is cured in an autoclave by the simultaneous application of heat and pressure. Most autoclave processes also use vacuum to assist in the removal of trapped air or other volatiles (Fig. 6.24). The vacuum and autoclave pressure cycles are adjusted to permit maximum removal of air without incurring an excessive resin flow. Vacuum is usually applied only in the initial stages of the curing cycle, while autoclave pressure is maintained during the entire heating and cooling cycles. Curing pressures are normally in a range of 50–200 psi (3–12 MPa). Compared with vacuum bag moulding, this process yields laminates with closer control of thickness and lower void content.

The heat transfer design of autoclaves is not straightforward. Both

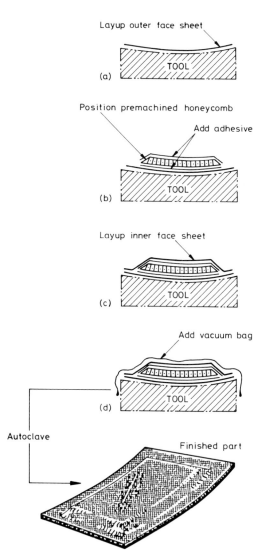

Layup outer face sheet

TOOL

(a)

Position premachined honeycomb

Add adhesive

TOOL

(b)

Layup inner face sheet

TOOL

(c)

Add vacuum bag

TOOL

(d)

Autoclave

Finished part

6.25   Steps in the manufacture of a honeycomb panel
by co-curing.

convective and radiative heat transfer mechanisms occur and thermal resistance is a function of both the mould and the layup itself. Cycle times, certainly in the heating phase of the process, can be dominated by this latter effect.[31] An advantage of autoclave type techniques is that components which may require secondary bonding can be co-cured in a single operation. Figure 6.25 shows the possible steps in the fabrication of a honeycomb panel in a single step.

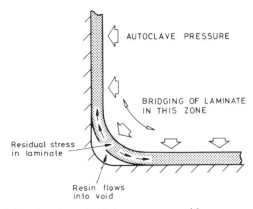

6.26   Laminate bridging across mould corner.

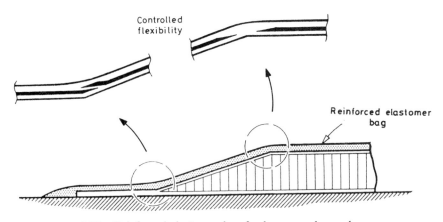

6.27   Reinforced elastomer bag for honeycomb panel.

A key factor in autoclave moulding is the behaviour of the bag while the consolidation and curing processes are underway. Although the external pressure on the bag is uniform, that which is applied to the component can vary. There are a number of reasons for this, but the two most significant are insufficient elongation to allow the bag to conform to the mould shape and problems with the breather ply which could prevent the development of the full pressure onto the surface of the component. One particular problem is bridging across mould corners (Fig. 6.26).[32] Here the normal pressure transmitted to the laminate prevents inplane slippage of plies and the laminate is not in contact with the mould in the corner zone. A solution to this problem is to use a reinforced elastomer as the bag material, the disposition of reinforcement being determined by the special requirements of each area of the moulding. For example, in the honeycomb panel in Fig. 6.25 the tailored

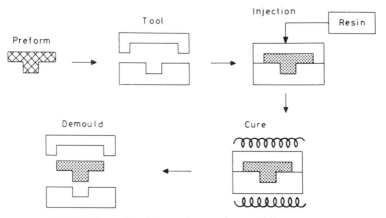

6.28   Schematic of the resin transfer moulding process.

stiffness as provided in the bag shown in Fig. 6.27 will help to consolidate materials at the change in geometry.

Resin injection processes

The moulding processes described so far begin with an open mould or tool which is subsequently closed to surround, heat and compress the charge of fibre and resin. In terms of production methods, they are highly discontinuous and there is usually significant manual intervention in each production cycle. Robotic placement and automatic metering of resin represent ways forward to provide an element of automation, but this does not address the basic drawbacks, in terms of production rate, of filling an open mould. Injection processes, on the other hand, offer potential in terms of overcoming some of the inherent difficulties.

Resin-transfer moulding

Resin-transfer moulding (RTM) is a technique whereby a fibre preform is inserted into the mould which is then closed and resin is mixed and injected. RTM is shown schematically in Fig. 6.28.[33] During the impregnation stages the advancing resin front expels the air and the resin-impregnated reinforcement is allowed to cure prior to removing the part. The low injection pressures used for RTM, typically less than 7 bar, permit the use of low stiffness shell moulds which are a low cost, lightweight alternative to the conventional machined metal moulds used for matched tool pressing techniques.[34] Component integration can be achieved including the manufacture of stiff double-skinned components[35] and the reduction in the corresponding

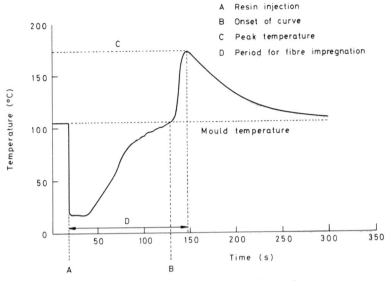

6.29   Temperature profile of a RTM cycle.

number of fabrication operations can lead to further savings in overall costs. Reinforcements can take a number of forms including random mats, woven roving and unidirectional material. There is also potential for net-shape manufacture, eliminating expensive trimming and post-forming operations.

Resin transfer is accomplished using a reciprocating pump or a simple pressure pot in which a gas under pressure, usually air, is applied directly on to the pre-mixed resin in the storage tank. Where necessary, for example with two-part resin systems, mixing prior to injection, usually via a static mixer, is carried out. Surface finish and mechanical properties can be improved if the cavity is evacuated and the resulting vacuum used to draw pre-mixed resin into the mould.[36]

Figure 6.29 shows a typical temperature profile with an RTM moulding cycle.[37] The key events such as resin injection, cure and exotherm can be observed. The details of the temperature profile can be shown to be dependent on resin formulation, preheating and the thermal capacity and conductivity of the tool. Higher thermal conductivity metal tools result in lower peak exotherm temperature and low thermal conductivity GRP materials cause greater temperature variations throughout the mould. In addition, a GRP mould requires a longer time to reach the preset temperature following impregnation and cure which can result in slower reaction rates.

Control of the process to increase production rate is complex because of the number of interrelated parameters. For example, preheating the resin prior to injection could reduce cycle time, but may require a reduction in mould temperature to prevent premature cure during the resin injection phase. It is

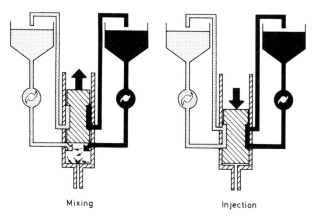

Mixing   Injection

6.30   Principle of reaction injection moulding.

probable that the thermal history of the mould gate has a controlling influence on the mould cycle time and the time available for fibre wetting.

Tooling must be designed to minimize deflection and this is governed by applied pressures.[34] The pressure cycle for RTM comprises three distinct phases: impregnation, hydrostatic and cure. The impregnation phase influences the structural design of the mould in addition to affecting the subsequent equilibrium hydrostatic phase. The cure phase consists of pressure peaks before and during exotherm which are caused by thermal expansion effects of resin and laminate. These, however, are found not to be influential in mould design.

Reaction injection moulding

Reaction injection moulding (RIM) and reinforced reaction injection moulding (RRIM) also use relatively low pressures (approximately 0.5–1 MPa) to produce large mouldings, principally in polyurethane elastomers. The principal difference from resin-transfer moulding is that, instead of using a pre-catalysed resin with a relatively slow cure, the RIM process brings two fast-reacting components together and mixes them just prior to injection into the mould. The mixed system can be tailored to cure in the mould within 30–60 s, thus giving rise to component cycle times of the order 1–2 min. The basis of the process is illustrated in Fig. 6.30.[38] The equipment comprises two reservoirs storing, typically, a diisocyanate and a polyol, which polymerize when intimately mixed. During the mixing/injection stage the two reactants are pumped through an impingement chamber in the mixing head, and thereafter to the mould cavity. The impingement chamber produces highly turbulent flow conditions and hence rapid mixing of the two constituents, and this is followed by rapid gelation and cure in the mould. In the second stage, all the mixed polymer is ejected from the mixing head and the two reactants are separated and continuously recirculated ready for the next moulding cycle. As a result, the mixing head is self-cleaning and any cured polymer in the connecting pipe from the mixing head to the mould cavity is removed as a piece of flash attached to the moulding.

A variety of urethane formulations is available to give polymers with a range of hardness from soft and rubbery to hard and brittle. The mouldings can be reinforced to some degree with short glass fibres by including them in one or both of the liquid reactants to form a slurry. The limitation on fibre length and hence reinforcing efficiency, and the quantity of fibre that can be incorporated, is set by the design of the mixing head and its ability to handle the material without risk of blockage or of poor flow of the fibres within the moulding resulting in a non-uniform distribution of reinforcement.

Integrated manufacturing systems

A drawback of many moulding processes is the number of stages necessary in part manufacture. Whilst the consolidation and curing sections of the process can be automated, the preliminary operations concerned with material handling are much more difficult to mechanize. Integrated manufacturing systems for the production of comparatively small batch sizes of structural components and which employ robotic devices for handling and layup of prepreg type material within a dedicated work cell provide scope for the development for faster rate fabrication systems.[39] The types of assembly operation which could be incorporated within such systems include:

- Cutting sheet to predetermined shape. This would be designed to minimize wastage through scrap.
- Layup within the tool.
- Post-machining.
- Non-destructive testing.

Each of these poses its own particular problem.[40] For the layup procedure, pre-cut pieces of material need to be laid accurately on the mould tool with precise butting of adjacent sections. Because of curvatures and general complex geometries there will be a variety of shapes and sizes to be handled and this, combined with the flexibility and variation of tack of the material, means that the design of grippers for placement devices needs to be carefully designed. Ideally the laying machine must be capable of a number of functions among which include de-reeling of material, removal of backing paper and any cover film, inspection for faults and cutting and trimming to shape. Figure 6.31 shows an outline of a conceptual plant where this type of technology for handling materials is incorporated in a plant layout – in this case for a helicopter spar blade.[41] It consists of three main elements:

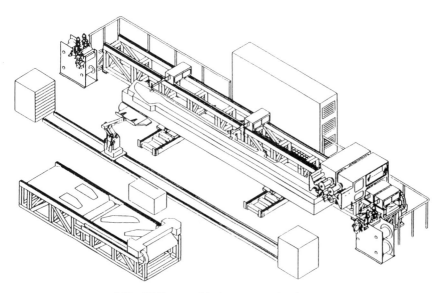

6.31   Helicopter blade spar production cell.

- A profiling and tape placement machine for full length plies.
- An X–Y profiling machine with an ultrasonic knife head to cut out plies for infill packs.
- A transfer system to present and lay-in infill packs to the tool on demand in a predetermined sequence.

Computer-aided design, computer-aided manufacturing (CADCAM) techniques are linked to the production unit so that profiles for new designs and machine control instructions can be modified to suit particular requirements. The system can cope with any blade design which fits within its operating envelope and can readily be extended to accommodate a wide range of materials.

Processing of thermoplastic composites

Most methods for manufacturing composites can be used to a greater or lesser extent for both thermoset and thermoplastic matrix systems. Figure 6.32, for example, shows some filament wound thermoplastic components. Implementation of a given method, however, is quite different for the two types of material system. Curing of thermosets entails chemical change and usually results in the conversion of a viscous liquid to a solid phase. Even for prepregs the temperature achieved during cure is sufficient to permit resin flow. Thermoplastics processing, on the other hand, involves physical transformations and these are effected by melting and freezing operations. Matrix flow

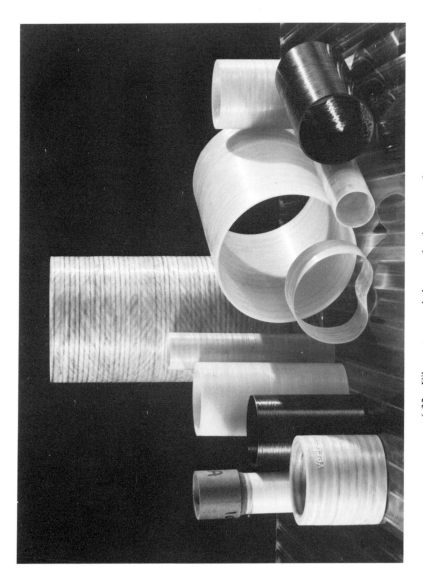

6.32   Filament wound thermoplastics composites.

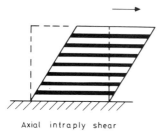
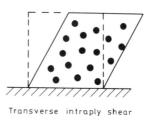

Axial intraply shear   Transverse intraply shear

6.33   Shearing deformation in thermoplastics
composites.

does occur but this is governed by the principles of polymer rheology and as a
result there are fundamental differences in the way the materials behave during
the production cycle.

Thermoplastic composites for structural applications need to be processed
at relatively high temperatures and thermal stability can set an upper limit to a
number of parameters which control production rate. Degradation can be
accelerated in the presence of oxygen and water and in these cases an inert
atmosphere or a vacuum needs to be used. As with cure in thermosets, the
equivalent in thermoplastics – cooling to achieve solidification – is often an
important factor in determining production rate.

When first reviewing the application of these materials initial consideration
can focus on processes which are commonly used for thermoplastics where
ease of forming is a noted advantage of the materials. In fibre reinforced
systems, particularly in continuous systems, matrix deformation is highly
constrained and limited to shear mechanisms.[42] Two such mechanisms are
shown in Fig. 6.33. As well as the intraply shear deformation shown, it is also
possible for plies to move with respect to one another in the interlaminar sense.
Consideration of the means by which deformation occurs during forming is
important as the analysis of the rheology of the material can be used to
establish limits for process conditions. For example, if the matrix viscosity is
too low and the timescales long, shaping pressures may cause resin to be
squeezed out of the laminate, whereas too high a viscosity coupled with a short
duration can result in the fibres being unable to adjust to the flow pattern and
they may then become distorted (Fig. 6.34).

Another property which is sensitive to processing history is crystallization
behaviour. Rates of crystallization can be different from those for unreinforced
matrix because of possible constraining effects of the fibres, but as with neat
resin, they are dependent on cooling rate. This leads to the possibility of
crystallinity gradients within the moulding. As yet these effects are only found
to be of secondary importance in terms of mechanical property behaviour.[43]

Overall, the position with the processing of structural thermoplastics
composites continues to be developmental. There are numerous examples of

Resin migration   Fibre kinking   Interply slip

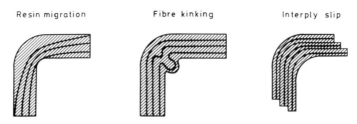

6.34   Distorsion of fibre distribution on forming.

successful application but further work is necessary to demonstrate that their full potential can be fulfilled.

Manufacturing defects

Defects and flaws are present in any engineering material, but with composites there are a number of issues that do not feature when considering systems that are homogeneous and isotropic. Perhaps the most significant form of manufacturing defect which can be encountered with composites is misalignment of the reinforcing fibre. For unidirectional material a variation in orientation of, say, 10° can result in a significant reduction in strength – up to a factor of 0.5 in extreme cases (see Chapter 3). A typical laminate will consist of many thousands of individual filaments and it is not possible for each to be fully tensioned and aligned. Some deviation is inevitable and acknowledgement of this must be made during design. Misalignment also occurs in systems made from woven cloth or preimpregnated fibre sheet. In the latter case it can be due to problems in individual laminae or when the layers are stacked to give a laminate. Also a factor, and one more difficult to control, is the movement of fibres that may occur in fabrication when, for example, a dry fibre preform is subjected to handling, moulding pressures and the effects of resin impregnation.

Voids are always present in fibre reinforced composites to a certain extent. They are due to air trapped between fibres, the presence of solvents or other volatiles or incorporated into the resin during mixing. Figure 6.35 shows a cross-section indicating void content in a typical laminate, in this case a filament wound GRP laminate. Volume void fractions can range between from less than 1% up to 5%. Voids may act as stress raisers which can result in lower transverse, flexural, shear and compression strengths, corrosion resistance and electrical properties. De-aerating the resin is one option to minimize air entrainment prior to processing, but this is only possible where quantities of material to be handled are small and the resin itself is not affected by the effect of vacuum, e.g. the preferential loss of one constituent. The key to control in a process is regular monitoring of the void content within a quality

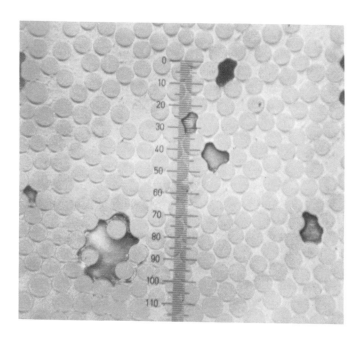

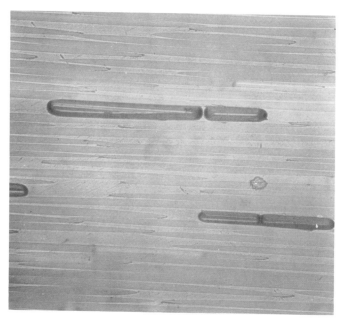

6.35   Voids in GRP laminate.

programme so that those aspects of the production method that influence voidage can be identified.

Another factor which requires careful attention is the achievement of an adequate bond between fibre and matrix. Without effective load transfer between constituents the material will not deliver its anticipated range of properties. Considerable effort has been devoted to the optimization of coupling objects for bonding purposes but this will be of limited value if there is surface contamination, moisture, excessive temperature or insufficient resin for fibre wetting. Conditions of raw material storage can be particularly important in this respect. As well as providing a mechanism for bonding, surface coatings are also used to provide a measure of protection to fibre surfaces. Despite this, tensile strengths can be significantly reduced through mechanical handling damage. The level and frequency of mechanical processing should be a minimum conducive to the requirements of the manufacture technique.

Composite design usually assumes a given fibre loading and that this is uniformly distributed throughout the section. During the initial stages of forming a composite the volume fraction of fibres can be below the values required and temperature to increase the flow properties of the resin, and pressure, are needed to increase the amount of fibre relative to the matrix. In vacuum bag and autoclave moulding bleeder plies are often used around the preimpregnated fibre to absorb any excess resin which is squeezed from the body of the laminate. If errors do occur, for example if gelling occurs prematurely restricting the levels of consolidation possible, both properties and dimensional control could be adversely affected. A non-uniform distribution of reinforcement could have significant influence on bending properties, especially where variation with respect to a neutral axis could reduce rigidity. Also, formation of 'weld lines' within a material could provide local areas of reduced stiffness and strength.

Temperature effects arising through curing must feature when considering the potential for manufacture-induced flaws. Many resin systems experience exotherm if laminate sections are too thick or curing temperatures too high. In addition to discoloration, local embrittlement and disbonds may occur. The dissimilar thermal expansion of constituents and lamina layers can also provide a source of difficulty. Stresses under such circumstances (see Chapter 5) can be of sufficient magnitude to generate internal cracking and problems of dimensional tolerance. In the latter case distortion and dimensional changes can be large. Hand lamination of thick flat plates, for example, can be particularly troublesome. Resin shrinkage on cure can result in significant bowing and a solution commonly adopted is to apply successive laminate layers on opposite faces. Whilst this tends to compensate for distortion it also sets up a distribution of residual stresses in the section. Figure 6.36 shows a similar example, in this case a 'T' beam stiffened panel.[44] This effect is due to

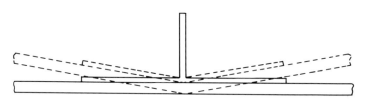

6.36   Thermal distortion of a 'T' beam stiffened panel.

the spring forward phenomenon, so-called as it causes an angle section moulded on a 90° tool to distort such that the included angle is less than the 90° intended. This distortion occurs as a result of the large difference in thermal expansion coefficients in the plane of the material and through its thickness.[45]

Clearly an ideal design/manufacture arrangement would be one where flaws were minimized or, at least, fully accounted for in the design analysis. In practical circumstances this ideal is rarely achievable and the uncertainties this generates should be reflected in the choice of a value for design margin. Of necessity, it is likely that factors to account for this type of variability will be higher for composites than for their metal counterparts. Coupled with this there must be a set of acceptance criteria against which a component can be inspected. This would need to address each of the types of defect likely to be encountered and provide guidance on the maximum level, e.g. size, distribution, position, which can be sustained in service. Each component and application will require its own set of criteria – (see Table 6.2 for allowable defects in chemical process plant),[30,46] and these will be derived from a consideration of technical and commercial requirements together with experience.

Machining

Composite materials can be more difficult to machine than metals because they are anisotropic, non-homogeneous and their reinforcing fibres tend to be abrasive. During machining, defects are introduced into the workpiece, and tools wear rapidly. Traditional machining techniques such as drilling or sawing can be used with modified tool design and operating conditions,[47] as can some of the more sophisticated processes such as laser and ultrasonic machining and electro discharge techniques.[48]

Drilling and machining processes

Drilling is the most widely employed composite machining technique, as numerous holes must be drilled in order to install mechanical fasteners. To

Table 6.2. Permissible limits for hand laminate defects

Defect	Inner (process) surface (thermoset linings only)	Outer (non-process) surface
Blisters	None	Maximum 6 mm diameter, 1.5 mm high
Chips	None	Maximum 6 mm, provided it does not penetrate the reinforcing laminate
Cracks	None	None
Crazing	None	Slight
Dry spots	None	Maximum 10 per m^2 with total not greater than 100 mm^2 in area
Entrapped air	None at surface. If in laminate, not greater than 1.5 mm diameter and not more than 2/100 mm^2	3 mm diameter maximum; no more than 3% of area
Exposed glass	None	None
Exposed cut edges	None	None
Foreign matter	None	None if it affects the properties of the laminate
Pits	Maximum 3 mm diameter, 0.5 mm deep, number shall not exceed $1/10^4$ mm^2	Maximum 3 mm diameter, 1.5 mm deep
Scores	Maximum 0.2 mm deep	Maximum 0.5 mm deep
Surface porosity	None	None
Wrinkles	Maximum deviation 20% of wall thickness but not exceeding 3 mm	Maximum deviation 20% of wall thickness but not exceeding 4.5 mm
Sharp discontinuity	Maximum 0.5 mm	Maximum 1 mm

machine fibre reinforced composites, the material from which the tool is made must be carefully selected. As glass and carbon fibres tend to be abrasive drill bits made from conventional high speed steel can fail after just a few holes. Tungsten carbide tools possess extended life particularly when they are coated with polycrystalline diamond (PCD).[49] Typical values of cutting speeds and feed rates are shown in Table 6.3.[47] An additional consideration when selecting drill material is whether or not the composite is bonded to a metal component. In these cases the drilling arrangement must be suitable for both materials. Aramid composites can also pose special problems due to the nature of the fibre which can fray and protrude from the surface, resulting in a poor quality hole.

Several types of damage can be introduced during drilling operations: matrix cracking, fibre pullout and fuzzing, interlaminar cracks and delamination, in addition to geometrical defects commonly found in metal drilling. Delamination effects can be particularly deleterious and there are a number of mechanisms by which these can arise. The two most significant are delaminations at the top layer through peeling of the laminate, and high thermal stresses and delaminations near the exit side, produced when the drill acts as a

Table 6.3. Drilling data for composite materials

Material	Tool material	Hole diameter (mm)	Material thickness (mm)	Cutting speed (m/min)	Feed rate (mm/rev)
Unidirectional CFRP	Carbide	5–8	0–13	43	0.03–0.05
	PCD	5–8	13–19	34	0.03
			0–13	61	0.05–0.09
			13–19	52	0.05–0.09
Multidirectional CFRP	Carbide	5–8	0–13	61	0.03–0.05
	PCD	5–8	13–19	43	0.03
			0–13	69	0.05–0.09
			13–19	61	0.05–0.09
Glass–epoxy GRP	High strength steel	3	10	33	0.05
Carbon–epoxy CFRP	Carbide	3	10	33	0.05
Boron–epoxy	PCD	6	2	91–182	25 mm/min
	PCD	6	25	91–182	25 mm/min
Kevlar–epoxy	Carbide	5.6	–	158	0.05

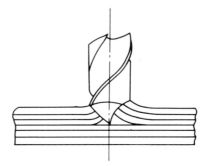

Peeling of top surface laminate

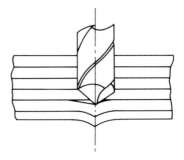

Separation of bottom surface laminate

6.37   Types of delamination damage with drilling.

punch forcing the thin, uncut layer from the main body of the laminate. Figure 6.37 shows typical types of delamination.[50] Delaminations can be minimized by reducing feed rates near the exit and using backup plates to provide support and to prevent deformations leading to exit side separations.

Much of the energy input during machining, e.g. turning, routing or grinding, is expended in the form of heat. This is manifested as temperature rises in the workpiece, tool and the material which is removed. With glass reinforced systems the heat balance is dominated by the thermal conductivity of the tool and this determines cutting temperatures and wear rates.[51] The nature of the cutting process is essentially one of fracture as opposed to plastic deformation as is the case with metals. The details tend to be governed by material and fibre orientation.

Lasers and waterjet cutting processes

Lasers provide an alternative to machining composite materials. A high energy infrared beam is focused on a very small spot, usually 0.1–1.0 mm in diameter, and causes melting, vaporization or chemical degradation throughout the depth of the material. Fluids and degradation products are removed by

Table 6.4. Laser cutting of composite materials

Material	Thickness (mm)	Cutting speed (mm/s)	Power (W)
GRP	3.2	5	250
	1.6	250	1 200
	1.6	87	450
Boron–epoxy	8.0	27	15 000
Aramid–epoxy	2	16–133	500
CFRP	0.5	38	400

a gas jet coaxial with the laser beam. Two types of laser are typically used; either a solid state Nd : YAG device or a CO_2 gas laser.[48]

In terms of quality of cut it has been shown that the best results are obtained when the thermal properties of the reinforcing fibres are closest to those of the matrix.[52,53] As a result aramids tend to be cut most easily, followed by glass and carbon. The influence of fibre orientation is most pronounced with carbon reinforced materials where the heat is transferred away from the cutting point in the fibre direction. The details of a laser cut surface show a number of common characteristics.[54] Initially there is a charred layer adjacent to which is a zone where fibres protrude from the matrix. The extent of local degradation can be considered analogous to the heat-affected zone in metals. Additionally, there may be some surface degradation due to heating from hot gases generated in the process. The transverse cutting speed which is achievable for a given system is dependent on laser details such as power and spot diameter and laminate parameters, for example, thickness and material properties. Table 6.4 shows typical speeds for laser cutting of composites.

Waterjet cutting can be seen as an alternative to laser techniques. The principle is to produce a thin waterjet with very high pressures and high velocities, and, upon impact, material is removed by localized shearing. Pressures up to 400 MPa are used, and waterjet diameters are in the 0.08–0.5 mm range. The performance is significantly improved when abrasive particles are added. Figure 6.38 shows a typical arrangement for an abrasive waterjet nozzle.[55] Some of the advantages of waterjet cutting include high cutting speeds, the absence of a heat-affected zone compared with traditional or laser machining techniques, and the elimination of dust by jet entrainment.[56] Several drawbacks of waterjet machining have been identified. At high cutting speeds, delaminations are introduced into the workpiece, and with composites containing aramid fibres fraying can occur, and in some cases, moisture absorption during waterjet cutting can lead to delamination under load.[57] In addition to cutting, variants of the waterjet method can be used for processes such as drilling, milling and turning.

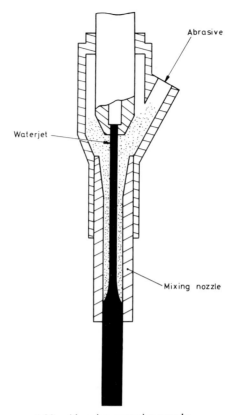

6.38   Abrasive waterjet nozzle.

Electro machining processes

For electrically conductive materials, electro-discharge machining (EDM) is possible for intricate and complex shapes. Here the workpiece is immersed in a dielectric fluid and material is removed by erosion caused by electrical discharge between the electrode and the workpiece. Due to the requirement of material conductivity the process is limited to carbon fibre reinforced polymers and metal matrix systems.[58] In the former case care must be taken to ensure temperatures are not excessive since, if they are, the material local to the cutting surface will be degraded. Figure 6.39 shows an outline of EDM equipment.[59]

An alternative which may be employed with non-conductive materials is electro-chemical spark machining.[60] In this process, two electrodes are placed in a tank filled with an electrolyte. The tool is an electrode as in EDM and when a DC current is applied, hydrogen gas bubbles form at the surface of the cathode and sparking occurs across these bubbles but not between electrodes.

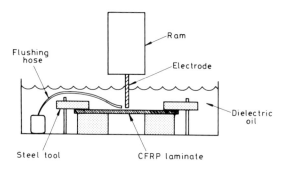

6.39   Outline of EDM equipment.

The electrical discharge generates large amounts of heat. The workpiece is held close to the electrode so that material removal can take place in the region where sparking occurs. The mechanisms for material removal are melting and vaporization.

Ultrasonic machining

Ultrasonic machining is a process in which the tool is vibrated at high frequencies.[61] Interactions between abrasive particles contained in a liquid slurry and the workpiece remove material by erosion as the tool vibrates. Tool displacements are small, typically a few thousandths of an inch at frequencies of the order of 20 kHz. This technique offers the potential for the production of complex geometrical forms not feasible with more conventional processes.

Non-destructive testing

Non-destructive inspection of fibre reinforced composite materials differs from that of metallic materials for two reasons. Firstly, physical properties such as thermal conductivity, acoustic attenuation, electrical resistivity and elastic behaviour are significantly different. This presents difficulties with the applicability of the physics of some of the techniques. Secondly, and more fundamentally, metallic structures are fabricated from feedstock, e.g. plate, bar and section, of known property, composition and quality. Subsequent inspection usually focuses towards joints, particularly welded connections. Composites on the other hand are heterogeneous and the material of construction and the component are formed in the one operation. Inspection therefore must be concerned with the whole surface of the structure; a different scale of problem altogether. In addition the range of flaws which may need to be detected is much larger and may include:

- State of cure.
- Volume fraction.
- Orientation of fibres/layup.
- Fibre matrix/interface condition.
- Interlaminar cracks.
- Inclusion.
- Strength of bonded joints.

As a result the NDT of composites is more complex than for metals both in terms of method of application and interpretation of results.

Visual inspection

For translucent GRP composites visual inspection methods can be the most useful of the available techniques. Where there is access to both surfaces, use of a strong light source and observing transmission through the laminate thickness can be particularly effective. Porosity, poor impregnation, delamination and inclusions can all be detected as well as surface flaws such as cracking and gel coat blisters.

Ultrasonic methods

For structural composites as a whole, ultrasonic inspection methods are probably the most widely used form of NDT. An ultrasonic pulse is propagated through the specimen and the reflection, either from the back surface or from a defect, is received by a transducer. Ultrasonic waves in composites are highly attenuated due to the heterogeneous nature of the materials and response is strongly dependent on frequency; a typical range being 1–10 MHz. Generally a higher level of amplification is required compared with other materials. Information regarding the defect can be obtained from amplitude and transit time of reflections. For small components testing can be carried out in a water bath (Fig. 6.40), although coupling agents can also be used. Figure 6.41 shows the results of ultrasonic testing of thick-walled CFRP cylinders – approximately 30 mm outside diameter × 10 mm thick. The areas of delamination are those arising from residual stresses generated during fabrication. Considering defects of a smaller scale, ultrasonic attenuation can be correlated with void content (Fig. 6.42).[62]

Measurements of ultrasonic velocity can be useful for the measurement of dynamic modulus and density. For an isotropic material, the relationship between velocity and modulus is given by:

$$\rho v^2 = \frac{E(1 - v)}{(1 + v)(1 - 2v)} \qquad [6.5]$$

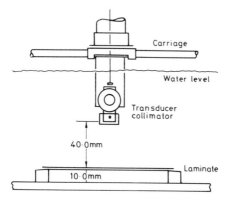

6.40 Ultrasonic scanning arrangement.

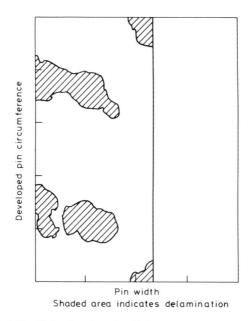

Pin width
Shaded area indicates delamination

6.41 Delaminations in thick-wall CFRP cylinder.

where v is the ultrasonic velocity, ρ the material density and E and v are modulus and Poisson's ratio respectively. The equivalent expression for propagation in the longitudinal direction of a unidirectional composite material is:

$$\rho v^2 = C_{11} \qquad [6.6]$$

where C_{11} is the longitudinal compliance. For off-axis material wave propagation is more complex due to attenuation at fibre and matrix interfaces.

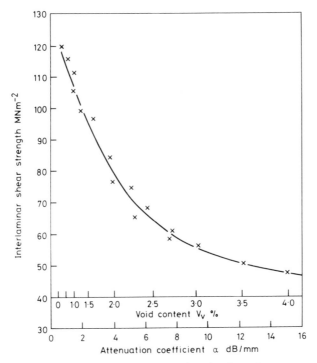

6.42 Relationship between interlaminar shear
strength, void content and attenuation coefficient.

Figure 6.43 shows a comparison of dynamic modulus values measured by ultrasonic and resonant techniques. Results are in reasonable agreement with expectations, the greatest deviation being for $\pm$ 45° to the material's axes.[63] In principle it is possible to correlate velocity measurements to assess issues such as fibre alignment and volume fraction, but the calculations are complex.[64]

Calibration of ultrasonic equipment using a characterized sample which is representative of the material to be tested provides a most effective means of determining deviations from a required specification. Variations in modulus would indicate, for example, possible variation in fibre content or, alternatively, the technique can be used to determine thickness as part of a quality control programme.

Radiography

X-radiography techniques are available and can be used to determine foreign inclusions, interlaminar cracks and voids although problems with low contrast may occur. This can be improved by the use of radio-opaque penetrants such as sulphur, trichlorethylene, carbon tetrachloride or methyliodide. In this way very fine cracks can be resolved. Resolutions can be

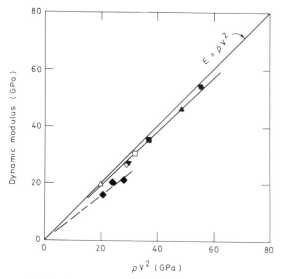

6.43 Comparison of ultrasonic and resonance
modulus measurement methods.

improved by using projection methods resulting in primary enlargements up
to 100 times. For glass reinforcement composites the absorption coefficient of
glass is approximately 20 times that of most resins and so X-ray methods can
be used to discern details of composition.[65] Film density measurements can
also be correlated to fibre content and distribution, orientation and, in the case
of woven materials, the quality of the weave.

Eddy current and microwave methods

For composites where the fibres are conductive, eddy current techniques can
be used to determine resin content and layup geometry. Thickness measurements can also be made through the use of a metal foil bonded to the
non-contact surface of the laminate. Accuracies of 0.1–0.2 mm have been
reported.[65]

Microwaves can be used for locating defects and for measuring thickness
and dielectric properties.[65] For GRP the dielectric constant of glass is much
greater than that for resins and therefore measurement of dielectric properties
of a laminate can yield information concerning fibre content. As dielectric
contact is strongly dependent on chemical composition, microwaves can also
be used to monitor the extent of cure and, because of interaction with free
water, moisture content. To be of significant value in the quantitative sense it is
important that materials to be examined are well characterized as there are
many interacting effects which can affect response. Comparative monitoring
of changes in a structure during service would present a useful application.

Mechanical methods

The relatively low modulus values of composites offers the possibility of NDT through the application of mechanical loads. For example, measuring deflections whilst applying a transverse pressure to an essentially flat component provides a measure of stiffness and if there are substantial areas of delamination, deformations will be greater than anticipated. This is due to reductions in flexural rigidity that arise if transfer of shear loading cannot occur between laminate layers on bending. With complex geometries a deformation signature will need to be taken at the outset of service and this can be used as a reference for subsequent measurements. More subtle changes of modulus occur in composites with the application load because of damage accumulation. However, these are generally small, certainly for loads within an envelope of design allowables, and the technique can only have application for a limited range of laminate types. Of course, any loads applied as part of an NDT test will need to be considered as part of the design.

Measurements of damping characteristics can be used to detect and locate damage through observing changes in natural frequency. The concept is potentially attractive as access is required to only a small area of the surface of the structure in order to impart the load impulse. The range of mechanisms by which energy is dissipated is varied and this means that results must be interpreted with care. Ideally, the experimental technique should be tailored to suit the material, structure and type of damage to be detected.[63] A practical development of the technique is tap testing. The basis of the method is to characterize the frequency spectrum of the component and to use this as a basis for comparative assessments later during service life.[66]

Thermographic methods

Thermographic methods whereby the diffusion of heat over a surface is monitored has a significant advantage over techniques that require the use of a contact transducer, such as ultrasonics. Large areas of a component can be evaluated rapidly with the minimum of preparation. Infrared detection systems are used, with or without subsequent image enhancement, and a temperature resolution of around 0.2 °C is achievable.[63] Heat input is usually provided by an arrangement of powerful flash guns. Thermography is particularly effective in determining the presence of delaminations, say in an aluminium honeycomb structure where heat transfer into the aluminium is prevented by the presence of the flaw (Fig. 6.44). The analysis of carbon fibre reinforced materials can prove problematic due to the comparatively high thermal conductivity and its anisotropy. Practical difficulties are also apparent because of the effects of ambient temperature changes or variability of surface emissivity.

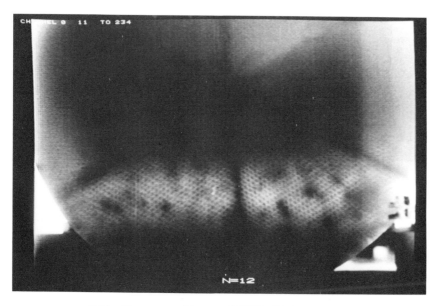

6.44 Thermographic test of CFRP/Al honeycomb
structure showing areas of delamination.

A second thermographic technique makes use of stress generated thermal
fields which arise as a result of cyclic loading.[67] Stress fields can be related to
surface thermal patterns and changes can be attributed to the development of
damage. Test frequencies are between 1 and 50 Hz, the value depending on the
material. There are threshold frequencies below which temperature changes
cannot be discerned.

Acoustic emission

With the application of load onto a composite material there are many
mechanisms by which acoustic emissions are generated. Examples are resin
cracking, fibre fracture, debonding and interlaminar cracking. Figure 6.45
shows the generation of acoustic emissions with applied stress for a variety of
glass reinforced materials.[63] A feature of the emissions, which is known as the
Kaiser effect, is that once loaded to a given stress level a material should emit
no further noise on unloading or on subsequent loading to the previous
maximum load.[68] In assessing structural response, failure of the component to
follow this rule could be an indication of damage. This analysis can be
quantified through the ratio of stress level at which emissions begin again to
that which was applied originally. This is often known as the Felicity ratio and
should be greater than a prescribed value which is close to one. Time-
dependent emission can also be a sign of damage. In normal circumstances the

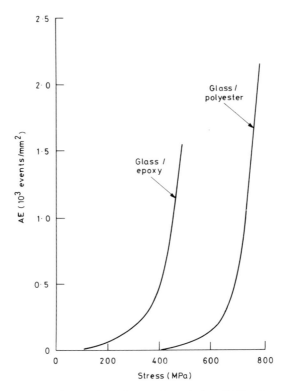

6.45   Generation of acoustic emission (AE) with
applied stress.

emission count falls rapidly when the specimen reaches, and is maintained at, a constant load. A sustained level of noise would indicate continued propagation of cracking. Because of the wide variety of possible sources of emission composites must be regarded as 'noisy' materials. As a result it is vital that a baseline set of data is generated which represents a normal material of acceptable quality. An increased level of output with respect to this data would then be cause for further examination. Figure 6.46 shows output from two pipe specimens; one in good condition and the other containing a defect.[69] The differences can clearly be seen.

Given a set of acoustic emission data it would be useful if the results could be studied in order to obtain information regarding the nature of the mechanisms which are generating the noise. Analysis of the amplitude distributions provides some guidance where peaks in the spectrum can be associated with specific types of event. For example, in GRP examination of data suggests that amplitudes in excess of 70 dB are due to fibre failure whereas those less than this value are due to matrix cracking phenomena.[70] Identification of failure of reinforcement would be important in assessing structural integrity and the

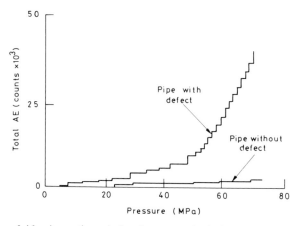

6.46 Acoustic emission for pressurized pipes with and without defects.

onset of failure. However, there remains significant uncertainty over whether or not it is possible to be rigorous in the application of this type of criteria.

It is evident that the use of acoustic emission entails the application of load to the structure of concern. To be of value, therefore, a rationale must be developed whereby loads are minimized and certainly below design values. For GRP process vessels such procedures have been developed[71] and the guidelines are of the following type:

- Total acoustic emission counts should not exceed a specified number.
- No more than a small number of counts should exceed a specified amplitude.
- The Felicity ratio should be close to one.

Testing is carried out using an array of transducers which are placed over the whole surface of the component, but with a concentration in critical areas.

References

1 Gandhi M V, Thompson B S and Fischer F 'Manufacture process driven design methodologies for components fabricated in composite materials', *Composites Manuf*, **1**, 32–40, 1990.
2 Peters S T, Humphries W D and Foral R F, *Filament Winding Composite Structure Fabrication*, SAMPE, Covina, California, 1991.
3 Rosatio D V and Grove C S, *Filament Winding*, Interscience, New York, 1964.
4 Goldsworthy W B, 'Composites fabrication', in *Encyclopedia of Polymer Science and Engineering*, **4**, Wiley-Interscience, New York, 1986.
5 Piola A, 'On site filament winding of 'Jumbo' tank products', 31st Annual Technical Conference, Reinforced Plastics/Composites, SPI, Washington, DC, Paper 23-B, 1976.

6 Short R J and Kozloff A, 'Patented design will enable 5.28 million gallon RP tank to be air, highway, rail and ship transportable', 36th Annual Technical Conference, Reinforced Plastics/Composites, SPI, Washington, DC, Paper 3-D, 1981.

7 Hayes C, 'Filament winding' in *Encyclopedia of Polymer Engineering and Science*, 2nd Edn, Vol 7, p 35, John Wiley and Sons, 1987.

8 Soden P D and Eckold G C, 'Design of GRP pressure vessels', in *Developments in GRP Technology*, Ed B Harris, Applied Science, London, 1983.

9 Wood A S, 'The majors are taking over in advanced composites', Modern Plastics International, April 1986.

10 Lynn V, 'What is filament winding?' Aerospace Design and Components, Sept. 1986.

11 Spencer B E, 'Application of the filament winding process', Proceedings of the 2nd Conference on Advanced Composites, Dearborn, Michigan, ASM International, Metals Park, Ohio, USA, 18–20 November, 1986.

12 Colluci F, 'The starship enterprise', *Aerospace Comp. Mat.*, **1**, 12, 1988.

13 Rowan J H C, 'Advanced filament winding: Evolution and revolution', *Metals Mat*, **4**, (5), 280–284, 1988.

14 Sternfield A, 'Filament winding has come a long way since the simple cylinder', Modern Plastics International, **16**, 1986.

15 Bowen D H, 'Filament winding in the 1980's', Fibre reinforced composites '84, University of Liverpool, 1984.

16 Kellog C W, US Patent No 4 267 007, 1981.

17 Eckold G C, Lloyd Thomas D G and Wells G M, 'Computer aided filament winding, design in composite materials', IMechE, London, March 1989.

18 Middleton V, Owen M J, Ellimen D G and Shearing M, 'Developments in non axisymmetric filament winding', Automated Composites 88, PRI, Noordwijk, September 1988.

19 Kirbery K W, Michaeli W, Menges G and Seifert A, 'Process simulation in filament winding', Automated Composites 88, PRI, Noordwijk, September 1988.

20 Divita G, Marchetti M, Moroni P and Perugini P, 'Designing complex shape filament wound structures', *Comp Manufac*, **3**, 53–58, 1992.

21 Spencer R A P, 'Developments in pultrusion', in *Developments in GRP Technology*, Ed B Harris, Applied Science, London, 1983.

22 Kelly A and Milenko S T (Eds), 'Fabrication of composites', in *Handbook of Composites, Vol. 4*, North-Holland, Amsterdam, 1983.

23 Lamb D W, Lo C Y, Quinn J A and Gibson A G, 'Understanding the factors controlling pultrusion', 2nd International Conference on Automated Composites, Noordwijk, 1988.

24 Devlin B J, Williams M D, Quinn J A and Gibson A G, 'Pultrusion of unidirectional composites with thermoplastics matrices', *Comp Manufac*, **2**, 203–207, 1991.

25 Astrom B T, 'Development and application of a process model for thermoplastic pultrusion', *Comp Manufac*, **3**, 189–197, 1992.

26 Shaw Stewart D E, 'Pullwinding', 2nd International Conference on Automated Composites, Noordwijk, 1988.

27 Schwartz M M, *Composites Material Handbook*, McGraw-Hill, New York, 1985.

28 Visconti I C and Langella A, 'Analytical modelling of pressure bag technology', *Comp Manufac*, **3**, 3–13, 1992.

29 G Lubin (Ed), *Handbook of Composites*, Van Nostrand, New York, 1982.

30 BS 4994, 'Design and construction of vessels and tanks in reinforced plastics', BSI, London, 1987.

31 Monaghan P F, Brogan M T and Oosthuizen P H, 'Heat transfer in an autoclave for processing thermoplastic composites', *Comp Manufac*, **2**, 233–242, 1991.

32 Musch G and Bishop W, 'Tooling with reinforced elastomeric materials', *Comp Manufac*, **3**, 101–111, 1992.

33 Karbhari V M, Slotte S U, Steenkamer D A and Wilkins D J, 'Effect of material process and equipment variables on the performance of resin transfer moulded parts', *Comp Manufac*, **3**, 143–152, 1992.

34 Kendall K N, Rudd C D, Owen M J and Middleton V, 'Characterization of the resin transfer moulding process', *Comp Manufac*, **3**, 235–249, 1992.

35 Harrison A R, Ginlino J and Berthet G, 'A composite tailgate for the P100 Sierra Pickup: resin transfer moulding with a class A surface finish', Proceedings of the IMechE, Autotech 89, 1989.

36 Howard J S and Harris B, 'The effect of vacuum assistance in resin transfer moulding', *Comp Manufac*, **1**, 161–166, 1990.

37 Owen M J, Middleton V, Rudd Cd, Scott F N and Hutcheon K F, 'Materials behaviour in resin transfer moulding (RTM) for volume manufacture', 2nd International Conference on Automated Composites, Noordwijk, 1988.

38 Becker W E, *Reaction Injection Moulding*, Van Nostrand Reinhold, New York, 1979.

39 Johnson D G and Hill J J, 'Flexible manufacturing of composite aerospace structures', in *Fibre Reinforced Composites*, IMechE, Liverpool, 113–116, 1986.

40 Evans G J, 'The processing of pre impregnated composite materials', 2nd International Conference on Automated Composites, Noordwijk, 1988.

41 Holt D, 'Mechanized manufacture of composite main rotor blade spars', in *Fibre Reinforced Composites*, IMechE., Liverpool, 125–132, 1986.

42 Cozswell F N, 'The experience of thermoplastic structural composites during processing', *Comp Manufac*, **2**, 208–216, 1991.

43 Blundell D J and Willmonth F M, 'Crystalline morphology of the matrix of PEEK – carbon fibre aromatic polymer composites, Part 3: Prediction of cooling rates during processing', *SAMPE*, **17**, (2), 50–58, 1986.

44 Zahlan N, 'Design with thermoplastic based composites: thermal effect consideration', *J Comp Manufac*, **3**, 70–74, 1992.

45 Zahlan N, O'Neil J M, 'Design and fabrication of composite components : the spring forward phenomena', *Composites*, **9**, 77–81, 1989.

46 ASTM C-582, *Standard Specification for Contact Moulded Reinforced Thermosetting Plastic Laminates for Corrosion Resistant Equipment*, American Society for Testing and Materials, Philadelphia, PA, 1987.

47 Abrate S and Walton D, 'Machining of composite materials Part 1: Traditional methods', *Comp Manufac*, **2**, 75–83, 1992.

48 Abrate S and Walton D, 'Machining of composite materials Part 2: Non traditional methods', *Comp Manufac*, **2**, 85–94, 1992.

49 Mackey BA, 'How to drill precision holes in reinforced plastics in a hurry', *Plastics Engineering*, **36**, (2), 22–24, 1980.

50 Ho-Cheng H and Dahran C K H, 'Delaminations during drilling in composite laminates', *J Eng Ind*, **112**, 236–239, 1990.

51 Sakuma K and Seto M, 'Tool wear in cutting glass fibre reinforced plastics (the

relationship between cutting temperature and tool wear', *JSME Bulletin*, **24**, (190), 748–755, 1981.

52 Tagliaferri V, Di Ilio A and Crivelli Visconti I, 'Laser cutting of fibre-reinforced polyesters', *Composites*, **16**, (4), 317–325, 1985.

53 Mello M D, 'Laser cutting of non-metallic composites', *Proc SPIE – Laser Processing: Fundamentals, Applications, and Systems Engineering*, **668**, 288–290, 1986.

54 De Ilio A, Tagliaferri V and Veniali F, 'Machining parameters and cut quality in laser cutting of aramid fibre reinforced plastics', *Mat Manufac Proc*, **5**, (4), 591–608, 1990.

55 Hashish M, 'Cutting with abrasive waterjets', *Mech Eng*, **106**, 60–69, 1984.

56 Howarth S G and Strong A B, 'Edge effects with waterjet and laser beam cutting of advanced composite materials', 35th International SAMPE Symposium, 1990.

57 Mortimer J, 'New technology brings quality to manufacture', *Industrial Robot*, **14**, 103–104, 1987.

58 Lau W S, Wang M and Lee W B, 'Electrical discharge machining of carbon fibre composite materials', *Int J Machine Tools Manufac*, **30**, 297–308, 1990.

59 Cope R D and Brown J C, 'An investigation of electrical discharge machining of graphite epoxy composites', *Composites Manufac*, **1**, 167–171, 1990.

60 Jain V K, Tandon S and Kumar P, 'Experimental investigations into electrochemical spark machining of composites', *J Eng Ind*, **112**, 437–442, 1990.

61 Friend C A, Clyne R W and Valentine G G, 'Machining graphite composite materials', in *Composite Materials in Engineering Design*, Ed BR Noton, American Society for Testing and Materials, Philadelphia, PA, 1973.

62 Stone D E W and Clarke B, 'Ultrasonic attenuation as a measure of void content in carbon fibre reinforced plastics', *Non Destructive Testing*, **8**, 137–145, 1975.

63 Harris B and Phillips M G, 'NDE of the quality and integrity of reinforced plastics', *Developments in GRP Technology*, Ed B Harris, Applied Science, London, 1983.

64 Reynolds W N and Wilkinson S J, 'The propagation of ultrasonic waves in CFRP laminates', *Ultrasonics*, **12**, 109–114, 1974.

65 Torp S, Farli O and Malmo J, Proceedings of the 32nd Annual Technical Conference, SPI, Washington DC, 9-A, p 5, Sept. 1977.

66 Cawley P and Adams R D, 'Vibration techniques for nondestructive testing of fibre composite structures', *J Comp Mat*, **13**, 161–175, 1979.

67 McLaughlin P V, McAssey E V and Dietrich R C, 'Nondestructive examinations of fibre composites structures by thermal fluid techniques', *NDT Int*, **12**, 56–67, 1980.

68 Phillips M G and Harris B, 'Acoustic emission monitoring of GRP plant', Reinforced Plastic Constructed Equipment in the Chemical Process Industry, IMechE/ICE, Manchester, Paper 7.

69 Phillips M G, Ackerman F J and Harris B, 'A study of acoustic emission amplitude distributions from composites, using simple statistical techniques', *Ultrasonics International '81*, Conference Proceedings, IPC Science and Technology Press, London, 1981.

70 Fowler T J, 'Acoustic emission testing of fibre reinforced plastics', Convention of ASCE, San Francisco, Paper 3092, 1977.

71 CARP/Howard Adams C, 'Recommended practice for acoustic emission testing of fiberglass tanks/vessels', Society of the Plastics Industry Inc., New York, 1982.

7

METAL AND CERAMIC MATRIX COMPOSITES

Considerable efforts have been made to increase the capability of polymer matrix composites in terms of temperature range and performance. However, these efforts will perhaps always be faced with increasing levels of difficulty with each incremental improvement because of the intrinsic characteristics of an organic material. Two matrix materials which have potential to overcome temperature limitations are metals and ceramics.[1-3]

Metal matrices offer the advantage of higher operating temperatures, in the region of 500 °C, with the additional advantages of effectively nil moisture absorption and outgassing, good thermal conductivity in all directions, good transverse and shear moduli and strengths, and high thermal stability. Although many combinations of fibre and metal have been explored, most commercial and potential systems are based on boron, carbon, silicon carbide and alumina fibres in aluminium or an aluminium alloy, a titanium alloy or magnesium matrix. In addition to continuous reinforced short, randomly or pseudo-aligned fibres or whiskers or whisker/particulate reinforcement can also be employed.

The principal disadvantage of metal matrix composites is their higher specific gravity compared with polymer composites. The higher temperature fibres, apart from carbon, have specific gravities between 2.5 and 4.0, and the aluminium and titanium alloys have values of 2.77 and 4.43 respectively. Also, fabrication tends to be much more difficult than is the case for organic matrix materials. A variety of techniques based on diffusion bonding, hot moulding, conventional moulding, powder metallurgy and vapour deposition among others are available, but as yet are still in the development stage. Important considerations yet to be fully resolved include mechanical damage of the fibres, intermetallic diffusion between fibre and matrix, dissolution of the fibre in the matrix, chemical reactions and poor wetting, all of which result in products with properties that are often well below the theoretically attainable.

Actual and potential applications for metal matrix composites include equipment panels for satellites, struts, tanks for propellants, bearings, aircraft components for which good hot-wet properties are required, and underbonnet and aero engine components.

Ceramic matrix composites have been under study for many years with the first development programmes undertaken in the early 1960s. The earliest work incorporated refractory metal wires into traditional ceramic matrices to improve their thermal and mechanical shock behaviour, but the composites had a high density and disappointing behaviour in air at elevated temperatures because of oxidation of the fibres. In the late 1960s and early 1970s a range of high performance carbon and silicon carbide fibre reinforced glass and glass–ceramic systems were developed. The fabrication technology was based on a hot-pressing, powder route and a number of components were fabricated. Room temperature properties of the composites were comparable with those of conventional polymer matrix composites. In inert atmospheres the carbon fibre reinforced glasses and glass–ceramics retained these properties to temperatures at which the matrices began to soften between 500 °C and 800 °C depending on the matrix, but in air the fibres rapidly oxidized at 350 °C–400 °C. The silicon carbide fibres were produced on a tungsten substrate, which resulted in relatively low specific properties because of the high fibre densities. Subsequent development of ceramic fibres of low specific gravity, e.g. Nicalon SiC, created a renewed interest in fibre reinforced glasses and glass–ceramics, the emphasis being on glass–ceramic matrices rather than engineering ceramics because of their simpler fabrication properties. In terms of potential application, carbon fibre reinforced glasses could be employed as alternatives to high temperature polymer matrix composites for use in air up to 350 °C–400 °C. Their fabrication is potentially no more difficult than that of some of the higher temperature polymer systems and they offer a significantly higher temperature capability combined with resistance to chemical and environmental effects. Possible applications exist in underbonnet components of automobiles, and in space applications where the absence of absorbed moisture could be a significant advantage. In inert atmospheres carbon fibre reinforced glass–ceramics could be used to much higher temperatures, ~ 800–1000 °C, although this condition is clearly a severe restriction on material potential. Silicon carbide reinforced glass–ceramics are more expensive, because of the higher cost of fibres, but can be used for prolonged periods to temperatures approaching 1000 °C in air with little reduction in properties. The prime application for these materials is as components of high temperature engines.

Carbon/carbon composites are another type of high temperature material (though strictly carbon is not a ceramic). They have useful high temperature properties in certain circumstances and they have a limited number of important, commercial uses, e.g. aircraft and racing car disc brakes.[4]

Metal matrix composites

Metal matrix composite (MMC) materials have been under development for around 20 years, initially using continuous filament reinforcement for

Table 7.1. Typical metal matrix materials

Constituent phase	Material
Matrix	Ti, Mg, Ni, Al (alloys)
Fibres	
Metal	Stainless steel, Bo, Mo, W
Ceramic (non-oxide)	SiC, C, B, $Ti_4 B_2$
Ceramic (oxide)	Al_2O_3, glass

Table 7.2. Typical properties of metal matrix composites

Matrix	Fibre	Fibre content (vol %)	Service temp. (°C)	Density (g/cm^3)	Ultimate tensile strength (MPa)	Tensile modulus (GPa)
Aluminium	Boron	50	350	2.62	1482	221
	B_4 coated boron	50	350	2.62	1517	221
	SiC (CVD)	50	350	2.96	1724	214
	SiC whiskers	20	350	2.80	483	110
	Al_2O_3 tow	60	350	3.45	586	262
	Carbon	40	350	2.37	455	131
Magnesium	Boron	50	300	2.18	1310	193
	Carbon	40	300	1.82	565	110
	Al_2O_3	50	300	2.80	517	200
Titanium	SiC-coated boron	35	650	3.76	758	207
	SiC (CVD)	35	650	4.00	862	193
Iron, nickel, base alloys	SiC	50	800	5.41	1655	311
	Tungsten wire	35	1150	11.7	1793	304

applications in aerospace, but more recently with particulate reinforced materials accompanied by a low cost fabrication route. Typical examples of the materials of construction and material properties are given in Tables 7.1 and 7.2.[5]

The main types of MMC are as follows:[1]

- **Dispersion-strengthened** – this type of composite consists of a micro-structure of an elemental matrix within which fine particles are uniformly dispersed. The particle diameter ranges from about 0.001 to 0.1 μm, and the volume fraction of particles ranges from 1 to 15%.
- **Particle-reinforced** – this material is characterized by dispersed particles of greater than 1.0 μm diameter with a volume fraction of 5 to 40%.
- **Fibre (whisker)-reinforced** – the reinforcing phase in fibre composite materials spans from 0.1 to 250 μm in length to continuous fibres, and includes the entire range of volume fractions, from a few percent to greater than 70%.

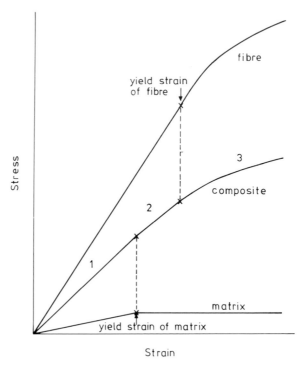

7.1   Stress/strain curves of typical matrix metal, fibre
and metal matrix composite.

In common with other composite materials, values of specific strength and stiffness are often cited as major advantages for MMC, but additional benefits with the metal matrix include:

- Low sensitivity to temperature changes or thermal shock.
- High surface durability and low sensitivity to surface flaws.
- High electrical and thermal conductivity.
- High vacuum environment resistance.

In terms of disadvantages MMCs tend to have lower toughness characteristics than the parent metal matrix and may also be prone to the effects of thermal fatigue.

In many respects the analysis of metal matrix composites with its attendant problems of anisotropy and microscopic heterogeneity follows that which has been established for polymer systems. The rules governing the transformation of stiffness to give properties of laminated structures are essentially the same. A significant difference, however, lies in stress/strain behaviour. Unlike most polymer systems a significant amount of plastic deformation may occur due to the ductility of the metal matrix. A typical stress/strain plot is shown in Fig. 7.1.[1] In the general case there are three regimes in stress/strain behaviour,

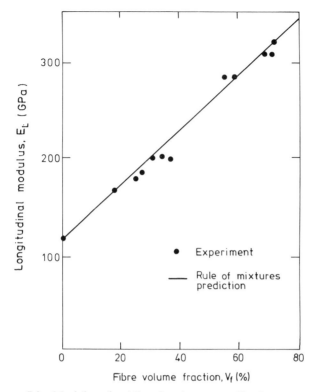

7.2   Modulus of unidirectional tungsten fibre/copper
composite.

the details of which are dependent on the particular material of concern. In (1) both matrix and fibre deform elastically, in (2) plastic deformation occurs in the matrix and in (3) both phases undergo inelastic strain.

For unidirectional reinforcement materials the rule of mixtures can be used to good effect for the determination of basic modulus and yield strength data in the primary material directions. Figures 7.2 and 7.3 show the results of some early work on such calculations for tungsten fibre copper composites.[6] With discontinuous materials there is marked deviation from behaviour anticipated from a continuous mechanics approach, particularly for strength where experimental values can exceed those obtained from simple theory by some margin. The rationale for this behaviour remains under investigation, but it is considered that the prime reasons are those due to high dislocation density and small subgrain sizes which arise through the difference in thermal expansivity between constituent materials.[6,7]

For fracture strengths the principles of mechanics are still valid, i.e. a load share approach as defined by the rule of mixtures, although subtleties in material behaviour can have a dramatic effect. Figure 7.4 shows results for two

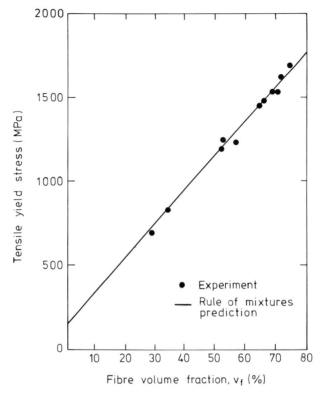

7.3   Tensile yield stress of unidirectional tungsten
fibre/copper composite.

SiC reinforced material systems.[8] The first has a high purity aluminium matrix whilst the second is an aluminium alloy. A rule of mixtures approach is approximately valid for one showing an increase in strength with volume fraction, whilst for the other no strengthening is apparent. This disparity in behaviour is due to the formation of reactive products during processing which results in a large thickness of brittle material on the fibre surface. This causes the fibres to fail at very low levels of strain.

The principles adopted for the properties of polymer components can also be used for off-axis strengths. Figures 7.5 and 7.6 show experimental results for aluminium matrix materials compared with the maximum stress criteria.[9]

For short fibre composites the expressions which govern load transfer through shear mechanisms can be applied yielding the conventional formula for strength, σ_L, in terms of a critical fibre length, l_c.[10]

$$\sigma_L = v_f \sigma_{fb} \left(1 - \frac{l_c}{2l} \right) + v_m \sigma'_m \quad \text{for } l \geq l_c$$

[7.1]

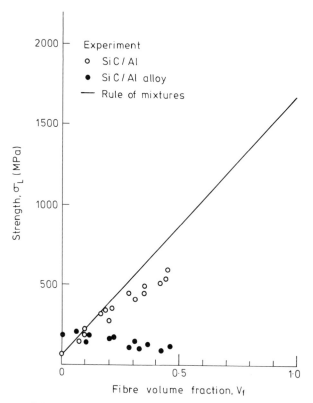

7.4   Longitudinal tensile strength unidirectional SiC
fibre – aluminium and aluminium alloy.

$$\sigma_L = v_f \sigma_{fb} \frac{l}{2l_c} + v_m \sigma_m \qquad \text{for } l < l_c$$

where v_f and v_m are volume fractions of fibre and matrix respectively, σ_m is the strength of the matrix metal, σ'_m is the flow stress of the matrix at the failure strain of the fibre and σ_{fb} is the maximum stress which can be carried by the fibre. In practical materials where there will be variable fibre lengths these equations can be employed by making use of the appropriate distributions.

As has already been cited, thermal fatigue can be a key factor in determining the success or failure of a metal matrix component in a given application. This arises because of the high residual stresses that are generated during fabrication as a result of thermal expansion mismatch. A simple analysis based on compatibility of strain yields the following expressions for residual stress:[11]

$$\sigma_m = \frac{v_f E_m E_f \Delta \alpha \Delta T}{E_c}$$

[7.2]

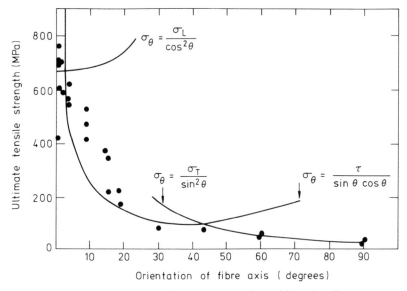

7.5   Off-axis tensile fracture stress of a unidirectionally
aligned 50% v_f silica/aluminium composite.

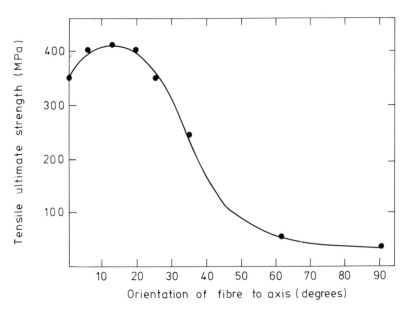

7.6   Tensile fracture stress of $\pm\ \theta$ stainless steel
fibre/aluminium composite.

$$\sigma_f = \frac{v_m E_m E_f \Delta\alpha\Delta T}{E_c}$$

where $\Delta\alpha = \alpha_m - \alpha_f$, E_c is the modulus of the composite, E_m and E_f are modulus of matrix and fibre respectively and ΔT is the temperature change.

Fibre/matrix interactions are in reality more complex than indicated by this rather simple elastic analysis. Under constrained conditions the matrix will behave as an elastic/plastic/creep material and this has to be accommodated in any model for constitutive behaviour. Also, during processing, reaction zones occur at the fibre/matrix interface which have different properties. Calculation of stress distributions in and around these regions requires a more complex model formulation.[12-16] The thermal expansion coefficients themselves can be calculated to a first approximation by conventional methods for composites.

Being relatively new, the behaviour of metal matrix systems across the complete spectrum of design and operating conditions has not been fully characterized. However, a brief resumé of certain aspects of performance is given below:

- **Fatigue.** In common with most composite materials fatigue performance of metal matrix systems is superior to that of the parent matrix. This arises because of the ability to accommodate the progression of microcracks and damage through mechanical interactions between constituent phases.
- **Wear.** Wear rates reduce with increasing volume percent of reinforcement and in the case of particulates with increasing particle size.
- **Oxidation.** The behaviour of metal matrix composites in hot oxidative atmospheres is complex. Ceramic fibres tend to be inert at the temperatures of concern, but metals are susceptible to oxidation. Even though efforts are made to ensure fibres are fully coated by matrix this is difficult to achieve in practice. Oxidation can be initiated at fibre ends and can then proceed down the fibre/matrix interface. The chemical reactions which occur result in volume changes which initiate cracking.
- **Corrosion.** Particular attention needs to be given in those material systems where there is a difference in electrochemical potential between constituents. Galvanic corrosion can be a significant effect in such circumstances.
- **Creep.** Owing to the use of the materials at high temperature creep can become a significant design issue. Generally creep effects are restrained with reinforced materials as compared with the matrix.

In terms of manufacture method, distinction can be drawn between process routes for particulate and fibre reinforced materials.[17] Particle reinforced composites may be cast to shape using an intermediary casting feedstock or for higher performance components may be produced by forging. By contrast,

fibre reinforced material can be produced by diffusion bonding for monofilament or, more recently, by plasma spray processing. These both tend to be
high cost processes. Lower cost multifilament tows can be pre-infiltrated with
liquid metal to form a composite wire which may be processed by hot pressing
into shape. A more attractive option is the cast to shape technology in which
reinforced castings are produced by pressurized feeding of liquid metal into a
mould cavity containing a fibre preform, followed by solidification of the
casting in the usual way. This confers good dimensional control and the
process economics are potentially attractive. The process involves the steps of
fibre preforming, preform preheating, preform infiltration and solidification.
Properties may be improved by subsequent heat treatment. Fabrication of
metal matrix preforms is an emerging technology which resembles conventional polymer preforming in some respects. For tow fibres, filament winding
techniques may be used to produce unidirectional pre-preg sheets which may
then be stacked to produce laminates. Alternatively, filament winding may be
used to generate preforms directly. Conversely, short fibres may be preformed
using dispersed slurry methods to form planar random distributions, but
volume fractions are lower because of packing difficulties. The maximum
volume fraction of reinforcement available in this system is approximately
25%. Because of the near random fibre distribution, maximum directional
properties are substantially lower than the equivalent volume fraction in a
unidirectional composite. A method for fabrication of short fibre preforms in a
cylindrical configuration is shown in Fig. 7.7.

Preforms may also conveniently be assembled from woven or braided fibre
cloths although this can lead to relatively poor properties. Recent advances in
fibre processing technology include the lost yarn preforming technique, in
which a consumable textile yarn of viscose is used to produce knitted
preforms. Unfilled areas may be progressively reinforced by injecting aqueous
slurries of short fibres with a binder, so that transverse strength is improved.
On heating the preform, the viscose yarn is consumed, leaving a structured
material. Some kind of binder is required to prevent disintegration of the
preform. Binder selection is important since residues can have a significant
effect on the fibre/matrix interface, and may influence fibre degradation during
processing. Typical organic binders include polyethyl glycol and polyethylene
oxide. These are volatilized at relatively low temperatures and leave a minimal
level of residues.

Various methods exist for applying pressure during infiltration (Fig. 7.8 and
7.9).[18] These range from high pressure processes, such as squeeze casting,
through intermediate techniques, such as pressure die casting, to low pressure
methods. The squeeze casting process is well adapted as a route for the
manufacture of fibre reinforced material. This is because most preforms of
practical interest may resist liquid metal infiltration. This resistance has
contributions from capillary repulsion at the melt/preform interface and from

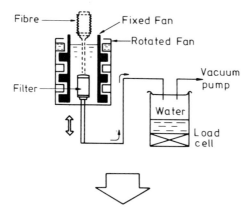

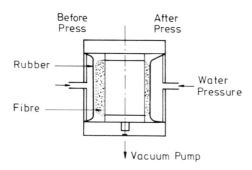

7.7   Outline of the preform process.

viscous flow effects. The pressure available with the squeeze casting process is generally acknowledged to be considerably higher than that necessary to effect infiltration. The infiltration rates need to be kept as low as possible (subject to adequate filling of the die cavity) to minimize loads on the preform which can cause delaminations or, in the worst case, complete disintegration. In squeeze casting of the metal matrix preforms, the reinforcement must be preheated to prevent premature solidification since a cold preform can rapidly chill an advancing melt front so that effective infiltration is blocked. However, the mechanical properties of the composite may be impaired if processing occurs at high temperatures which lead to degradation of the fibre as a result of interactions with the matrix. Because of these interaction effects, optimized properties may be obtained in composites with pure matrix materials instead of alloys. The pressures available with pressure die casting are typically of the order of 40 MPa. However, conventional pressure die casting is not well adapted to infiltration because metal velocities are too high for effective infiltration.

Process	Outline	Characteristics
New Die-Cast		Casting Pressure 50 ~ 300 kgf/cm^2 Filling Speed 0.3 ~ 0.8 m/sec
Squeeze -Cast		Casting Pressure > 1000 kgf/cm^2
Die-Cast		Casting Pressure 550 ~ 1000 Filling Speed 25 ~ 40 m/sec
Low Pressure Die-Cast	Air	Casting Pressure < 0.3 kgf/cm^2 Filling Speed 0.1 ~ 0.2 m/sec
Gavity Die-Cast	Riser Gate	Casting Pressure < 0.3 kgf/cm^2 Filling Speed 0.1 ~ 0.2 m/sec

7.8   Casting method comparisons for MMCs.

Ceramic matrix composites

Monolithic ceramics are well known for their refractory behaviour, but their use for engineering structures is severely limited by their poor thermal shock resistance and low fracture toughness. Despite attempts to improve fracture properties, e.g. transformation toughening of zirconia,[19] the materials are still relatively brittle compared with other engineering materials with fracture strains typically within the range 0.1–0.2%.[20] Because of the effects of brittle behaviour, the concept of strength as a material property no longer has the same meaning as for conventional engineering materials and this requires a somewhat different approach when considering aspects of design.

Fibre reinforcement of ceramic matrices offers the potential of tougher materials together with an increased strain at fracture.[21] The main mechanisms by which these improvements occur are due to matrix microcracking and

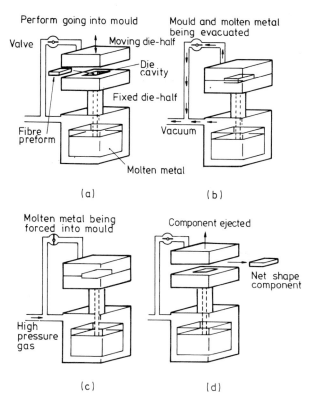

7.9   Diagrammatic representation of liquid pressure
forming.

subsequent load transfer within the material. These energy-absorbing effects
give rise to a form of pseudo-plasticity and result in nonlinear elastic
behaviour. Figure 7.10 shows the stress/strain behaviour of a glass–ceramic
composite containing SiC fibres.[22] During matrix microcracking, there is a
nonlinear region of reduced modulus associated with extension of fibres. A
maximum stress occurs corresponding to fibre failure and this may be followed
by fibre 'pull out'. In terms of design, the important characteristics of this
behaviour are as follows:

- Comparatively high work of fracture due to load-carrying ability beyond
  matrix cracking.
- Increased failure strain and tolerance of overstressing.
- Reduced sensitivity to flaws within the material.

These effects are advantageous in a number of design situations, not only for
simple static load cases, but also in areas of high stress concentration, thermal
shock, impact and abrasive wear. Table 7.3 shows typical values of fracture

Table 7.3. Toughness properties for a range of materials

Material	Work of fracture (kJ/m^2)	Fracture toughness (MPa m$^{1/2}$)
Mild steel	200	200
Al Alloy	150	50
Wood	6	7.7
Glass	0.01	0.3
Ceramic (typical)	0.1	4.0
Carbon fibre/borosilicate glass	5	–
SiC/fibre/borosilicate glass	18–26	50
SiC whisker/Al$_2$O$_3$	–	9.5
Sic whisker/Si$_3$N$_4$	–	6.5
Al$_2$O$_3$ fibre/glass	1	7.0
GRP*	150	–
CFRP*	80	–

* Parallel to fibre direction.

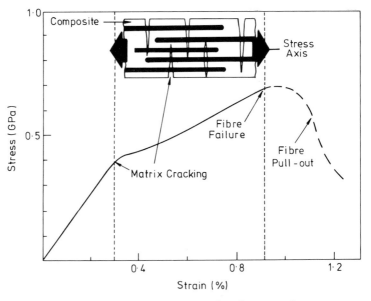

7.10   Stress/strain behaviour for glass ceramic
containing unidirectional SiC fibres.

properties[23] for a range of materials. As can be seen, significant improvements can be achieved by using reinforcement either in fibre or whisker form. In the latter case mechanical behaviour is broadly similar to monolithic ceramics except with higher toughness and less scatter in strength.[24,25]

Examples of some ceramic matrix composites currently under consideration include:

- Low temperatures (up to 400 °C) – carbon fibre reinforced borosilicate glass.
- Intermediate temperatures (up to 1100 °C) – silicon carbide reinforced glass ceramics (lithium alumino silicate, cordierite and mullite).
- High temperatures (up to 1500 °C) – oxidation-resistant carbon/carbon (not yet commercially available).

Whilst the benefits of fibre reinforcement are significant, it is important that the resulting properties are put into perspective, particularly with regard to those obtained from other engineering materials such as metals and polymer-based composites. As can be seen from Table 7.3 fibre reinforcement of ceramics offers significant benefits, but for the purposes of design they must still be considered as comparatively brittle materials. Generally the degree of toughness enhancement increases with reinforcement content, for example Fig. 7.11 shows results for a silicon carbide reinforced silicon nitride.[26] A similar trend can be applied to other properties and as with polymer and metal matrix systems the simple rule of mixtures approach provides a good approximation to directional properties. Figure 7.12 shows a comparison between experimentally determined modulus values[27] with those obtained from the simple calculation method. The correlation between theory and experiment is good. The deviation that is present has been attributed to the effects of porosity and fibre misalignment. For strength properties, however, there are distinct differences in behaviour between ceramic and other matrix types. A significant factor that must be considered is the comparatively low strain to failure of ceramic matrices. Figure 7.13 shows/stress strain curves for epoxy and glass matrices reinforced with silicon carbide fibres.[28] With a tensile strain of approximately 0.3% the slope of the curve for the glass composite decreases significantly, whereas that for the epoxy system is essentially linear to failure. This behaviour is due to multiple matrix microcracking in the ceramic matrix prior to fibre failure.

Using the simple strain compatibility model employed to derive the rule of mixtures equation for modulus the stress at which matrix cracking occurs is given by:

$$(\sigma_c^*)_m = \sigma_m^*[1 + v_f(E_f/E_m - 1)] \qquad [7.3]$$

where $(\sigma_c^*)_m$ is the composite stress at which the matrix will crack, σ_m^* is the ultimate strength of the unreinforced matrix, v_f is fibre volume fraction and E_f and E_m are modulus values of fibre and matrix respectively.

The ultimate strength of the composite can also be calculated by a rule of mixtures approach:

$$\sigma_c^* = v_f\sigma_f^* \qquad [7.4]$$

where σ_c^* is the ultimate strength of the composite and σ_f^* that for the fibre.

In terms of design, the ramifications of matrix cracking require careful

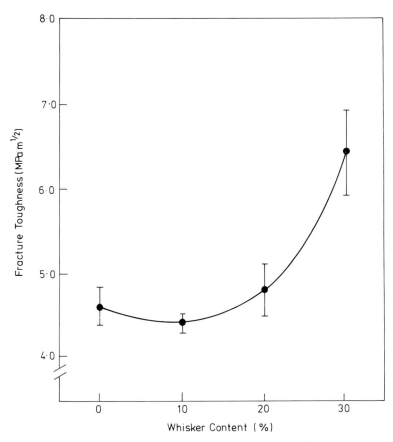

7.11 Fracture toughness as a function of whisker content (SiC whisker/$Si_3 N_4$).

thought as, although there is still considerable strength and toughness retention, problems may arise due to propagation of cracks through repeated loading, and the possibility of degradation of exposed fibres through environmental attack or thermal fatigue loading.

Experimentally it is noted that equation 7.3 generally underestimates the matrix cracking stress. In unreinforced ceramics the propagation of a crack from a single flaw may be sufficient to cause total failure, whereas in composites the presence of fibres can inhibit crack growth. The theory describing the development of cracking in fibre composites (Aveston, Cooper, Kelly – ACK)[29,30] is relatively well established and uses energy considerations to derive expressions from which the matrix cracking strain may be calculated:

$$(\varepsilon_c^*)_m = \left\{ 12 \frac{\tau_i \gamma_m v_f^2 E_f}{r E_c E_m^2 (1 - v_f)} \right\}^{1/3} \qquad [7.5]$$

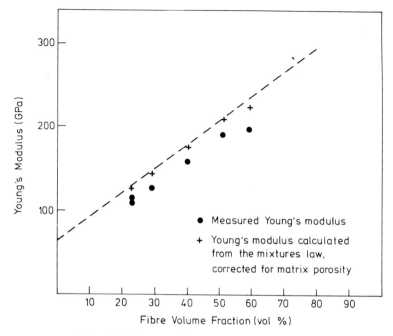

7.12 Modulus as a function of volume fraction (carbon fibre/glass).

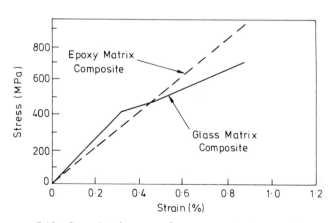

7.13 Stress/strain curves for epoxy and glass matrices (SiC reinforcement).

where $(\varepsilon_c^*)_m$ is the failure strain of the matrix, τ_i the shear strength of the matrix, E_c the modulus of the composite, γ_m the fracture surface energy of the matrix and r the fibre radius.

As can be seen from equation 7.5 the matrix failure strain can be enhanced by increasing fibre volume fraction and reducing fibre diameter. Examination

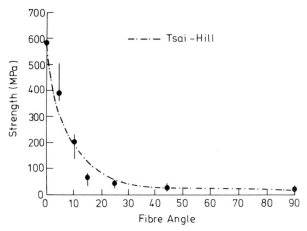

7.14   Tsai–Hill failure criterion (SiC/glass).

of the situation in terms of a fracture mechanics approach[31,32] leads to a similar expression for matrix cracking stress:

$$(\sigma_c^*)_m = \frac{12\tau_i\gamma_m}{r}\left\{\frac{v_f^2}{(1-v_f)}\frac{E_f}{E_cE_m^2}(1-v^2)^2\right\}^{1/3}\cdot\frac{E_c}{1-v_m^2} \qquad [7.6]$$

The results from these expressions give good agreement with experimental data.[33] For lamina with fibres orientated off-axis the often used maximum stress and distortional energy failure criteria can be employed giving strength values in the major directions (see Fig. 7.14).[34,35]

Owing to the brittle failure modes of ceramics and the inherent variability in strength values it is with ceramic matrix composites that probabilistic methods come to the fore in composite design. Figure 7.15 shows examples of Weibull plots for fibre types.[36] The effects of probabilistic design can be most apparent with structures under bending. This is because only a relatively small volume of material is subjected to high tensile stress. For a beam subjected to three-point bending it can be shown that the relationship between tensile and bending strengths is as follows:

$$\frac{(\sigma_c^*)_B}{(\sigma_c^*)_T} = [2\,(m+1)^2]^{1/m} \qquad [7.7]$$

where $(\sigma_c^*)_B$ and $(\sigma_c^*)_T$ are the composite strengths in bend and tension respectively and m is the Weibull modulus.

For a value of m of 20 (measured for SiC/glass), this ratio is 1.4, i.e. the bend strength is 1.4 times the tensile strength. This type of approach has been shown to be satisfactory for monolithic ceramics and is likely to be given reasonable results for discontinuous reinforcements.[37] In the case of unidirectional

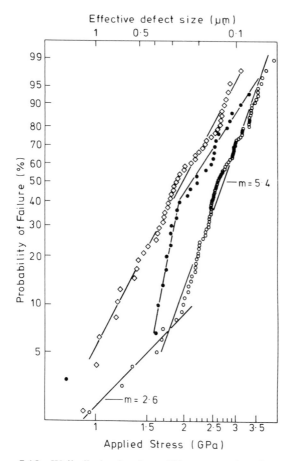

7.15 Weibull plot for three different samples of SiC.

composites, however, it is found that the bend strength is considerably greater than that predicted from this equation. Recent work[38] has shown that this behaviour may be rationalized by considering the composite as a bundle of fibres within the matrix. From this it can be deduced that the stress volume integral of a unidirectional material depends only on the stress distribution along the length of the beam and not through the thickness. This results in a modified expression for bend/tensile strength ratio:

$$\frac{(\sigma_c^*)_B}{(\sigma_c^*)_T} = 2(m + 1)^{1/m} \qquad [7.8]$$

In this case the Weibull modulus relates to that of the fibres and not the composite. Adopting a typical value, 5, the strength ratio becomes 1.6 which gives closer agreement with experiment.

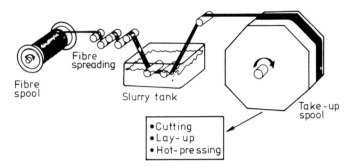

7.16   Schematic illustration of slurry method for
preparing long-fibre reinforced ceramics.

In a way similar to metal matrix materials the manufacture processes for ceramic composites can be divided into two types; those suitable for particulates and whisker reinforcement and those that can be used for long fibre composites. In the former case the technology established for monolithic ceramics can be used whereas for the continuous reinforced systems new techniques have had to be developed. This is due to the need to minimize fibre damage during processing and the fact that at sintering temperatures many fibres suffer degradation. An example of one such process is the manufacture of glass and glass–ceramic composites by the slurry impregnation route. This involves three stages: the production of a pre-preg tape of fibre and unconsolidated matrix material; cutting and lay-up of the pre-preg tape to an appropriate pattern for the required shape of component and stacking sequence of plies; and hot pressing to produce the final consolidated composite component. Figure 7.16 shows diagrammatically some key features of the process.[39] After being cut to appropriate sizes the laminae are laid up in a die for hot pressing, in a method similar to conventional polymer composite pre-preg technology, albeit at higher temperatures.

Current developments for the manufacture of ceramic composites centre around the infiltration of a fibre preform. The infiltrating matrix can take the form of a powder slurry, a melt, a liquid solution or a mixture of gases which react *in situ* (CVI – chemical vapour infiltration), Fig. 7.17.[2] Here a porous fibre preform is set inside an infiltration chamber which is fed with a gaseous precurser, e.g. in the case of SiC the precurser is CH_3SiCl_3 and hydrogen. A technique of particular interest is polymer infiltration and pyrolysis. Fibre preforms can be infiltrated with liquid polymers, either molten or in solution, which are then pyrolized to leave a ceramic deposit. For example, pyrolysis of polycarbosilanes and polysilazanes can be utilized to form matrices of SiC and Si_3N_4, respectively. The advantage of pyrolysis techniques is the low process temperature, typically $\leq 1000\,°C$. Moreover, it is easier to infiltrate a fibre preform with a homogeneous liquid than with a powder slurry.

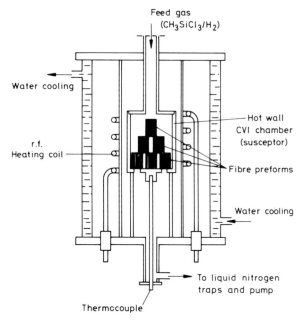

Feed gas
(CH_3SiCl_3/H_2)

Water cooling

r.f.
Heating coil

Hot wall
CVI chamber
(susceptor)

Fibre preforms

Water cooling

To liquid nitrogen
traps and pump

Thermocouple

7.17   Hot wall deposition chamber for processing
       SiC–ceramic–matrix composites.

In some circumstances a combined approach may be appropriate. Both
polymer pyrolysis and sol–gel techniques can be combined with CVI or
powder methods. For example, a first infiltration can be performed with a
polymer or sol carrying a dispersed ceramic powder while subsequent
infiltrations could be with pure polymer or sol. The final infiltration could be
CVI filling the smallest pores and providing an oxidation-resistant, high
purity surface coating.

References

1 Taya Minoru and Asenault R J, *Metal Matrix Composites*, Pergamon Press,
  Oxford, 1989.
2 Warren R, *Ceramic Matrix Composites*, Blackie, Glasgow, 1992.
3 Hancox NL and Phillips D C, 'Fibre composites for intermediate and high
  temperature applications', Materials Engineering Conference, London, 1985.
4 Fisher R, *Ceramic Composites and Coatings*, IBC, London, 1988.
5 Kelly A, *Concise Encyclopedia of Composite Materials*, Pergamon, Oxford, 1989.
6 McDanels D L, Jech R W and Weeton J W, 'Analysis of stress-strain behaviour of
  tungsten-fiber-reinforced copper composites', *Trans Metal Soc*, AIME, **233**,
  636–642, 1965.

7 Nardone V C and Prewo K M, 'On the strength of discontinuous silicon carbide reinforced composites', *Scripta Meta*, **20**, 43–48, 1986.

8 Kagawa Y and Choi B H, 'Relation between tensile strength and fracture toughness in brittle fiber reinforced metals', in *Composites '86, Recent advances in Japan and the United States*, Ed K Kawata *et al.*, Japan Society for Composite Materials, 1986, p. 537–543.

9 Jackson P W and Cratchley D, 'The effect of fibre orientation on the tensile strength of fibre-reinforced metals', *J. Mech Phys Solids*, **14**, 49–64, 1966.

10 Kelly A and Tyson W R, 'Tensile properties of fibre-reinforced metals: copper/tungsten and copper/molybdenum', *J Mech Phys Solids*, **13**, 329–350, 1965.

11 Koss D A and Copley S M, 'Thermally induced residual stresses in eutectic composites', *Metal Trans*, **2**, 1557–1560, 1971.

12 Hecker S S, Hamilton C H and Ebert L J, 'Elastoplastic analysis of residual stresses and axial loading in composite cylinders', *J Mat*, **5**, (4), 868–900, 1970.

13 Uemura M, Iyama H and Yamaguchi, J, 'Thermal residual stresses in filament wound carbon-fiber-reinforced composites', *J Thermal Stress*, **2**, 393–412, 1979.

14 Iesan D, 'Thermal stresses in composite cylinders', *J Thermal Stress*, **3**, 495–508, 1980.

15 Christensen R M and Lo H, 'Solutions for effective shear properties in three phase sphere and cylinder models', *J Mech Phys Solids*, **27**, 315–330, 1979.

16 Mikata Y and Taya M, 'Stress field in a coated continuous fiber composite subjected to thermomechanical loadings', *J Comp Mater*, **19**, 554–578, 1985.

17 Young R, AEA Technology, Private communication, 1992.

18 Hayashi T, Ushio H and Ebisawa M, 'The properties of hybrid fibre reinforced metal and its application to engine block', SAE, The Engineering Society for Advancing Mobility, Land Sea and Space.

19 Lankford J, 'Deformation of transformation toughened zirconia', in *Proceedings of the Symposium on Advanced Structural Ceramics*, Ed P F Becher, M V Swain and S Somiya, Materials Research Society, Boston, USA, p. 61, 1986.

20 Davidge R W, 'Fibre reinforced composites', *Composites*, **18**, 92, 1987.

21 Dawson D M, Preston R F, Briggs A and Davidge R W. 'Ceramic fibre/glass matrix composites with outstanding mechanical performance', AEAE Report R12025, Harwell Laboratory, 1986.

22 Prewo K and Brennan J J, 'Silicon carbide yarn reinforced glass matrix composites', *J Mat Sci*, **17**, 1201–1206, 1982.

23 Bailey J E, 'Principles and problems in the fibre reinforcement of brittle/ceramic materials', Conference on Ceramic Composites and Coatings, London, 1988.

24 Duffy S F, Manderscheid J M and Palko J L, 'Analysis of whisker-toughened ceramic components – A design engineer's viewpoint', *Cer Bull*, **68**, 2078, 1989.

25 Nemeth N N, Manderscheid J M and Gyekenyesi J P, 'Designing ceramic components with the CARES computer program', *Cer Bull*, **68**, 2064, 1989.

26 Hasselman D P H, 'Tailoring of the thermal transport properties and thermal shock resistance of structural ceramics', in *Tailoring Multiphase and Composite Ceramics*, Ed R E Tressler, Material Science Research Series, **20**, 731–754, 1986.

27 Phillips D C, 'The fracture energy of carbon-fibre reinforced glass', *J Mat Sci*, **7**, 1175–1191, 1972.

28 Prewo K N, 'Tension and flexural strength of silicon carbide fibre-reinforced glass ceramics', *J Mater Sci*, **21**, 3590, 1986.

29 Aveston J, Cooper G A and Kelly A, 'Single and Multiple Fracture', in *Proceedings of the Conference on the Properties of Fibre Reinforced Composites*, IPC, Guildford, 1971.

30 Hale D K and Kelly A, 'Strength of fibrous composite materials', *Ann Rev Met Sci*, **2**, 405–462, 1975.

31 Marshall D B, Cox B N and Evans A G, 'Tensile fracture of brittle matrix composites: influence of fibre strength', *Acta Metall*, **33**, 2013, 1985.

32 McCartney L N, 'Mechanics of matrix cracking in brittle matrix fibre reinforced composites', *Proc R Soc (London) Ser A*, **409**, 329, 1987.

33 Cooper G A, 'The structure and mechanical properties of composite materials', *Rev Phys Technol*, **2**, 49–91, 1971.

34 Phillips D C, Sambell R A J and Bowen D H, 'The mechanical properties of carbon fibre reinforced Pyrex glass', *J Mat Sci*, **7**, 1454–1464, 1972.

35 Briggs A and Davidge R W, 'Borosilicate glass reinforced with continuous silicon carbide fibres: A new engineering ceramic', *Mat Sci Eng*, **109**, 363, 1989.

36 Warren R and Anderson C H, 'Silicon carbide fibres and their potential for use in composites', *Composites*, **15**, 16, 1984.

37 Davidge R W, Mechanical Behaviour of Ceramics, Cambridge University Press, Cambridge, 1979.

38 Davidge R W and Briggs A, 'The tensile failure of brittle matrix composites reinforced with unidirectional continuous fibres', *J Mat Sci*, **24**, 1989.

39 Kelly A and Mileiko S T, *Fabrication of composites*, Elsevier, Amsterdam, 1983.

8

EXAMPLES OF COMPOSITE APPLICATIONS

The range of applications for which composites are candidate materials of construction is vast. The issues of weight and corrosion resistance are common to almost every industry and whilst there is no materials system which can offer a panacea to all problems, there has been, and continues to be, opportunities for composites to provide significant improvements in component performance. In the following sections the uses of composites are discussed in terms of the different application areas where the materials are currently being used or under investigation for the future. Sections are concluded with an example component, each of which varies with respect to materials of construction, motivating factors that determined the choice of composites in the first instance, and degree of maturity in the product development cycle.

Aircraft

While the utilization of composites, in tonnage terms, for aircraft components constitutes a relatively small percentage of total use, the materials often find their most sophisticated applications in this industry. In aerospace the demands placed upon materials can be greater than in other areas, often requiring a combination of light weight, high strength, high stiffness and good fatigue resistance.

Military aircraft were the first to use GRP composites in significant quantities. The first applications were in radomes and then in secondary structures and internal components. The modulus of glass, however, is low compared with that of metals and it was not until the advent of boron and carbon reinforcements that significant interest in terms of primary structures developed. The situation in the present day, where use of composites is extensive, has been the result of a gradual direct substitution of metal components followed by the development of integrated composite designs as confidence has increased. Figures 8.1[1] and 8.2[2] show two examples. In the first, the Airbus 320, a whole range of components is made from composites, including the fin and tailplane. This alone has led to a weight-saving of 800 kg

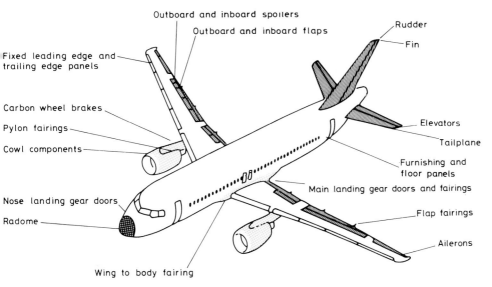

Outboard and inboard spoilers

Outboard and inboard flaps

Rudder

Fin

Fixed leading edge and trailing edge panels

Carbon wheel brakes

Pylon fairings

Cowl components

Elevators

Tailplane

Furnishing and floor panels

Main landing gear doors and fairings

Nose landing gear doors

Radome

Flap fairings

Ailerons

Wing to body fairing

8.1   Application of composites in the Airbus 320.

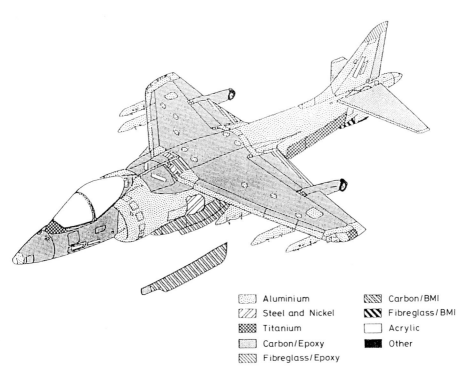

	Aluminium		Carbon/BMI
	Steel and Nickel		Fibreglass/BMI
	Titanium		Acrylic
	Carbon/Epoxy		Other
	Fibreglass/Epoxy		

8.2   Application of composites in the Harrier AV-8B.

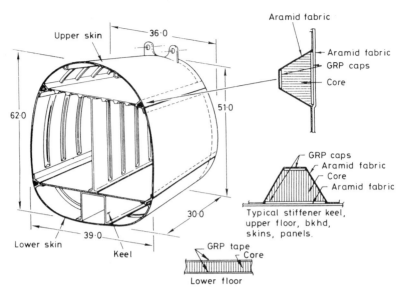

8.3   Test section of proposed fuselage design
(units-inches).

over its equivalent in aluminium alloy. The second example is a military
aircraft – the Harrier AV-8B. Primary structural applications are the wing
torque box and control surfaces, horizontal tail and forward fuselage.
Secondary structures are the gun and ammunition packs, strakes, ventral fin,
rudder, engine bay doors, nose cones and fairings. Twenty five per cent of the
airframe weight is fabricated from composite materials. For the European
Fighter Aircraft (EFA) currently under development, the projected target for
composites utilization is 35% involving the main wing, the forward fuselage,
and the fin and rudder.[3] For any structure or subassembly it is likely that a
combination of materials will be used, each applied so that its individual set of
properties can be used to best advantage. As an example, Fig. 8.3 shows a scale
test section of a proposed fuselage design where a number of combinations of
materials are employed. It incorporates carbon and Kevlar reinforced epoxy,
aluminium honeycomb and unidirectional and woven reinforcements.[4]

Parts integration is also a key factor. The imaginative design of tooling and
the taking advantage of the flexibility available in composites design allows
parts lists, compared with the equivalent metal fabrication, to be much
reduced. Figure 8.4 shows the design of an integrated composite horizontal
stabilizer which offers an overall 15% weight-saving.[5] Although the basic
material cost is more expensive, this is more than offset by reduction in
substructure and assembly costs due to the smaller number of component
parts. Overall the production cost savings are projected to be 18% over an
equivalent all-metal stabilizer. Taking this concept to the limit, designs are

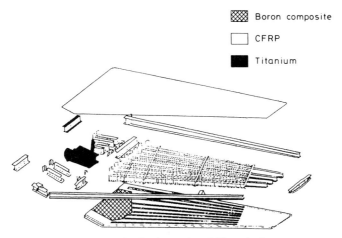

Boron composite

CFRP

Titanium

8.4   Composite horizontal stabilizer design.

now being proposed for an entire filament-wound aircraft fuselage. Two methods of providing the required stiffness characteristics have been studied. The first employs a filament wound isogrid network. Here skins are initially wound at $\pm 45/90$ and then split. A geodesic reinforcement is then bonded to the two skin sections. The second method uses a honeycomb stiffened structure with CFRP skins. Lightweight composite tooling is used which is capable of outward expansion and allows consolidation of the part during cure in the female mould.[6]

The excellent fatigue performance of composites is used to good effect with propeller designs.[7] Using a combination of a unidirectional carbon spar and glass cloth reinforced skins at $\pm 45°$ for torsional resistance produces a very effective component (Fig. 8.5). Complex aerodynamic shapes are more easily manufactured with composites than with metal and this provides added design flexibility. A tough polyurethane paint coat is often applied at the end of the production process to protect the structure from debris damage and runway stone impact. In jet engines there is scope for composites, particularly in the cooler regions. Carbon/polyimides are proposed for both rotors and stators and the early problems associated with turbine blades have been largely overcome, although in the latter case there is strong competition as a result of the good performance of titanium. Ducts of various types are also fabricated from carbon fibres, and Kevlar have application for the ring that surrounds the engine which is required to be designed to contain debris in the event of turbine blade failure.

Helicopter airframes also provide opportunities for composites. In a Franco-German programme a new aircraft is proposed which includes 80% composite; 24% CFRP; 42% CFRP honeycomb and 11% Kevlar honeycomb.[18] Design features are as follows:

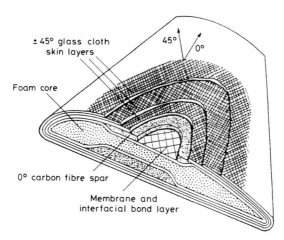

±45° glass cloth skin layers

Foam core

0° carbon fibre spar

Membrane and interfacial bond layer

45° 0°

8.5   Composite propeller design.

- Frames and beams of Kevlar 49 carbon laminates.
- Panels of carbon and Kevlar sandwich construction with a Nomex honeycomb core.
- Carbon/Kevlar sandwich structures for the underfloor for high energy adsorption.
- Landing gear frames of CFRP laminates.
- Carbon sandwich structures for the tail boom for stiffness and strength.
- Vertical and horizontal stabilizers of carbon and Kevlar laminates.

In a similar way to propellers, composites have revolutionized helicopter rotor blade design.[9] As the blade rotates, pitch changes, which are necessary to balance lift forces, cause very high levels of fatigue loading. Composite main rotor blades that utilize unidirectional CFRP in the spar design have virtually unlimited life (Fig. 8.6). Furthermore, with advances in aerodynamic design it is found that blades of complex form are required for optimum performance (Fig. 8.7). These are now universally made using composites – fabrication costs for a similar metal design would be prohibitive. Future blade developments are likely to focus on aeroelastic designs where the blade structure and material properties control motion in order to modify aerodynamic performance, or reduce stresses or vibration. To achieve this in a passive manner is likely to require radially distributed aerofoil sections in combination with highly asymmetric and complex hybrid laminate constructions.

Conventional rotor hubs, i.e. the structure that connects the blades and the main body of the aircraft, are very complex units which have a multiplicity of bearings, seals and lubricators to allow blade movement whilst ensuring proper load transfer. Novel designs using elastomeric composite materials with high levels of elastic deformation have resulted in concepts which are

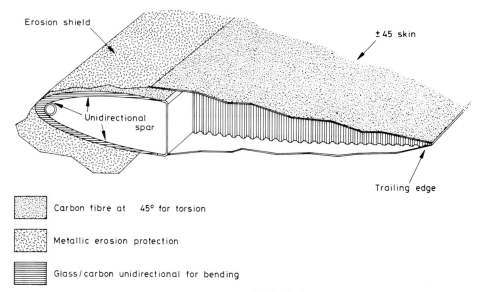

Erosion shield

± 45 skin

Unidirectional spar

Trailing edge

Carbon fibre at 45° for torsion

Metallic erosion protection

Glass/carbon unidirectional for bending

8.6 Helicopter rotor blade design.

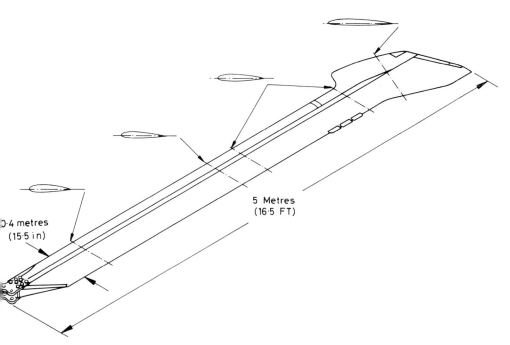

5 Metres
(16·5 FT)

0·4 metres
(15·5 in)

8.7 Helicopter rotor blade optimized for aerodynamic performance.

essentially bearingless.[10] The advantages of this are manifold, including reduced maintenance and drag, reduced parts count, lower weight and improved damage tolerance and lifetimes. The centrifugal load from the elastomeric bearings is carried by a hub plate which is also composite.

For all aircraft structures air certification is a major issue and the difference in behaviour between composites and metals has required reconsideration of the prevailing guidelines.[11] The static structural loading cases and the minimum ultimate factor of safety to be considered are the same for conventional metallic structures, i.e. the design ultimate load is 1.5 times the design limit load (the maximum expected service load). However, owing to their anisotropic elastic behaviour, composites can be sensitive to individual design features and therefore regulatory authorities normally require both structural analysis and test results for each design case. For analysis, design allowables are established in a similar way to metals, viz.:

- Failsafe/redundant designs require a value above which at least 90% of the population of values falls with a confidence of 95%.
- For a single load path design, the value above which at least 99% of the population is expected to fall with a confidence of 95%.

A typical route for obtaining the required evidence to substantiate a safe design would be as follows:

- Structural analysis identifying the key design features.
- Confirmation of analysis by loading a fully instrumented structure.
- Determination of allowable values for each significant structural feature with allowance for variability and environmental effects.
- Test of the complete structure to a minimum of the design ultimate loads and check that allowables are not exceeded.

For fatigue loading two approaches are used. In the first, a 'safe life' is defined where no significant damage is likely to occur. A component is not allowed to operate beyond this life. The second method is based on a 'failsafe' approach where the component can remain in service accompanied by an inspection regime which will detect any flaw before it becomes of critical importance. The choice of method depends on the aircraft of concern. For large transports a failsafe approach may be taken, whereas helicopters may be based on a safe life methodology.

When testing a large structure either statically or in fatigue there are a number of factors that need to be considered. Owing to cost the simplest way is to test at room temperature without imposing environmental conditions and using a factor on loads to cater for the effect of the environment, cycling and variability. Individual factors need to be determined for each case but they are often in the range 1–2. As these terms tend to be dominated by degradation of matrix-based properties, tension members and, indeed, any metal components can be unduly penalized. A second approach is to carry out rigorous testing on sub-assemblies and then carry out a static room temperature test on the completed structure up to ultimate load. If the strain value determined in the structure test exceeds values determined previously, the system is deemed to have failed. In addition to the mechanical design loads and the operating environment, any structure must also be seen to conform to performance standards for irregular loadings, for example impact (bird strike, turbine disc burst, low velocity impacts the results of which are not visible), flammability and lightning strikes.

Design example: sine wave spar

Proposed designs for composite wings often consist of a number of spars with laminated skins. A multispar design means that load on the individual spars is low and buckling tends to be the dominant design criterion. A structure of sine wave configuration is ideal for this application.[12]

The design of the sine wave configuration itself is influenced by a number of factors:

- The size and spacing of fasteners through the flanges.
- The width of the web.
- The critical buckling load.
- The ease of processing.

Different geometries are possible but it is found that a wave configuration based on arcs which are not tangential, but separated by a small flat region is the optimum of the alternatives. Of the other options tangential arcs pose tooling difficulties and a true sinewave has insufficient buckling stability.

A typical spar cross-section is shown in Fig. 8.8. The webs of the spars have three layers; two CFRP cloth plies with fibres oriented at $\pm45°$ to transmit shear loads, and one unidirectional CFRP ply in the centre to provide vertical stiffness. The $\pm45°$ layers are folded over to form the flange. Additional reinforcements are applied each side of the web, orientated at $90°$, to transmit the load from the fasteners into the web. The structure is completed by capping plies on the tops of each flange. The stability of the web as a function of the number of laminate layers is given in Fig. 8.9. Matched metal tooling has been developed for the sine wave spars (Fig. 8.10), as, in detail, the spars are not symmetrical about a centre-line and the tools must therefore be capable of splitting into six parts.

Space

In terms of the innovative use of new materials space applications in many ways provide more scope than the aircraft industry. For satellites the timescale from concept to part manufacture can be as little as two years and for the short

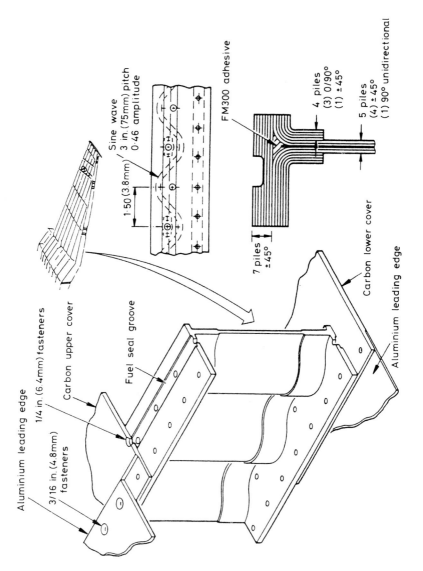

Sine wave
3 in. (75mm) pitch
0·46 amplitude

1·50 (3.8mm)

7 piles
±45°

FM300 adhesive

4 piles
(3) 0/90°
(1) ±45°

5 piles
(4) ± 45°
(1) 90° unidirectional

Carbon lower cover

Aluminium leading edge

Aluminium leading edge

1/4 in. (6.4mm) fasteners

Carbon upper cover

3/16 in. (4.8mm) fasteners

Fuel seal groove

8.8   Typical spar cross-section.

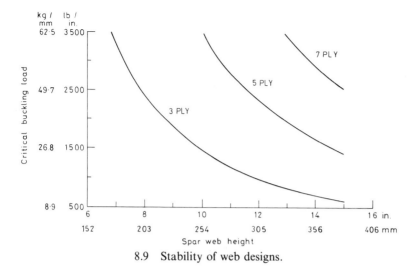

8.9   Stability of web designs.

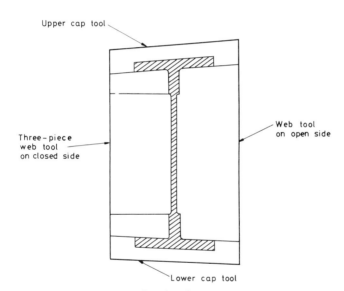

8.10   Tooling for sine wave spar.

product runs normally involved, the materials element in the final cost is often relatively low. Also in many applications no other material is suitable either for reasons of mass or thermal control.

The mechanical design requirements for satellite components are usually determined by launch conditions. Typically for conventional rocket launch, loading would be 7 g in the axial direction with 1 g in the lateral direction, whereas for a space shuttle launch the loading is more uniform with approximately 5 g in both directions. The natural frequency of the structure,

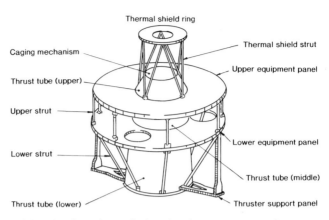

8.11   Design of a typical carbon/epoxy structure for a
communications satellite.

and hence its stiffness, will often control the mechanical aspects of the design. The potential of vibration-induced coupling with the launcher is a significant concern. Once in orbit, mechanical loads are comparatively low. Environmental conditions under some circumstances can be arduous and severe thermal cycling can feature, as well as the effects of high vacuum and erosion through atomic oxygen or micrometeroid impacts.

Where thermal insulation is important, for example in local bracketry, GRP is used. The material is also employed for some antenna reflectors, but for these structures metal meshes must be incorporated to provide a radio frequency conductive surface. Carbon fibre systems, however, are those materials most often associated with space applications. The potential for very high stiffness and excellent thermal stability over a wide temperature range make them ideal. Fairings, manipulator arms, antennae reflectors, solar array panels and optical platforms and benches are all examples of CFRP space structures.

While CFRP has been readily used for these subsidiary components which are primarily stiffness-critical, it is only comparatively recently that they have had applications for primary structures. In the past the need for a combination of stiffness and strength, and for thermal and electrical conductivity have favoured metals. However, the pressures for weight reduction have been unrelenting and there are now some satellites, for example Fig. 8.11, which have been built with a predominantly composite structure sub-system.[13]

Design example: collapsible tube mast

Long booms or masts are frequently required on spacecraft for a variety of purposes, such as supports for solar arrays, as antennas, or to locate

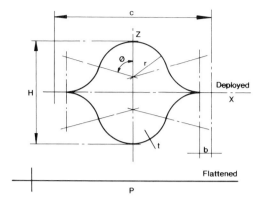

8.12   Tube section characteristics.

experiments remote from the spacecraft environment. Prior to, and during, launch the masts must be stored in a compact form on board the spacecraft. There may also be a need to retract and redeploy masts when in orbit, for example to minimize the angular momentum during changes in the attitude of the spacecraft.[14]

A variety of designs of extensible or foldable masts have been developed. Some are based on a series of telescopic tubes, others make use of the de Haviland tube principle consisting of a thin walled tube of elastic material, split longitudinally, which is opened out flat and coiled for launch. When uncoiled, it reverts to a circular cross-section and in some designs, provision is made for the edges of the split to interlock mechanically to increase the rigidity.

The collapsible tube mast (CTM) is a closed tube of lenticular cross-section that can be flattened and then reeled on a drum for storage in a relatively small volume. Collapsible masts made from thin metal sheet were explored in the United States space programme in the 1960s.[15-17] A design for a lenticular collapsible tube mast is shown in cross-section in Fig. 8.12. It consists of two channel sections, each composed of circular arcs and joined at the flanges by welding or adhesive bonding.

The tube shape and size are characterized by four parameters:

• The shape radius (r).
• The shape angle (ϕ).
• The flange width (b).
• The wall thickness (t).

Variation of the shape angle between 0 and 90° produces a family of cross-sections, with small values of ϕ yielding flat tubes, and large values giving rise to a more bulbous shape. At small angles, the bending stiffness EI_z is higher than EI_x and in practice the tube is often required to have

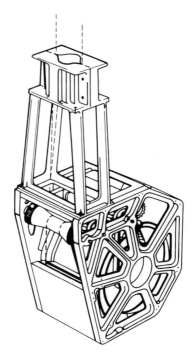

8.13   CTM storage box and deployment mechanism.

approximately equal stiffness about the x and z axes. This occurs for $\phi = 1.4$ radians, but to provide equal buckling resistance about the x and z axes a slight compromise is necessary and this is given by a value of 1.3 radians ($\sim 75°$).

In order to maintain the flattening strain to within acceptable limits (0.4%), the wall thickness for a shape radius (r) of 20 mm needs to be approximately 0.16 mm. To keep the overall stress level low, strains should be less than 0.25% and this gives the radius of the coiling drum $R = 64$ mm.[18] The values of r and t, together with the external volume of the storage compartment that can be accommodated in the spacecraft, define the number of turns and hence the length of mast that can be carried. As an example, approximately 60 m of mast can be carried in a storage volume of approximately 24 litres using the above dimensions. The width of the flanges is determined partly by the shear stress between the tube halves, partly by the need to minimize the overall storage volume, and partly by the requirement that sufficient width must be provided to enable the deployment mechanism to engage.

The deployment mechanism is shown in Fig. 8.13. The stresses induced by flattening and coiling, together with consideration of longitudinal stiffness for the mast dictate a material with orthotropic properties, and the need to fabricate masts some tens of metres in length suggests a plain weave fabric as

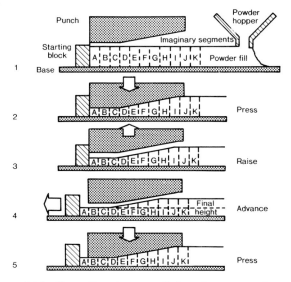

8.14   Operations in sequential motion compaction.

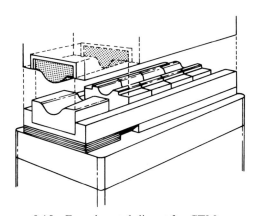

8.15   Experimental die set for CTM.

the most convenient feedstock material. In considering methods for fabricating long lengths of mast, truly continuous processes such as pultrusion and roll-forming must be rejected because of the risk of damaging or distorting the very thin fabrics required. An alternative technique is sequential moulding where high longitudinal stresses inherent in, for example, the pultrusion process are avoided. The principle of the method, as developed for the production of continuous profiles from metal or ceramic powders,[19] is shown in Fig. 8.14. Adaptation to the formation of the mast half-profile is straightforward using the open-ended die-set illustrated in Fig. 8.15.

Production of the mast takes place in two stages. In the first stage,

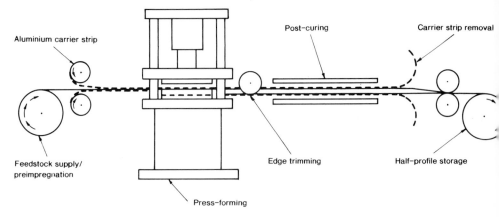

8.16   Schematic of CTM forming stage.

illustrated schematically in Fig. 8.16, preimpregnated fabric from a storage reel is sandwiched between two carrier strips of aluminium foil, 0.24 mm thick, which are treated, on their inner surfaces, with a release agent to prevent adhesion to the feedstock. After passage through the press, excess width is trimmed from the flanges of the moulded profile by a pair of guillotine blades mounted symmetrically about the apex of the profile. When a thermosetting matrix is used, a partial post-cure is given prior to removal of the carrier strips, whereupon the half-profile is flattened and reeled on to a storage drum. In the second stage, two identical half-profiles are bonded together at the flanges to form the mast. The completed tube mast structure mounted within a demonstration Perspex carrier drum is shown in Figure 8.17.

Process plant

Corrosion resistance, strength and ease of fabrication make composites, particularly glass reinforced plastics, admirably suitable for process plant applications. Large, complex GRP components are readily produced in small quantities and the installed cost of GRP plant is highly competitive with that manufactured from other more traditional materials, e.g. stainless steel, rubber-lined steels, etc. Storage tanks, pressure vessels, pipework, stacks and ducting systems are available from a number of organizations and there are numerous examples of successful operations over many years.[20–23]

The majority of process equipment are basically thin, cylindrical shells. Thicknesses are typically in the range 5–50 mm, the term 'thin' being used only because the thickness of the component is usually considerably less than its diameter which would be in the range 1–10 mm. Pipes may be up to several metres in diameter and characteristics and design problems apply equally to pipes and vessels. The length or height of vessels is usually of the same order of

8.17 CTM mast system.

8.18   Example of GRP storage vessel (photograph courtesy of PDE, Motherwell Plastics Fabrications).

magnitude as their diameter, although some vessels are quite tall, e.g. a vessel 4 m in diameter and 20 m high is not unusual. Typical examples are shown in Fig. 8.18 and 8.19.

Tanks for storing liquid at atmospheric pressure may have open tops but most tanks and all vessels will have end closures. These may be of flat, conical or domed form. Invariably a number of branches are required in the cylinder and ends to allow pipework to be connected to the vessel, openings and covers are required for access and inspection, and arrangements have to be made to support the vessel and for the attachment of internal and external fittings such as trays and ladders. Lifting lugs will be fitted and the vessel may be made in flanged sections to facilitate manufacture and erection. A variety of other features can be incorporated in the design.

Vessels are frequently required to contain hazardous, corrosive and toxic substances; operating temperatures may range from 0 to 110 °C and operating pressures are usually in the range 0–10 bar absolute. Figure 8.20 shows the limitations for pressure and size applicable in the current British Standard. Vessels have been manufactured outside these limitations but special consideration is required with respect to the design, manufacture and installation of such items. There is no size restriction on tanks subjected to the static head of liquid contents only.

In chemical process plant applications the provision of a corrosion-resistant lining is an essential feature of the design and it is often this consideration which leads to the selection of GRP in the first instance. A chemical-resistant lining may be obtained by one of two methods. The first of these is to offer a resin-rich surface of proven chemical resistance to the process medium. This surface layer would be reinforced with a C-glass surfacing mat, synthetic fibres or other suitable material with a thickness between 0.25 and 0.5 mm and backed with layers of reinforcement with a high resin content. Once this chemical barrier is complete, further layers of reinforcement may be added to satisfy the structural requirements of the particular item of plant. Alternatively, a thermoplastic lining may be employed. When supported by GRP laminates the established chemical resistance of thermoplastics can be used at temperatures and pressures well beyond their normal operating ranges. In this method a continuous chemical barrier is manufactured from thermoplastic sheet which may be formed and welded to suit the dimensions of the vessel. Depending on the nature of the thermoplastic, external GRP reinforcement may be bonded to the liner by either a chemical or a mechanical mechanism. The essential difference between these two methods of manufacture lies in the fact that, in the former, GRP laminates are subject to the effects of a corrosive medium whilst under load, whereas in the latter the GRP acts as a structural material only. Figure 8.21 shows examples of typical laminate configurations.

There are a number of standards covering materials and design of GRP process equipment and the industry is perhaps the most advanced in the

8.19   Example of GRP pipework (photograph courtesy of PDE, Motherwell Plastics Fabrications).

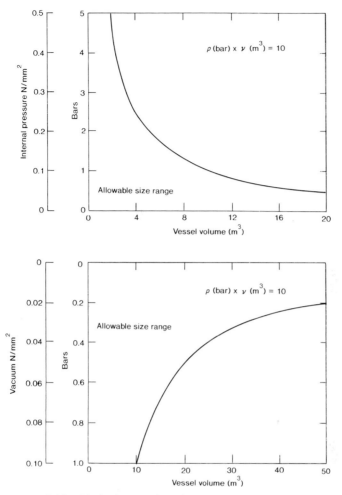

8.20   Limitation on size of vessel covered by the
design code BS 4994.

composites area in this respect.[24] The design of GRP process equipment when carried out in accordance with UK standards may be divided into three sections: the assessment of an allowable design strain; calculation of the applied unit loads; and the choice of an appropriate laminate construction.

Allowable Design Strain. Specifications BS 4994/6464 provide a carefully considered design method of the calculation of allowable design strains. The purpose of this procedure is to arrive at a value of strain at which the laminate will not undergo deterioration under the specified process conditions over the lifetime of the vessel. These calculations are carried out by the use of a number

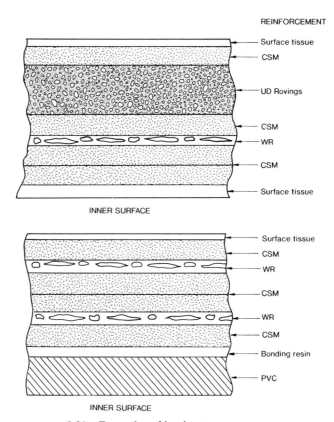

8.21   Examples of laminate arrangements.

of part factors which deal with the effect of loading and manufacturing conditions on the chemical and mechanical behaviour of GRP laminates.

These part factors may be described as follows:

- k_1 – method of manufacture (range 1.6–3.0).
- k_2 – long-term behaviour (range 1.2–2.0).
- k_3 – temperature (range 1.0–1.2).
- k_4 – cyclic loading (range 1.1—1.4).
- k_5 – curing procedure   (range 1.1–1.5).

The product of these factors and a further factor of safety of three results in an overall design factor, K, which is used to determine the allowable design strain, ε_L:

$$\varepsilon_L = \frac{u}{KX} \qquad [8.1]$$

where u is the laminate-allowable unit load and X is its extensibility or unit modulus. These properties have the units of load per unit width. The combined value K is usually in the range 8–15. There is a further over-riding upper limit to the design strain of 0.002 or 0.1 ε, (where ε is the fracture strain of the unreinforced resin in a simple tension test).

Although the above procedure is somewhat empirical and the relationships between the effects represented by the part factors k_1 and k_5 are not established experimentally, experience indicates that results obtained from it are approximately correct and lead to safe designs.

Applied Unit Loads. Within the standards equations are given for calculating the unit loads, Q, applied for common design cases, e.g. for spherical shells subject to internal pressure, p:

$$Q = pD_i/4 \qquad\qquad [8.2]$$

and for cylinders

$$Q = pD_i/2 \qquad\qquad [8.3]$$

The axial unit load in vertical cylindrical vessels and tanks due to combined loads is given as:

$$Q = pD_i/4 \pm 4M/\pi D_i + W/\pi D_i \qquad\qquad [8.4]$$

where M is an applied bending moment, e.g. that due to wind loading, W is an axial load, e.g. weight of the contents, p is applied pressure and D_i is the inside diameter of the vessel.

These and similar equations are simply expressions for the membrane unit load in thin shells. A typical equation given for end closures, i.e. for domed ends subject to internal pressure is:

$$Q = 0.55p\, D_i K_s \qquad\qquad [8.5]$$

where K_s is a shape factor based on theoretical stress concentrations tabulated for particular shapes of end.

As already noted, the elastic moduli for GRP materials are lower than those of steel, and therefore in certain types of vessels shell-buckling may be a potential problem. The code gives equations for the maximum permissible compressive unit load in a cylinder:

$$Q = 0.58\, t_{LAM}\frac{X_{LAM}}{FD_i} \qquad\qquad [8.6]$$

where t_{LAM} is the thickness of the laminate.

The minimum permissible thickness for shells subject to external pressure, t_m, may also be calculated. For example, for spherical shells:

$$t_m = 2R_i\sqrt{\frac{pF}{E_{LAM}}} \qquad\qquad [8.7]$$

In these equations F is a safety factor usually given a value of four, and X_{LAM} and E_{LAM} represent the overall laminate stiffness and modulus.

Laminate construction. GRP vessels are manufactured from the successive application of individual layers of reinforcement and this allows for great flexibility in the design of shell laminates. As a result of this flexibility there are many possible combinations of reinforcement type which will meet the structural requirements of any one design case. This enables a designer to select the laminate construction which will be best suited to his or her manufacturing facilities and hence be most cost-effective. It is due to this that the concepts of unit load, unit strength and unit modulus are most useful, as although possible laminate constructions are dissimilar in detail, they may still satisfy the calculated load requirement.

The allowable unit load, U, for a laminate may be calculated from:

$$U = \varepsilon_{LAM} X_{LAM} \qquad [8.8]$$

where ε_{LAM} is the lowest of all the different values of ε_L, 0.002 and $0.1\varepsilon_r$. X_{LAM} is the summation of the stiffness of all layers in the laminate and is given by:

$$X_{LAM} = \Sigma w_x n_x X_x \qquad [8.9]$$

where w_x is the weight of glass per unit area for one layer of type x, n_x is the number of layers of type x and X_x is the extensibility (unit load/strain).

Although the basic BS 4994 procedure concentrates on determining the required weight of reinforcement rather than thickness, the thickness of the laminate is required for those calculations involving shell bending and buckling. For a given weight of glass per unit surface area, w, and a given percentage glass content by weight, M_g, the thickness (t) of any layer can be calculated from:

$$\frac{t}{w} = \frac{1}{\rho_r}\frac{\rho_r}{\rho_f} + \frac{1}{M_g} - 1 \qquad [8.10]$$

where ρ_r and ρ_f are the densities of the resin and glass fibre. Hence the thickness of the whole laminate, t_{LAM}, can be determined. Figure 8.22 shows the relationship given by equation 8.10 in graphical form.

The effective Young's modulus, E_{LAM}, for a laminate is given by:

$$E_{LAM} = X_{LAM}/t_{LAM} \qquad [8.11]$$

Although these standards are perhaps the most comprehensive design code for GRP vessels in use at the present time, there are some fundamental shortcomings in the documentation which should be appreciated, particularly when designing critical items of chemical plant:

• The designs are based almost entirely on short-term tensile test data. Ideally, a method of predicting long-term behaviour from short-term information should be considered.

• It is assumed that strains within a laminate construction are constant

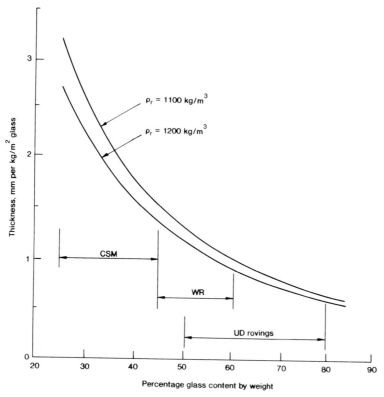

8.22   Thickness versus percentage glass content for
GRP layers.

throughout its thickness. Design cases that involve shell bending should be
given careful consideration.

- Within a laminate there may be individual layers with different thermal
  and elastic properties. No consideration is given to the interaction of these
  different layers under thermal load.

Design example: butterfly valve

The lives of cast iron components in the cooling water circuits of estuarine
power stations are often significantly reduced by a number of factors which
include graphitic corrosion, localized corrosion due to mixed metals, pitting of
surfaces, seizure of sealing faces during outages and brittle fracture. In order to
overcome these problems GRP is being considered as an alternative material
of construction and there are a number of units now in service. These are large
structures with bore diameters up to 1800 mm. Figure 8.23 shows details in

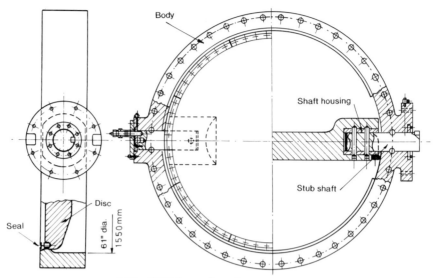

8.23   GRP butterfly valve component.

this case for a component 1524 mm (60 in.) diameter.[25-27] Typical design pressures are in the region of 0.5 MPa (5 bar g).

In this design the seal was made of nitrile rubber and was based on the geometry for the previous cast iron designs. The GRP disc thickness of 190 mm ($7\frac{1}{2}$ in.) was a compromise between that required for satisfactory disc deflection in the valve-closed position and that required for low hydraulic head loss. The disc was an all-woven roving GRP construction, except for the seal clamping ring and bolts, the stub shafts and their housings, all of which were constructed from stainless steel. A woven roving and isophthalic resin system was chosen, because of its relatively high modulus and proven durability and as it is a reasonably economic form of composite material.

The body was of a 'wafer' design, being clamped by long through-bolts between the adjacent pipe flanges. This design avoids the conventional flanges in the body, which could be subject to unknown system loads, and also provides a sufficient material thickness to limit the body's bending deflections due to pressure. The construction consisted of alternating layers of woven rovings and CSM, this type of construction having a higher shear strength than the all-woven roving design, but a lower modulus. Again an isophthalic polyester resin system was used. The surface mating with the rubber seal on the disc was the moulded gel coat, which therefore avoided the corrosion or seizing mechanism which occurs in the cast iron valve. The flange mating with the actuator gearbox must be capable of withstanding the motor stall torque of 27 kN m (20 000 lb/ft) and the bending moments due to the gearbox and actuator dead weight. Lifting lugs were omitted from the design as it was considered that the relatively low weight of the valve would enable it to be

Table 8.1. Butterfly valve material test results

	Glass (% by wt)	Tensile modulus (GPa)	Tensile strength (MPa)	Unit modulus $(\text{Nmm}^{-1}/\ \text{kg}\,\text{m}^{-2}\ \text{glass})$	Unit strength $(\text{Nmm}^{-1}/\ \text{kg}\,\text{m}^{-2}\ \text{glass})$	Shear strength (MPa)
(a) Valve body material*						
Design assumption	37	10	145	18 200	267	7
BS 4994	–	–	–	14 700	258	7
Measured average	40	10.57	159	17 110	255	9.4
(a) Valve blade material†						
Design assumption	48	14.5	215	19 000	280	7
BS 4994	–	–	–	16 200	300	7
Measured average	48	13.45	214	18 110	287	12.3

* Average results of eight separate test boards from both bodies.
† Average results of eight boards, four from each blade.

installed by a fork lift truck, rather than requiring a crane. The valve geometry was designed to material properties previously found to be typical for the two laminate types, based on the chosen resin and glass cloths. The valve body was constructed with alternating layers of woven roving and chopped strand mat. For this material, a modulus of 10 GPa and a strength of 145 N/mm^2 were assumed. The valve blade on the other hand was constructed from all-woven roving, because of its higher modulus, which was expected to be 14.5 GPa. Its anticipated strength was 215 N/mm^2. Table 8.1 shows details of material properties.

The stress analysis of the valve disc assuming constant thickness and homogeneous material properties is given below.[28] The equations assume a constant thickness disc, supported at two diametrically opposite points on the perimeter and under constant pressure loading. These equations were used to give the theoretical bending moments and deflections for the butterfly-valve disc. The relationships derived were as follows:

$$\frac{M_r}{pa^2} = \frac{(1-\rho^2)}{(3+v)}\left\{\frac{(3+v)^2}{16} + \cos 2\theta + \frac{3-v}{4}\cos 4\theta\, \rho^2 + \frac{2-v}{3}\cos 6\theta\, \rho^4\right\}$$

[8.12]

$$\frac{M_t}{pa^2} = \frac{1}{12(3+v)}\left\{\frac{3(3+v)}{4}[(3+v) - \rho^2(1+3v)] - 12\cos 2\theta(1+v\rho^2)\right\}$$

$$- \frac{1}{12(3+v)}\{3\cos 4\theta\, \rho^2\,[(3-v) - (1-3v)\rho^2]$$

$$\times 4\cos 6\theta\, \rho^2\,[(2-v) - (1-2v)\rho^2]\}$$

where M_r and M_t are radial and tangential bending moments per unit length respectively, p the water pressure on disc, v the Poisson's ratio, a the disc radius, r, θ the polar co-ordinates, and ρ is the ratio r/a.

The disc transverse deflection, w, is given by the expression:

$$\frac{2w\,D(3+v)}{pa^4} = \left(2\log_e 2 - 1\right) + \frac{1+v}{1-v}\left(2\log_e 2 - \frac{\pi^2}{12}\right)$$

$$- \rho^2 \cos 2\theta \left(\frac{1}{2} + \frac{1}{2}\cdot\frac{1+v}{1-v} - \frac{\rho^2}{6}\right) \qquad [8.13]$$

$$- \rho^4 \cos 4\theta \left(\frac{1}{12} + \frac{1}{24}\frac{1+v}{1-v} - \frac{\rho^2}{20}\right)$$

$$- \rho^6 \cos 6\theta \left(\frac{1}{30} + \frac{1}{90}\frac{1+v}{1-v} - \frac{\rho^2}{42}\right)$$

where

$$D = \frac{Et^3}{12(1-v^2)}$$

It was found that, on areas of the disc away from stress concentrations, there was agreement between the measured and theoretical stresses to within $\pm 14\%$. Also, there was agreement between the measured and theoretical disc deflections to within $\pm 5\%$ (Fig. 8.24). Thus, even though the theoretical analysis was based on homogeneous material properties, and a constant thickness disc, it may be used with confidence at the design stage for thickness calculations.

Valves of this size are now in service and during the reported inspection three years after installation the components were found to be fully satisfactory. At the time of manufacture the cost GRP alternative was found to be marginally more expensive (16%) than its traditional counterpart, coated cast iron, but these were found to be difficult to source. The other option using a corrosion-resistant metallic design would have been approximately 30% more expensive than GRP.

Medical

Lightweight, attractive stiffness characteristics and bio-compatibility has meant that composite materials are finding application in a number of areas in the medical sector. External components such as artificial limbs are those which are the most similar to conventional engineering structures and here high specific properties, fatigue resistance and flexibility of manufacture can

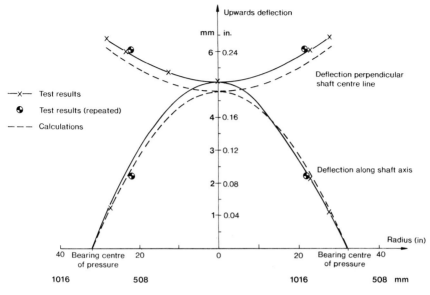

8.24   Disc deflections at 0.5 MPa (60 psi).

all be used to advantage. Figure 8.25 shows an early design of an artificial limb where the main load-bearing structures are press-moulded CFRP.[29] More recent developments incorporate a wider use of carbon fibre to include load-carrying links in the joint mechanisms, foot keels with sprung energy return, and hybrid designs incorporating elastomers to dampen shock loads.[30]

The chemical inertness of carbon fibre has led to a number of surgical applications where the material is used in conjunction or instead of metallic or polymeric materials. The behaviour of artificial joints which conventionally consist of an ultra high molecular weight polyethylene (UHMWPE) component articulating against a polished steel part can be improved by enhancing the wear characteristics of the polymer. The UHMWPE can be reinforced by a random distribution of short (3 mm) carbon fibres to provide the desired tribological properties.[31] Carbon composites are being implanted into cartilage to promote biological resurfacing of damaged areas. The open weave structure of the material promotes cell growth along and between individual fibres ultimately resulting in a suitable repair. In a related application carbon fibre tows, either used individually or in plaits, are being employed in the repair of damaged ligament. Loops of material are passed through holes drilled in the adjacent bone structures and then their length adjusted to achieve the correct tension for the particular patient. The mechanical properties of bone repair materials, often a self-curing polymethyl methacrylate, can be enhanced by the addition of carbon fibres. Tensile,

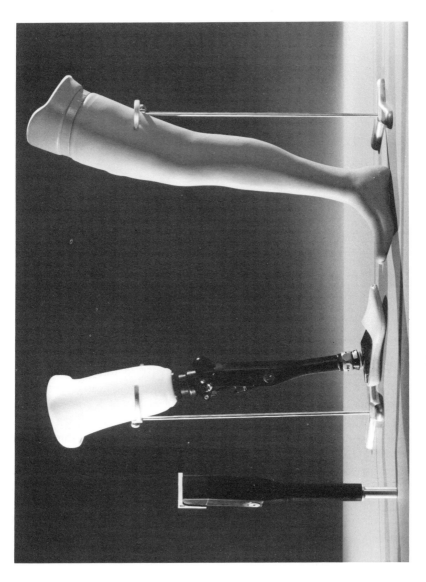

8.25　CFRP prosthesis design.

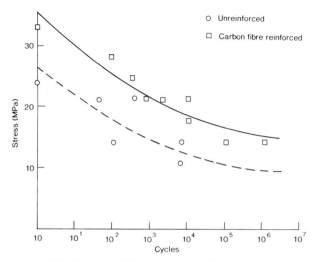

8.26   Results of fatigue tests on bone cements.

compressive and shear strengths as well as creep and fatigue performance are all improved and this could lead to wider clinical use of the material. Figure 8.26 shows some fatigue results on unreinforced and carbon reinforced bone cements.[32] The enhancement due to the carbon fibre can clearly be seen.

Design example: intramedullary nails

Intramedullary nails, used to repair fractured femurs, are currently made from stainless steel or a titanium alloy. The nail is slightly curved (typically a 1 cm bow over a 30 cm length) and hollow. Some designs have a longitudinal slit and holes at either end in which to locate fixing screws. Ideally, the nail should have similar mechanical properties to those of cortical bone, be capable of some bending deformation and have sufficient strength to allow the insertion of fixing screws. The mechanical properties for current materials are not ideally suited to this application, their stiffness being an order of magnitude higher than that of bone. This difference in rigidity means that there is not a uniform load distribution in the region of bone and implant and this can impede healing.[33,34]

Bone is a natural composite material and its properties vary with direction and water content. Typical bone has the following longitudinal and transverse properties: modulus 17.0 and 11.5 GPa; Poisson's ratio 0.64 and 0.58; ultimate tensile strength 148 and 49 MPa; and ultimate compressive strength 193 and 133 MPa respectively. Measured, the bending and shear stiffness of a human femur are found to be $1.7 \times 10^2 < EI < 3.1 \times 10^2$ N m^2 and $1.5 \times 10^2 < GJ < 2.1 \times 10^2$ N m^2. The variation is due to the changing

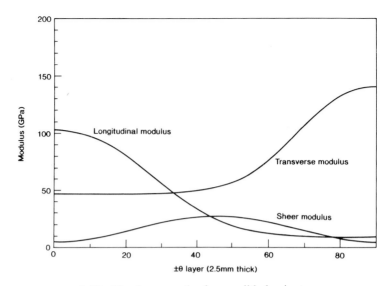

8.27   Elastic properties for possible laminate
configurations.

geometry of the femur. The objective of any design, therefore, is to match these
stiffness characteristics as closely as possible. Typical geometrical constraints
of a nail are inside and outside diameters of 5 and 12 mm respectively.[35,36]

A good way to manufacture a composite component with a substantial
degree of axial symmetry is by filament winding. The proposed layup of the
carbon fibre was, starting at the inside wall, $a + \theta$ layer of 1.25 mm thickness,
$a - \theta$ layer of 1.25 mm thickness and a 1 mm thick hoop layer. The latter was
included to assist in compressing the $\pm \theta$ layers and to improve transverse
strength and the ability of the structure to take fixing screws. The following
properties were assumed for a 60 vol.% unidirectional carbon fibre epoxy
composite: longitudinal modulus, E_1, 140 GPa, transverse modulus, E_2,
9 GPa, in-plane shear modulus, G_{12}, 5.5 GPa, and Poisson's ratio, ν_{12}, 0.18,
and a laminate analysis carried out to determine the elastic properties of a
$[\pm \theta, 90]_s$ lay-up. The results are shown in Fig. 8.27.[37] It is clear that it is not
possible to obtain both E_1 and E_2 in the desired range together with an
appropriate value of G_{12} to match the properties of bone. For simplicity of
manufacture, however, it was decided not to choose a more elaborate winding
pattern. That selected was $[\pm 35, 90]_s$ as this gave substantially similar
longitudinal and transverse moduli, higher than that of bone, but lower than
those of stainless steel or titanium. Figure 8.28 shows the final component.
These are currently undergoing clinical trials.

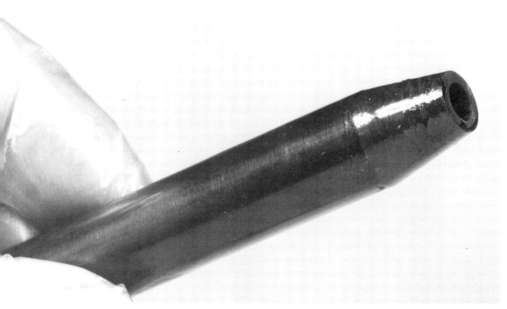

8.28   Intramedullary nail.

Automotive

Although there is a wide usage of plastics in the automotive industry, applications are somewhat restricted to non-safety critical components. The majority of current use is in thermoplastic injection moulded trim items. There is increasing interest, however, in the use of plastics for structural components, especially in the areas of fibre reinforced materials. Prototype components such as steering wheels, steering columns and seats have been developed, but, as yet, have not received wide acceptance.

There are basic differences in the approach adopted when considering the application of composites in the aerospace and automobile industries. This is primarily due to the production requirements of the two activities. In aerospace and defence, the design of the structure is optimized to provide the required performance, and the manufacturing process is subsequently selected on the basis that it is capable of achieving the desired design. In contrast, the automotive industry requires a rate of manufacture which is capable of producing components within a specified unit cost. Therefore, manufacturing processes which are suitable for volume output are the primary consideration, and design of a component must be tailored for that fabrication process. The fibre with the greatest potential for automobile structural applications, based

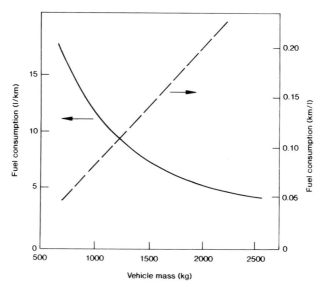

8.29   Relationship between fuel consumption and
vehicle weight.

on optimal combination of cost and performance, is probably E-glass fibre. Likewise, the less costly resin systems are likely to dominate, such as polyester and vinylester, at least in the near term. High performance materials such as epoxies and carbon fibre will find only specialized applications even though their ultimate properties may be superior.

The discovery of a lightweight material, with the potential to increase fuel economy, is one of the driving forces behind the interest in composites for automotive applications.[38] Figure 8.29 shows an empirically derived relationship between total vehicle weight and fuel consumption. The consequences of redesign on a given component are not only limited to the component itself. Secondary weight savings can be considerable, for example, substitution of a lighter body structure results in a reduced braking system (Fig. 8.30), tyres, engine and fuel tank, etc.

The potential use of composites in structural applications in automobiles can be designated into two categories:

- The direct replacement of existing components.
- The integration of multiple steel components into one composite component or structure.

The second category, involving parts integration, offers the most potential as it can result in a significant reduction in the number of individual manufacturing operations. In the shorter term, however, most applications of composites

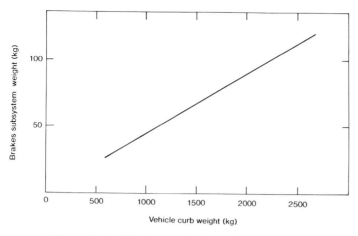

8.30 Relationship between brakes sub-system weight vs vehicle carb weight.

will be in single component replacement programmes, due, in part, to existing levels of capital investment in conventional materials.

The most significant examples of a direct substitution component currently in production are driveshafts.[39] These are widely used in high performance cars and have now found application for conventional vehicles. One example is in the special drive line configuration of the Ford Econoline van. These vehicles would normally have a two-piece steel driveshaft which incorporates a connecting centre bearing. The total length of the drive line dictates that a two-piece steel shaft must be used to prevent excessive vibration. The composite alternative provides satisfactory dynamic characteristics since the lower weight combined with high stiffness satisfies the bending-frequency requirements. The driveshaft is fabricated by filament winding on a continuous machine operating at speeds up to 2 m/min and includes longitudinal carbon fibres (parallel to shaft axis) to give the required bending stiffness, and $\pm 45°$ glass fibres to provide the torsional strength. The resin used is a vinyl ester which gives the appropriate combination of properties, processability and cost. This particular driveshaft is economically feasible largely because of the systems savings resulting from the elimination of the centre bearing.

Reductions in unsprung weight have added attraction due to enhanced performance characteristics, specifically improved ride.[40] GRP leaf springs that can replace steel in light vans and heavy lorries are the most advanced products in terms of market penetration. The feasibility of a number of components ranging from wheels to suspension arms has been assessed (Fig. 8.31).

As already cited the successful application of structural composites to large integrated automotive structures will be more dependent on the ability to use

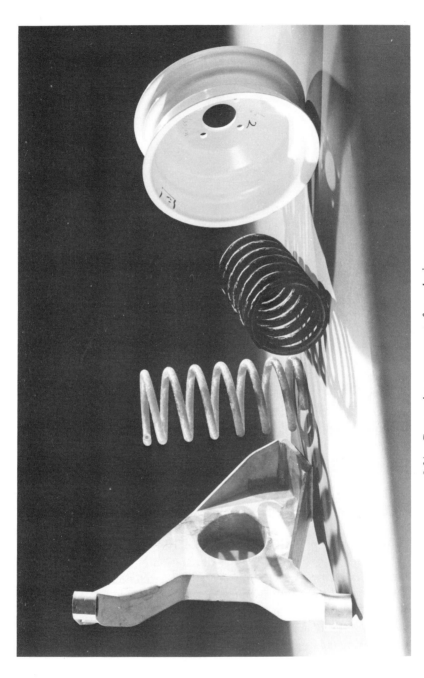

8.31   Composite components for reducing unsprung weight.

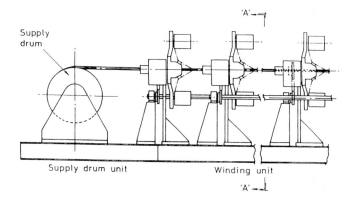

Section 'A' showing spools

8.32　Multi-head spiral winding machine.

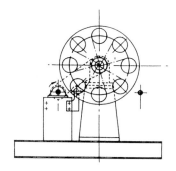

rapid and economic fabrication processes than on any other single factor. The fabrication process must also be capable of close control of composite properties to achieve lightweight, efficient structures. Figure 8.32 shows a scheme for a multi-head spiral winding machine to make tubular stock for the volume production of composite coil springs (Fig. 8.33).[41] The stock material for the spring consists of fibres at $\pm 45°$ wound over a feasible core. Prior to curing, the tubular sections are wound over a segmented spring former (Fig. 8.34).

Composite body panels have been used for many years in the speciality car market. Here the comparatively low cost of tooling and flexibility of design are the main attractions.[42,43] In order to demonstrate the potential of composites a number of demonstration vehicles have been built in CFRP and other materials. Figure 8.35 shows an example and the associated weight savings are given in Table 8.2. In this case savings of the order of 55–65% were achieved. These are clearly extreme values and it is unlikely that CFRP will gain wide acceptance for body parts due to cost. Glass reinforced systems and more

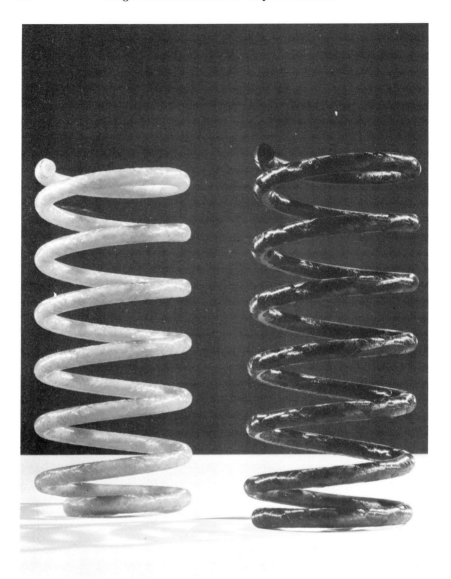

8.33   Composite coil springs.

modest weight reduction targets of the order of 15–20% are likely to be the norm in the immediate future.

Underbonnet applications are also of interest, not only for low stress components such as casings and housings, but also for structures within the engine itself where the benefit of weight savings can be magnified by up to an order of magnitude due to the reciprocation of mass. Gudgeon pins, valve

Table 8.2. Major weight savings in a CFRP vehicle

Component	Weight (kg)		
	Steel	CFRP	Reduction
Body	192.3	72.7	115.0
Front end	43.2	13.6	29.5
Frame	128.6	93.6	35.0
Wheels (5)	41.7	22.3	19.4
Bonnet	22.3	7.8	14.7
Decklid	19.5	6.5	13.1
Doors (4)	64.1	25.2	38.9
Bumpers (2)	55.9	20.0	35.9
Driveshaft	9.6	6.8	2.8
Total vehicle	1705	1138	566

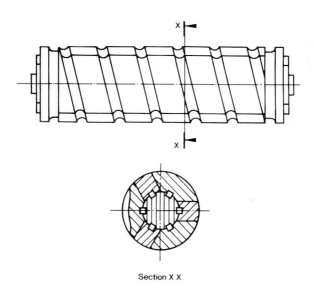

Section X X

8.34   Segmented spring former.

springs and connecting rods are all being evaluated and there have even been concepts put forward for the complete composites engine. These are all some way from commercial exploitation but developments are continuing apace.

Design example: flexible driveshaft

For many driveshaft applications, universal joints must be used to accommodate shaft misalignment and significant advantage could be gained if a component could be developed which maintained torsional stiffness but was flexible in the longitudinal direction. Such a system is not possible with steel or

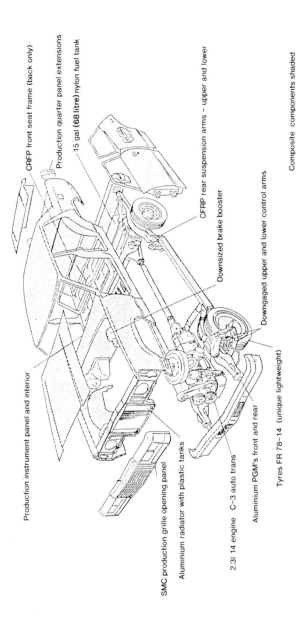

CRFP front seat frame (back only)

Production quarter panel extensions

15 gal (68 litre) nylon fuel tank

CFRP rear suspension arms – upper and lower

Downsized brake booster

Downgaged upper and lower control arms

Composite components shaded

Production instrument panel and interior

SMC production grille opening panel

Aluminium radiator with plastic tanks

2.3l I4 engine   C–3 auto trans

Aluminium PGM's front and rear

Tyres FR 78–14 (unique lightweight)

8.35   Exploded schematic view of CFRP concept vehicle.

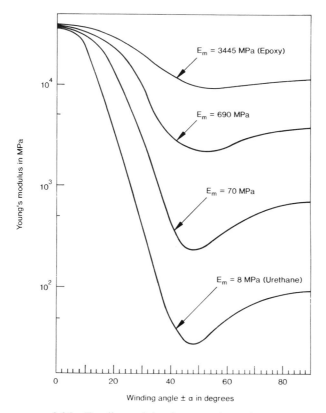

E_m = 3445 MPa (Epoxy)

E_m = 690 MPa

E_m = 70 MPa

E_m = 8 MPa (Urethane)

Young's modulus in MPa

Winding angle ± α in degrees

8.36   Tensile modulus for several matrix types
($v_f = 50\%$).

with a conventional epoxy composite, but there is scope if an elastomeric matrix is used.[44] Figures 8.36 and 8.37 show the variation in modulus values for laminae with various resin systems ranging from a stiff epoxy to a flexible urethane system. S-glass is the reinforcement system. As can be seen, the shear modulus for angles usually used for the transmission of torque (about ±45°) is effectively constant, but the longitudinal modulus can be reduced by around two orders of magnitude.

The design specification for the driveshaft is given in Table 8.3. The criteria for fatigue loading were established after a series of road tests. Calculations were performed using standard formulae for torque transmission and the results are also given in Table 8.3. For comparative purposes designs were carried out for both flexible (urethane) and rigid (epoxy) composite alternatives. Both designs embraced a ±45° laminate configuration. Figure 8.38 shows the results of finite element calculations on the two tubes and the benefits of the more flexible system in reducing the axial stress due to misalignment is evident.

Table 8.3. Design requirements for flexible driveshafts

Design specification	
Horsepower	96
Torque	123 N m (1090 lb in.) at 5600 rpm
Length	1117 mm (44 in.)
Maximum diameter	76.2 mm (3 in.)
Critical speed	150 Hz
Fatigue conditions	
Torque	$\pm$ 536 N m (4800 lb in.)
Lateral deflection	$\pm$ 15.75 mm (0.62 in.)
Bending angle	2.5°
Frequency	1 Hz
Life	200 000 cycles

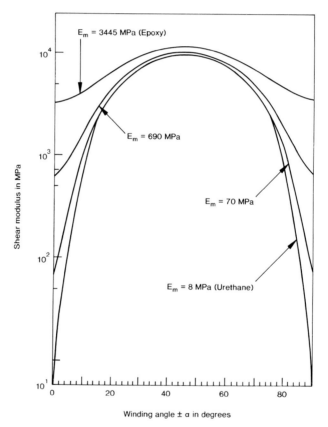

8.37   Shear modulus for several matrix types
$(v_f = 50\%)$.

Table 8.4. Fatigue test results

Test parameters	Before test	After test	Before test	After test
	Epoxy-glass		Urethane-glass	
Number of cycles	–	5800	–	231 000
Bending stiffness (N/m)	0.36	–	0.06	0.06
Torsional stiffness (N/m)	70.4	–	83.2	81.6
Torsional strength (N m)	2380	–	2268	2168

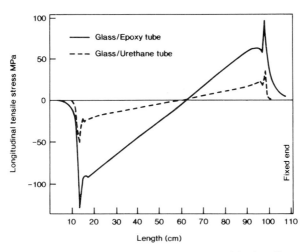

8.38   Longitudinal tensile stress induced by bending.

The results of fatigue testing are shown in Table 8.4. The carbon/epoxy shaft failed after only 5800 cycles, whereas the flexible component went on to exceed the specification with little or no loss in properties, thereby demonstrating the concept. The next step is to confirm its capability to operate fully satisfactorily in an automotive environment and to quantify its performance with respect to temperature (both elevated, ambient and due to hysteresis damping), creep, and chemical resistance.

Offshore

Composite materials are now being widely considered as materials of construction for a number of offshore structural components where their properties of light weight and corrosion resistance offer both design and operational advantage. The use of composites, however, is not new to the petroleum industry. Glass reinforced materials in particular are extensively employed and have become established in a number of applications such as flow and gathering lines, casings, surface injection and salt water disposal lines, CO_2 handling and tank battery piping as well as the relining of existing gas and liquid piping. Examples have been documented where composites have out-performed their conventional counterparts and demonstrated considerable durability under arduous service conditions,[45-48] for example, downhole tubing which was withdrawn from a depth of 990 m (3250 ft) after 23 years and found to be in excellent condition and subsequently returned to service. More recently composites are being assessed in terms of application for structural components where both loading and environmental conditions can be severe. This is especially the case for offshore installations where the mechanics of platform design are such that weight reduction in a given component can lead to substantial further saving in the supporting structure. Typical examples are risers, tubulars, tethers for tension leg platforms, process equipment, firewater mains and topsides structures.

GRP pipework has been used for many years in a number of chemical plant installations and its use offshore is primarily just an extension of current technology. In principle the same is true of casing except there are added design requirements specifically in terms of axial loading and external collapse. GRP casing has performed successfully in sizes from $4\frac{1}{2}$ in. to $9\frac{5}{8}$ in. (115–245 mm). Typically, these tubulars have tensile load ratings up to 68 t (150 000 lb), internal pressures of 20 MPa (3000 psi) and external collapse pressures of 45 MPa (6700 psi). Documented experience is available for GRP casing which has seen service in wells up to 3000 m (10 000 ft) deep.[46]

Details of the API (American Petroleum Institute) standards which relate to offshore GRP tubular products are given in Table 8.5. For line pipe, components are rated against service pressure. There are two methods of specification:[49] cyclic pressure and static pressure. In the former case, the test procedures establish the life of the product under cyclic pressure by testing to 15×10^6 cycles with subsequent extrapolation by a further decade (Fig. 8.39), and in the second case 20-year service life under hydrostatic loading is determined. This entails testing to 10^4 h with extrapolation to design life. It is usual in this latter case to downrate the design stress obtained by the test by a factor of two (Fig. 8.40). Additionally, short-term failure pressure, cyclic performance at design level and resistance to impact loads are obtained. Quality control tests are detailed covering dimensional checks, hydrostatic pressure tests, burst pressure, degree of cure and visual inspection. With specific regard to GRP casing, minimum quoted mechanical standards are given in Table 8.6. These are compared with selected steel casing ($9\frac{5}{8}$ in. OD; 245 mm) in Tables 8.7 and 8.8. As can be seen, with the exception of all but the high strength steel, GRP systems compare favourably and even in this case the selfweight carrying capability is similar. Furthermore, if the buoyancy effects are considered, the selfweight values would be considerably enhanced in favour of the composite alternative. Standards for tubing and casing remain

Table 8.5. Status of FRP specifications for offshore products

Product	Low-pressure line pipe (< 1000 psi)	High-pressure line pipe (> 6.8 MPa, 1000 psi)	Downhole tubing/casing
Specification	API 15 LR	API 15 HR	API 15 AR
Manufacturing process	Filament winding centrifugal casting	Filament winding	Filament winding
Material	Fibre glass reinforced epoxy, polyester or vinylester	Fibre glass reinforced epoxy or vinylester	Fibre glass reinforced epoxy or vinylester
Primary connection	Tapered box, epoxy bonded	Threaded	Threaded
Basic design criteria	Cyclic pressure test ASTM D2992 Proc. A	Static pressure test ASTM D2992 Proc. B	Across joint tensile internal/external pressure test
Design verification test	● Short term failure at 400% ● Cyclic pressure 750 cycles ● Impact resistance	● Short-term failure pressure ● Cyclic pressure at, 5000 cycles at 150% ● Static pressure 10 000 h at 200%	(In committee)
Quality control test	● Dimensional ● Hydrostatic test at 150% ● Short-term failure pressure ● Degree of cure or percent extractables ● Visual	● Dimensional, including thread ● Hydrostatic test at 150% ● Acoustic emission (optional) ● Degree of cure ● Short term failure pressure ● Glass content ● Visual	(In committee)

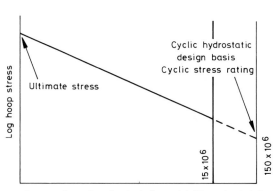

8.39   Cyclic pressure testing.

Table 8.6. Minimum mechanical requirements for composite casing

Test method	Required property value
Long-term static pressure Strength at 100 000 h ASTM D 2992 Procedure B	54 MPa (8 000 psi)
Short-term rupture strength (weep) ASTM D 1599	115 MPa (17 000 psi)
Cyclic pressure strength Long-term: 150×10^6 Cycles: ASTM D 2992 Procedure A	27 MPa (4 000 psi)
Cyclic pressure strength Short-term: 750 cycles ASTM D 2143	68 MPa (10 000 psi)
Ultimate axial tensile strength across the threaded joint ASTM D 2105	119 MPa (17 500 psi)
Axial tensile strength across the pipe wall ASTM D 2105	150 MPa (22 000 psi)
Modulus of elasticity in tension across the pipe wall ASTM D 2105	18 400 MPa (2.7×10^6 psi)

Table 8.7. Characteristics of $9\frac{5}{8}$ in. casing

Material	t mm (in.)	wt kg/m (lb/ft)	Axial strength MPa (psi)	Hoop strength MPa (psi)
FRP	13.84 (0.545)	19.22 (12.9)	150 (22 000)	115 (17 000)
	25.4 (1.0)	33.38 (22.4)	150 (22 000)	115 (17 000)
Steel:				
H40	7.92 (0.312)	48.3 (37.3)	272 (40 000)	272 (40 000)
K55	10.03 (0.395)	59.6 (40.0)	374 (55 000)	374 (55 000)
N80	13.84 (0.545)	79.7 (53.5)	544 (80 000)	544 (80 000)

Table 8.8. Strength of $9\frac{5}{8}$ in. casing

Material	t mm (in.)	Axial load t	Selfweight* m (ft)	Internal pressure MPa (psi)	Collapse pressure MPa (psi)
FRP	13.84 (0.545)	153	8000 (26 000)	13.0 (1910)	3.9 (570)
	25.4 (1.0)	267	8000 (26 000)	23.9 (3510)	24.4 (3590)
STEEL:					
H40	7.92 (0.312)	163	3400 (11 000)	17.6 (2590)	9.4† (1380)
K55	10.03 (0.395)	282	4700 (15 000)	30.7 (4510)	17.4† (2560)
N80	13.84 (0.545)	556	7000 (23 000)	61.6 (9060)	45† (6620)

* Length of string that can be supported (in air).
† Plastic collapse.

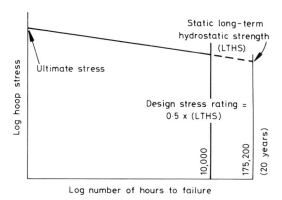

8.40   Static pressure testing.

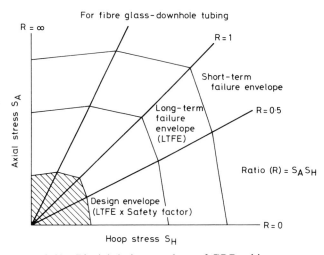

8.41   Biaxial design envelope of GRP tubing.

under development, the present documents being considered to be transitional. Because of the importance of axial loading in both these components the derivation of a tension/pressure envelope (Fig. 8.41), is being advocated.[47] This allows the performance under all likely loading conditions to be assessed and also recognizes the anisotropic nature of composite materials.

A significant difficulty with directional drilling arises because of the high frictional forces developed between the drillpipe and the well bore, and these can limit the maximum achievable depth and reach of the drilling operation.[50] One method of reducing these frictional forces is to employ a lightweight drillstring. Some progress in this area has been achieved with the development of aluminium, titanium and high strength steel drillpipe, but whilst these are

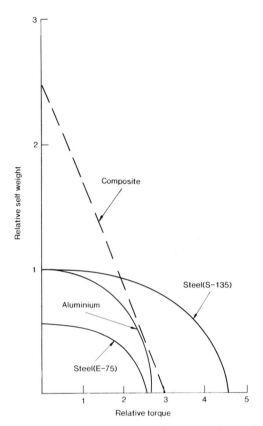

8.42   Relative design envelope for composite-steel and
aluminium drillpipe.

indeed lighter, their comparatively high cost cannot always be justified in terms of their improved capabilities.[51,52] The density of the composite materials is significantly lower than that of any of the alternative materials. This is magnified still further due to buoyancy effects as their density is close to that of typical drilling muds.

In addition to the primary objective of reducing torques and frictional loads, there are a number of other advantages to be gained through the use of a composite drillstring; rig capacity is increased, or alternatively rig loads reduced, and chemical resistance and fatigue life are improved.[53] Figure 8.42 shows the results of design calculations for a composite drillpipe design, in this case for an 'E-75 equivalent design', that is a design that matches or exceeds the load-carrying capability of E-75 grade steel. For the case shown, the laminate is of hybrid construction (combination of glass and carbon reinforcements) and is of similar thickness to its steel equivalent. Also shown in Fig. 8.42 are envelopes for steel (E-75 and S-135) and aluminium drillpipe.

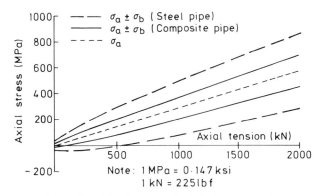

8.43 Combined stresses in drillpipe exposed to combined tension and bending ($r = 400\,\text{m}$).

The term 'relative selfweight' is the allowable suspended length normalized with respect to that of S-135 steel. The torque is normalized with respect to a typical design value. Clearly the performance of the composite design exceeds that of the alternatives by considerable margins. Drilling is, of course, a dynamic operation with significant interactions between the drillpipe and its surroundings resulting in cyclic loads being superimposed on the static forces. In curved portions of the well the drillpipe is subjected to a cyclic bending stress which is superimposed on the axial tensile stress due to selfweight. This bending stress is a function of both the well bore and the axial load applied, with a period equal to the rate of rotation. If the bending stresses are high they can lead to problems with fatigue and reduced lifetime for the pipe. These stresses can be calculated and are illustrated in Fig. 8.43 where the net axial stress (combined axial and bending) is plotted with respect to the tension force. The assumed radius of curvature for the calculation was 400 m (4.3°/100 ft). As can be seen, the bending stress amplitude, and therefore the stress concentration at the tool joint, is considerably lower for the composite drillpipe, thus reducing the fatigue loadings. The tool joint design makes use of conventional metal end fittings of box and pin form attached to the composite pipe. The fittings are attached to the pipe by a combination of mechanical keying, to resist both torsion and tension, and adhesive bonding. The adhesive serves to seal the joint as well as providing efficient load transfer between the fitting and the composite. In order to assess the potential of lightweight drillstrings in a drilling situation, calculations can be performed to determine the load distribution along the drillstring in a horizontal well. Typical results of this type of calculation are shown in Fig. 8.44. These calculations are based on the length of horizontal leg that will give rise to a tension of 1 t in the vertical leg of the drillstring (the vertical leg will buckle if it goes into compression) and limiting torque to a typical working value of 16 kN m (12 000 lb ft). On the

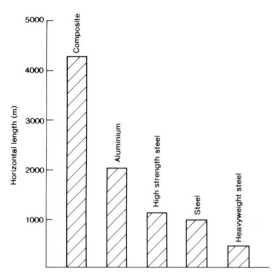

8.44   Relative horizontal reach capability for drillpipe.

basis of this example, a lightweight composite drillstring offers the potential of a fourfold reach advantage over steel.

A number of studies have been carried out on the use of GRP in topsides design, specifically in the accommodation module and process areas. In one case the study achieved an effective replacement of 66% of the primary and secondary steel components and 89% of architectural items for an example accommodation unit.[54] A global weight reduction of more than 20%, from 2000 to 1600 t, was indicated for only a small increase in anticipated cost. Behaviour in a fire situation is of vital importance for this type of application. The advent of cold cure phenolic resins which are inherently fire retardant has provided a boost in this area, although much debate remains concerning issues such as total fire loading, smoke emission and fume toxicity. In terms of physical behaviour composite constructions have much lower thermal conductivity than their steel equivalent and this means that wall temperatures can be much reduced (Fig. 8.45). This can be of important benefit in inhibiting the spread of a fire. This aspect of behaviour is used to good effect in offshore firewater mains. GRP is attractive as the debris from corrosion of steel systems (offshore firewater mains contain seawater) can block sprinkler systems and it has been shown that, under flowing condition, a GRP pipework system remains intact during simulated fire situations.

Subsea structures offer scope for employing GRP composites.[55] The main advantages are that corrosion resistance gives rise to simplified maintenance procedures and the light weight makes them easy to install. As an example, a conceptual study for a wellhead protection structure (Fig. 8.46), advocating

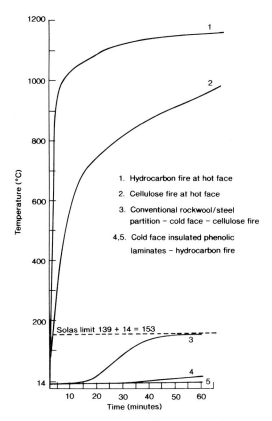

8.45   Performance of fire-rated partitions.

GRP sandwich panels indicated at weight of 9.7 t compared with over 100 t for its steel counterpart.

Design example: TLP components

Floating platforms such as tethered leg platforms (TLPs) are the structures which are the most sensitive to weight issues, and buoyancy calculations can be a dominant feature of the structural design. This effect can be most acutely demonstrated in the consideration of riser design. These are the lengths of tubing that convey process fluids from the platform to the drillhead on the seabed. A platform may have up to 50 risers of different diameter. For a given TLP design, and therefore given platform displacement, every tonne saved from the riser top tension (or tether pretension) is an extra tonne of potential payload. If buckling were the limiting design criterion, it would be possible to allow riser tension to fall to a minimum value only slightly greater than the apparent weight of the member concerned. In the case of TLPs this cannot be

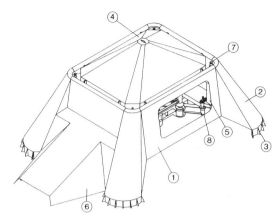

Structural elements

Item	Description	Material	Thickness
1	Inner shell	Sandwich	36 mm
2	Corner ramp	Sandwich	56 mm
3	Mudline penetrator	Laminate	30 mm
4	Top cover	Sandwich	46 mm
5	Inner corner	Sandwich	56 mm
6	Flowline protector	Sandwich	56 mm
7	Lifting/guide post	Steel	
8	Locking system	Steel	

Sandwich panels

Thickness	Outer/inner laminate	Core
56 mm	2 x 8 mm	40 mm
46 mm	2 x 8 mm	30 mm
36 mm	2 x 8 mm	20 mm

8.46   View of well-head protection structure.

allowed, since in order to prevent hydrodynamic interaction and possible contact, it is important to maintain compatible profiles between adjacent risers at all times. The principal design factors that control the detailed riser design are the potential blowout pressure (34 MPa, 5000 psi) and the mean axial top tension (typically 75 tonnes/riser).

Consideration of these loadings gives rise to a hybrid design with longitudinal carbon fibres and circumferential glass reinforcement.[56,57] Table 8.9 and Fig. 8.47 show the details of the design. The internal diameter (9 in.; 230 mm) was determined by the duty for this particular component; conveying two $3\frac{1}{2}$ in. (90 mm) steel tubing. The blow-out pressure determined the thickness of the circumferential layers whilst the thickness of the longitudinal layers are determined by the pressure end load and axial tension. The axial loads also determined the joint detail at the ends of each tube length. The end connectors themselves are of threaded steel design. Table 8.10 shows the mechanical properties of the design. The rubber liners on the internal and external surface of the tube are applied to ensure pressure tightness of the system.

A feature of the composite concept is the way its elasticity is used to advantage. With a steel riser the top ends are suspended with tensioners which

Table 8.9. Production riser physical characteristics

Internal diameter	9 in. (230 mm)
Joint lengths	24 m
Wall thickness:	
Outer liner (rubber)	1 mm ⎫
Longitudinal helically wound layer (carbon fibres/resin)	5.9 mm ⎬ 17.4 mm
Circumferential layer (glass fibres/resin)	9.5 m ⎪
Inner liner (rubber)	1 mm ⎭
Winding angles:	
Longitudinal helically wound layer	
Circumferential layer	$\pm 20°$
Apparent weight (with inserts, couplings and tubings)	90°
	45 kg/m

Table 8.10. Mechanical characteristics of composite risers

Axial modulus	E_a	51 300 MPa
Circumferential	E_h	38 600 MPa
Poisson's ratio	v_{ah}	0.22
Shear modulus	G_{ah}	11 350 MPa
Axial stiffnesses:		
Tension	ES	610 MN
Pressure	$ES/(1-2\,v)$	1 090 MN
Ultimate tension		450 tonnes
Burst pressure		105 MPa (15 000 psi)
Collapse pressure		38 MPa (5400 psi)

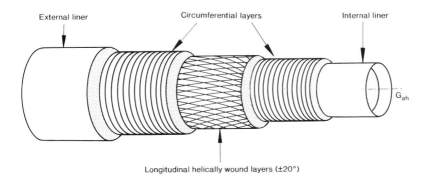

8.47   Composite riser design.

must have sufficient stroke to accommodate vertical movements. However, the low modulus and high strain capability of the composite hybrid (compared with the steel tethers securing the platform to the seabed) means that an active tension system is not necessary. A further subtlety in the design is the optimization of Poisson's ratio. Long cylinders under pressure can have large axial extensions due to end load. This is in part compensated for by Poisson's ratio contraction. If an all-CFRP design was used Poisson's ratio would be small, about 0.08, which could cause difficulties in terms of control of riser tension. The hybrid design gives rise to a Poisson's ratio of 0.22 which enables axial extension to be kept within convenient limits.

When a riser is attached to the seabed structure, there is a sudden change of stiffness and this joint may be subjected to large forces and bending moments generated as a result of platform movement. This can cause problems at the riser joint which can result in failure or over-stressing of internal tubing. To overcome these load concentrations it is now common to incorporate a stress joint within the riser assembly. This typically consists of a section of tapered wall thickness to provide a gradual transition in stiffness. Current designs are usually manufactured from single piece steel forgings, although there is work ongoing to examine the feasibility of titanium. Potential advantages of a composite design include:

- Lightweight.
- Improved performance, notably:
  - fatigue durability;
  - corrosion resistance;
  - high strength;
  - favourable elastic properties.
- Reduced installation costs.
- Flexibility of design leading to significant cost benefits.
- Simplification of assemblies.

Two of these factors are of particular interest for the stress joint application. Firstly, the combined stiffness/strength characteristics of composite materials are ideally suited to the design of a component with requirements for both high strength and compliant behaviour. This means that the stress joints can be shorter and will induce lower loads and bending moments into the subsea structure. Secondly, flexibility of manufacture means that small numbers can be produced cost effectively. In a given platform installation the majority of risers may be of standard diameter (9 $\frac{5}{8}$ in.; 245 mm), but there will also be requirements for non-standard sizes (up to 20 in. diameter; 508 mm) and these larger components can be easily fabricated in small numbers without excessive cost penalty.

In one proposed design composite, fibres are wound over a standard riser section with thickness increasing toward the bottom flange.[58] This provides a gradual increase in stiffness which results in a smooth transition in the curvature from the riser above to the wellhead below. After winding, the riser pipe remains an integral part of the structure. Other composite design options are available which are of lighter weight and therefore more structurally efficient, but the overwound design has the advantages of simple concept, a steel liner which simplifies the design, particularly for internal pressure, and maintenance of the use of the standard end connectors attached to the riser. Also, as the mandrel for winding the composite cylinder is a standard riser the

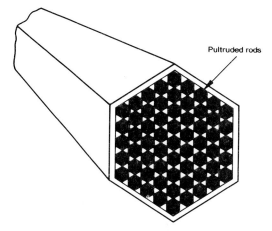

Pultruded rods

8.48   Conceptual design for CFRP rope.

costs are somewhat lower especially for small volumes. Connection to the seabed structure would be through a flange similar to existing arrangements.

For tethered platforms the weight of the tethers themselves can be significant. Hollow steel tendons, designed to be of neutral buoyancy, are conventionally used but as water depth increases instability due to external pressure comes into play. Solid tendon structures from composites would overcome this problem. Figure 8.48 shows a conceptual design of carbon fibre tethers fabricated from an assembly of pultruded CFRP rods.[59] One difficulty with this particular application, which is perhaps unique, is that the tendons for a single platform in very deep water ($> 10\,000$ ft; 3050 m) would require a major percentage of the world's annual capacity of carbon fibre!

Wind turbines

Glass reinforced plastics have been used extensively in the construction of blades for large wind turbines.[60] In some machines they are used to provide lightly stressed aerofoil profiles over a main load-bearing steel spar or as a cladding material for wood. In other designs, the GRP provides the main load bearing structure. Carbon fibre materials, although highly suitable in terms of mechanical loads, are not normally considered owing to cost. For large machines, blades fall into three groups: tape wound, filament wound and hand laid-up. Figure 8.49 shows an outline of each of the main types of design.[61,62]

The design requirements for a typical rotor blade are as follows:

- Low amplitude, high cycle fatigue properties (10^7–10^8 cycles).
- Stiffness and strength to accommodate 30-year extreme wind conditions.
- Adequate stiffness and structural damping to avoid resonance and to ensure aeroelastic stability.

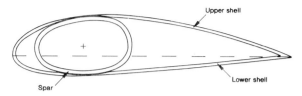

Tape wound blade

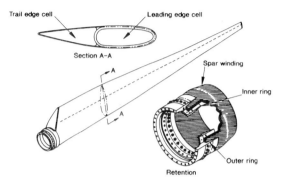

Filament wound blade

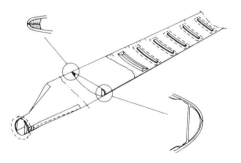

Hand layup blade

8.49   Wind turbine blade designs.

Figure 8.50 shows the layout of a typical medium sized (20 m, 630 kW) rotor blade.[63] The conceptual design of the rotor is a prismatic spar with aerodynamically shaped outer skins. The inner wing spar is steel while the outer is 'D'-shaped of GRP construction. The spar was wound using glass tape in such a manner that the fibres are disposed predominantly in the longitudinal direction. The use of steel for the inner spar and GRP in the outer is a compromise driven by the need to minimize weight at a large radius. The stresses in the outer wing originate from the aerodynamic loads, the centrifugal forces on the rotor and the forces generated as a result of the tilt of

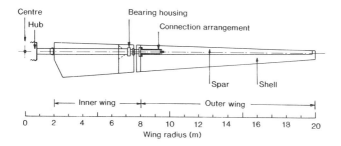

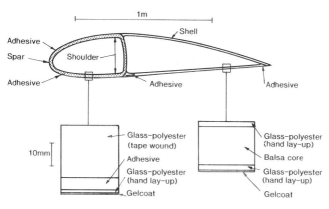

8.50 Typical rotor blade (20 m).

the blades from the rotor blade. The static stresses under extreme wind conditions (56 m/s) are calculated to be about 10 MPa near the wing tip to about 70 MPa at the connection between inner and outer radius. The dynamic stresses are estimated to be about 35 MPa for 20% of operational time, about 21 MPa for 40% of the time and 6 MPa for the remainder. The maximum deflection under storm conditions is about 1 m. The resonance frequency for a single outer rotor is about 2.2 Hz, but because of interaction between the three blades in the turbine, this decreases to 1.93 Hz.

To date a series of wind turbine structures has been constructed, many of them of similar basic design and some with blade configurations of diameters up to 100 m. These large systems are at the demonstration stage awaiting evaluation in terms of efficiency and cost per kilowatt/h.

Design example: passive aerodynamic control using composites

The main forces acting on the blade of a wind turbine are primarily aerodynamic and centrifugal. Depending on the type of machine, these may lead to blade bending or axial tension. To achieve power control via blade

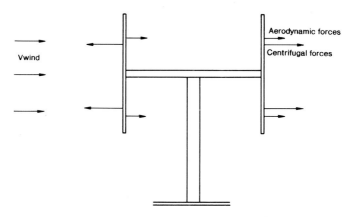

8.51   Forces on vertical axis wind turbine.

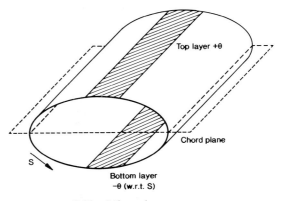

8.52   Mirror lay-up σ.

twisting therefore requires coupling between bending and twisting or between axial extension and twisting.[64]

In a vertical axis wind turbine (Fig. 8.51), with bending/twisting coupling, centrifugal forces give rise to twist whose magnitude remains constant as the rotor turns. Also generated are aerodynamic forces which act radially inwards on an upwind blade, but radially outwards on a downwind blade. These forces give a blade twist whose magnitude varies cyclically as the rotor turns. One means of achieving coupling of this type is using a 'mirror' type layup where the laminate angle is of opposite sign above and below the chord (Fig. 8.52). The magnitude of coupling is determined by the materials of construction, particularly the degree of anisotropy and the details of the aerofoil geometry. Figure 8.53 shows the twisting curvature/bending curvature ratio for a two-layer laminate (θ, ϕ). In general the maximum coupling is obtained when the thicker lamina is orientated at $20°$ to the longitudinal axis. Given the

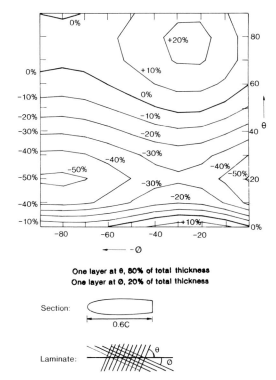

One layer at θ, 80% of total thickness
One layer at ∅, 20% of total thickness

Section:

0.6C

Laminate:

8.53   Contours of the ratio of twisting curvatures
(relative to bending curvature).

example of $\theta = 20$, $\phi = -20$ the ratio of curvatures is 0.5. For a blade of length 5 m, chord 1 m, the bending curvature is $1.53°/m$ for a strain of 0.2%. This is equivalent to a maximum end to end twist of 3.8°. Similar contour plots can be determined for stiffness properties. Figure 8.54 shows the effective bending stiffness as a percentage of the value obtainable when all the fibres are placed along the longitudinal axes. For a given laminate a judgement can be made as to whether or not the stiffness is sufficient to prevent dynamic instabilities. Figure 8.55 shows calculations indicating the effect of twist on turbine performance. The $-3°$ curve indicates a 3° twist due to centrifugal forces plus a cyclically varying twist of 2° due to the aerodynamic force. The influence of twist on performance is clearly indicated.

Other applications

In the preceding sections a whole range of applications together with design examples has been given – from satellites to subsea, from automotive to artificial legs. The list provided is by no means exhaustive; composites in one

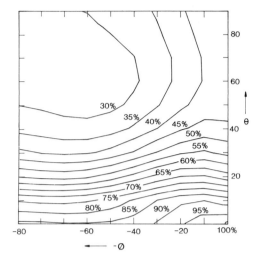

One layer at θ, 80% of total thickness
One layer at Ø, 20% of total thickness

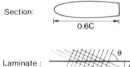

8.54   Contours of bending stiffness (relative to stiffness
at 0° laminate).

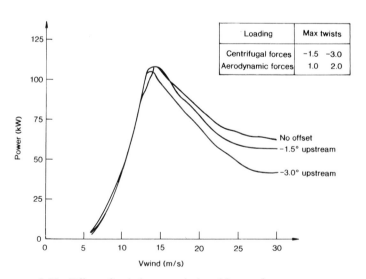

8.55   Effect of twisting on wind turbine performance.

form or another now pervade most sectors of industrial activity. Of all applications of composites the two applications that are perhaps most significant in terms of public awareness are in the sports and marine industries. Indeed, the leisure industry ranks as one of the biggest consumers of reinforced materials, both carbon and glass. Improvements in performance, whether real or perceived, through light weight and novel stiffness characteristics (or market hype!) have led to a growing consumer-led market. Composite golf clubs, fishing rods, racquets and skis are now established products within the range (just) of the average consumer. More are on the way, for example, bicycles, and no doubt others will be available as technology and fashion allow.

Marine craft has developed somewhat differently. From being one of the first areas of significant GRP use it has revolutionized the industry from a craft-based activity to one of mass production. All variety of boats and vessels are available ranging from the 'do it yourself' dinghies to the latest class of minesweeper with a hull of over 60 m in length. In a number of ways, experience from the marine industry has set the scene for other applications. Developments in fabrication technology and the design of chemical-resistant structures have been widely applied and continue to give confidence in long-term behaviour.

The building and civil industry is also a bulk user of GRP. Lightweight, corrosion-resistant accommodation and small housings and enclosures are universally found to give excellent service. What is less well known is that external wall cladding is commonly of GRP construction. Reduced weight can have a considerable effect on installation costs and ease of replacement means that the style of the building can easily be changed. Developments in this sector are linked to new materials offering fire protection service. CFRP is also finding application in the reinforcement of concrete structures. It is found that the bonding for very thin layers of unidirectional carbon laminates can considerably strengthen a steel-reinforced concrete beam. Figure 8.56 shows a load deflection curve for such a system. When this type of operation is done retrospectively for a repair or to increase load-carrying capacity, significant cost savings are possible owing to reduced scaffolding and lifting requirements.

Design example: CFRP bridge

To take advantage of any arising technology requires skill, commitment and not a little imagination. In some cases the engineer must be bold in concept and audacious in intent. Even if the proposal does not come to fruition, discussion of its basis and feasibility can prompt more imaginative thinking in what may be more mundane projects. The last design case in this chapter is one such example – the use of CFRP potentially to double, or triple, the span of

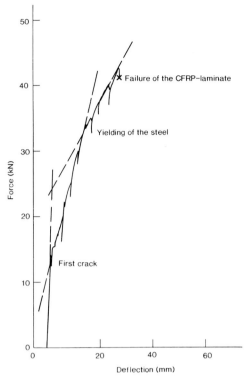

8.56   Experimental loading curve of a concrete beam
strengthened with a 0.2 mm thick GRP plate.

suspension bridges.[65] Bridges from conventional materials have a theoretical
maximum span of around 5000 m. The relationship between load and span is
given by the following expression:

$$\frac{w_s}{w_d + w_e} = \frac{L}{\left(\dfrac{\sigma_{all}}{\alpha\gamma}\right) - L}$$ [8.14]

where w_s, w_d and w_e are the dead load of the superstructure, deck load and
heave load respectively, L is the span, ρ is the density, α is a design coefficient
and σ_{all} is the allowable load for the hangers and cables. The limiting span L_{lim}
is given by:

$$L_{lim} = \frac{\sigma_{all}}{\alpha\rho}$$ [8.15]

Figure 8.57 shows the specific design load as a function of span ($\alpha = 1.66$)
using different materials of construction. Use of GRP doubles the potential
span whereas CFRP would triple it.

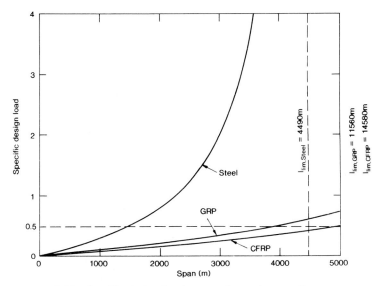

8.57 Specific design load versus the main span for a
suspension bridge.

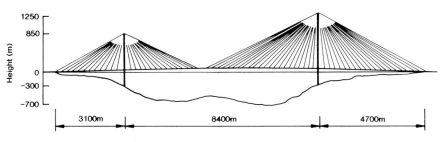

8.58 Proposal for a CFRP bridge across the Strait of
Gibraltar.

In a typical superstructure the cables represent about 70% of the
superstructure weight, so even limiting the concept to using unidirectional
materials for composite ropes will achieve the benefit to a significant degree.
Where could such a bridge be constructed? The Strait of Gibraltar has been
suggested (Fig. 8.58). Owing to water depth a span of at least 8400 m is
necessary; well beyond the scope of a steel structure. Will it be built? Certainly
not in the short or medium term. Difficulties in securing the necessary
financing arrangements for such a venture would surely be prohibitive without
the experience of smaller similar projects.

Whether it is ever built or not; whether it is outrageous or not; it is an
example whereby engineers and their colleagues – materials scientists,
chemists and designers – are using composites to push the limits.

POST SCRIPT: the opening of the world's first structural GRP bridge has been announced.[66] It has an overall length of 10 m suspended by Kevlar fibres from two structural GRP towers of height 17.5 m. Weight of the deck structure is quoted as a mere 150 kg/m.

Clearly there is a long way to go to match the grandest of ambitions, but the opportunities are there for the taking.

References

1 Ridell J C, 'Use of composite materials in aerospace applications', Proceedings of the BWEA - DEn Workshop, E7SU-N-109, Nov. 1987.
2 Riley B L, 'AV-8B/GR MK5 airframe composite applications', *Fibre Reinforced Composites*, I Mech E, Liverpool, 1986.
3 Bowen D H, 'Application of composites: An overview', in *Concise Encyclopedia of Composite Materials*, Ed A Kelley, Pergamon, Oxford, 1989.
4 Foreman C R, 'Design concepts for composite fuselage structure', in *Fibrous Composites in Structural Design*, Ed E M Lenoe, D W Oplinger and J J Burke, Plenum, New York, 1980.
5 Lubin G and Dastin S J, 'Aerospace applications of composites', in *Hand Book of Composites*, Ed Glubin, Van Nostrand, New York, 1982.
6 Peters S T, Humphrey W D and Foral R F, *Filament Winding: Composite Structure Fabrication*, SAMPE, Covina, California, 1991.
7 McCarthy R, 'Manufacture of composite propeller blades for commuter aircraft', ASE Technical Paper No 850875, SAE Technical Meeting, Kansas, 1985.
8 *Advanced Composites Engineering*, Nov. 1991, 6–8.
9 Holt D, 'Mechanized manufacture of composite main rotor blade spars', in *Fibre Reinforced Composites*, I Mech E, Liverpool, 1986.
10 Bashford D P, 'The novel use of composites in engineering applications', in *Fibre Reinforced Composites*, I Mech E, Liverpool, 1986.
11 Bristow J W, 'Certification of civil aircraft composite structures', in *Advanced Composites*, Ed I K Partridge, Elsevier, London, 1989.
12 Huttrop M L, 'Composite wind substructure technology on the AV-8B advanced aircraft', in *Fibrous Composites in Structural Design*, Ed E M Lenoe, D W Oplinger and J J Burke, Plenum, New York, 1980.
13 Kawashima T, Inove T and Seko H, Design and development of the graphite epoxy structure for the CS-3 satellites, *Proceedings of the ESA Noordwijk Symposium*, Netherlands, 1985, p 267.
14 Aguirre-Martinez M, Bowen D H, Davidson R, Lee R J and Thorpe T, 'The development of a continuous manufacturing method for a deployable satellite mast in CFRP', BPF Congress, 1986, pp 107-110.
15 Gessma L W, Dunn J and Demphe E E Jr, 'Evaluation of one type of foldable tube', NASA TM X-1187, 1965.
16 Renil BB, 'New closed tubular extendable boom', Proceedings of the 2nd Aerospace Mechanisms Symposium, Boeing Aircraft Co, August 1967.
17 NASA, 'Tubular spacecraft boom', NASA SP-8065, 1971.
18 Aguirre-Martinez M A, 'Developments in deployable masts to support flexible solar arrays', Procdings of the 4th European Symposium on Photovoltaic

Generators in Space, Cannes, 18-20 Sept. 1984 (ESA SP-210 Nov. 1984).

19 Emley F and Deibel C, *Progress Powder Metall*, **5**, 15, 1959.

20 Wheatley B D, 'Design of chemical plant process pipework in GRP', North Western Branch I Chem E/I Mech E Joint Symposium on Reinforced Plastic Constructed Equipment in the Chemical and Process Industry, Manchester, 1980.

21 Fowle D J, 'The fabrication of GRP pipes and piping components for chemical process plant', North Western Branch I Chem E/I Mech E Joint Symposium on Reinforced Plastic Constructed Equipment in the Chemical and Process Industry, Manchester, 1980.

22 Gray B D, 'Glassfibre reinforced plastics pipework for chemical plant', 4th International Conference on Plastic Pipe, Plastics and Rubber Institute, Brighton, 1979.

23 Gray B D, 'The design and construction of large vessels in GRP materials', North Western Branch I Chem E/I Mech E Joint Symposium on Reinforced Plastic Constructed Equipment in the Chemical Process Industry, Manchester 1980.

24 BS 4994, 'Specification for vessels and tanks in reinforced plastics', BSI, London 1973.

25 Bryan-Brown M H, 'Structural analysis of 60″ diameter valves in GRP material', CEGB Report SSD/SW/81/N29, July 1981.

26 Bruel and Kjaer, Technical Publication, October 1975.

27 Sully S, 'Non-metallic valves for auxiliary cooling systems', CEGB Report SSD/SW/82/N164, 1982.

28 Timoshenko S and Woinowsky-Krieger S, *Theory of Plates and Shells*, McGraw-Hill, New York, 1959.

29 Shorter J J, Carbon fibre uses and prospects', in *Proceedings of the 3rd PRI International Conference: Carbon Fibres* 3, New Jersey, Nayis Publications, 1986.

30 The Design Council, *Advanced Composites Engineering*, April 1992.

31 Saha S and Pal S, in *Concise Encyclopedia of Composite Materials*, Ed A Kelly, Pergamon Press, Oxford, 1989, pp243-248.

32 Pilliar R M and Blackwell R, 'Carbon fibre reinforced bone cement in orthopaedic surgery', *J Biomed Mat Res* **10**, 893-906, 1976.

33 Tayton K J J, 'The use of carbon fibre in human implants', *J Med (ed) Eng & Tech*, **7**, No 6, 271–272, 1983.

34 McKibbin B, *Recent Advances in Orthopaedics*, Churchill Livingstone, Edinburgh, pp 129–203.

35 Currey J D, *Handbook of Composites*, **4**, North-Holland, Amsterdam, 1983, Chapter IX.

36 Allen W C, Piotrowski G, Burstein A H and French V H, *Biomechanical Principles of Intramedullary Fixation*, Pitman Medical/Lippincott, Chapter 2.

37 Dowell J, Hancox N L, Lee R J and Wells G M. 'A carbon fibre reinforced epoxy resin intramedullary nail'. *J Materials Science Letters*, not yet published.

38 Dhoran C K H, 'Design of automotive components with advanced composites', *Proceedings of ICCM2 Conference*, Toronto, p 1446–1461, Ed B Norton et al, Metallurgical Society of AIME, Warrendale Pa, 1978.

39 Beardmore P, 'Composite structures for automobiles', *Composite Structures*, **5**, 163-176, 1986.

40 Wootton A J, Hendrey J C, Cruden A K and Hughes J D A, 'Structural automotive components in fibre reinforced plastics', 3rd *International Conference on Composite Structures*, Paisley, 1985.

41 Lee W A, 'A design study of material production for composite coil springs', *Composite Structures* **4**, Ed I H Marshall, Elsevier Applied Science, Paisley, 1987.

42 Beardmore P, and Johnson C F, The potential for composites in structural automotive applications', *Comp Sci Tech*, **26**, 251-281, 1986.

43 Pain B R K, Charlesworth D and Bennett A, *Development of a Cost Effective Composite Automotive Exterior Body Panel*, I Mech E, Liverpool, 257-260, 1986.

44 Hannibal A J and Avila J A, 'A torsionally stiff-bending soft driveshaft', 39th Annual Conference on Reinforced Plastics/Comp Inst, Soc Plastics Ind, 1984.

45 Puckett D B, 'Fibre reinforced pipe – a perspective for the future', Proceedings of the 38th Annual Conference Reinforced Plastics/Composites Inst, Society of Plastics Ind, Session 10-A, 1984.

46 Oswald K J and Moore R H, 'RTRP pipe systems demonstrate versatility and durability in oil field application', Proceedings of the 38th Annual Conference on Reinforced Plastics/Composites Inst, Society of Plastics Ind, Session 7-B, 1983.

47 Pickering F H, 'Design of high pressure fiberglass downhole tubing – a proposed new ASTM Specification', Proceedings of the 38th Annual Conference on Reinforced Plastics/Composites Inst, Society of Plastics Ind, Session 7-A, 1983.

48 Hicks A, 'Fiberglass reinforced plastic sucker rods', Proceedings of the 38th Annual Conference on Reinforced Plastics/Composites Inst, Society of the Plastics Ind, Session 7-D, 1983.

49 ASTM D2992, 'Hydrostatic design basis for fibreglass pipe and fittings', American Society for Testing and Materials, Philadelphia, PA, 1991.

50 Tolle G and Dellinger T, 'Mobil identifies extended reach advantage possibilities in the North Sea', *Oil Gas J*, **78**, May 1986.

51 *Reynolds Aluminium Drillpipe Engineering Data*, Edition No 9, Reynolds Metal Co. Richmond Va, 1987.

52 Tsukano Y and Ueno M, 'Development of lightweight steel drillpipe with 165 ksi yield strength', IADC/SPE, Drilling Conference, 1990.

53 Eckold G C, Bond A E and Halsey G, *Design of a Lightweight Drillstring Using Composite Materials*, SPE, Dallas, 1991.

54 Godfrey P R and Davis A G, 'The use of GRP materials in platform topsides construction and the regulatory implications', 9th International Conference, OMAE, Houston, 1990.

55 Steinsvaag O, Pettersen B and Janson J E, 'GRP structures in subsea applications – development for application in the Drauges field', 9th International Conference, OMAE, Houston, 1990.

56 Sparks C P, 'Lightweight composite production risers for a deep water tension leg platform', 5th International Conference, OMAE, 1986.

57 Sparks C P and Schmitt J, 'Optimized composite tubes for riser applications', 9th International Conference, OMAE, Houston, 1990.

58 Tapered Stress Joint, Pat Appl No P9205943.5.

59 Kiwi K S, Hahn H T and Williams J G, 'Application of composites in TLP tethers', 7th International Conference, OMAE, Houston, 1988.

60 Elliot G and Wootton A J, 'Wind turbines – the potential for composites', *Composite Structures*, Vol. 4, Ed I H Marshall, Elsevier Applied Science, Paisley, 1987.

61 Johansen B S, Lilholt H and Lystrup A, 'Wingblades of glass fibre reinforced polyester for a 630 kW windturbine', ICCM3, Paris Ed A R Bunsell et al, Pergamon, Oxford, Vol 2, pp 1355–1367, 1980.

62 Guemes J A and Avia F, 'Design manufacturing and tests of large wind turbine rotor blades', in *Composite Structures*, Vol. 4, Ed I H Marshall, Elsevier Applied Science, Paisley, 1987.

63 Gewehr H W, 'Large low cost composite wind turbine blades', NASA Conf Pub 2106, DOE Publication Conf – 7904111, 1979.

64 Karaolis N M, Musgrove P J and Jeronimidies G, 'Passive aerodynamic control using composite blades', in *Use of Composite Materials for Wind Turbines*, Proceedings of BWEA – D En Workshop, 1987.

65 Meier U R S, 'Future use of advanced composites in bridge construction engineering', in *Fibre Reinforced Composites*, I Mech E, Liverpool, 1986.

66 Raymond J A, Private communication, Scott Bader Co Ltd, 1992.

INDEX